TOP 100

Houses & Landscapes
100 最佳楼盘与景观

佳图文化 编

VI

中国林业出版社

图书在版编目（CIP）数据

100最佳楼盘与景观．6 / 王志 主编． -- 北京：中国林业出版社，2015.4

ISBN 978-7-5038-7851-0

Ⅰ．①1… Ⅱ．①王… Ⅲ．①建筑设计－图集②景观设计－图集 Ⅳ．
① TU206 ② TU986.2-64

中国版本图书馆CIP数据核字（2015）第 029727 号

中国林业出版社·建筑与家居出版分社
责任编辑：李 顺 唐 杨
出版咨询：（010）83143569

出 版：中国林业出版社（100009 北京西城区德内大街刘海胡同7号）
网 站：http://lycb.forestry.gov.cn/
印 刷：利丰雅高印刷（深圳）有限公司
发 行：中国林业出版社
电 话：（010）83143500
版 次：2015年4月第1版
印 次：2015年4月第1次
开 本：889mm×1194mm 1/8
印 张：64
字 数：400千字
定 价：798.00元

Preface 前 言

Living in the house and being taken good care of when becoming old is the dream of most ordinary people. And turning the houses into "beautiful housing" is the dream of the real estate developers and the designers. The "Meiju Jiang" is undertaking the humanistic spirits of "Chinese Dream" and the pursuit of high quality life. "2014 China Meiju Forum & Awards Ceremony", initiated by JTart Publishing & Media Group and sponsored by New House, has come to a perfect ending. "Meiju Jiang" sets 10 awards, including the Most Beautiful Apartments, the Most Beautiful Villas, the Most Stylish Apartments, the Most Beautiful Residential Landscapes, the Most Beautiful Commercial Properties, the Most Beautiful Tourist Resorts, the Most Beautiful Hotels, the Most Beautiful Cultural Buildings, the Most Beautiful Showflats and the Most Beautiful Spaces. This book has selected the most beautiful architectural works that have led fashion in 2014 and publish them together in this book.

As the first brand professional book introducing features of the residential buildings and landscape design, this book is the continuation and innovation of the former five books. This book shows the projects in categorization with styles as the main line and explains each project case in detail with excellent pictures and detailed descriptions. It comprehensively introduces every case from the aspects of the award-winning reasons, reasons for being shortlisted, project's general introduction, planning, construction, landscape and interior details. This book continues to use the compiling pattern of "theory + case" of Top Houses & Landscapes V, and the theoretical part focuses on the new development trend of "Beautiful Housing" design, discovering the new development trend of future residential buildings from the perspective of "Beauty". The book has succinct words, novel perspective, rigorous layout and rich & detailed contents. At the same time, it also represents the latest concept and idea of the architecture and landscape design, being a high-level professional book worthy of reading.

居有其屋，老有所养，这是大多数普通老百姓的梦想。而使其屋成为"美居"，则是所有地产开发商和设计师们的梦想。"美居梦"正承载着"中国梦"的人文精神和对生活高品质的追求。由《新楼盘》杂志主办、佳图出版传媒集团承办的美居奖，正是基于实现广大中国的"美居梦"而孜孜不倦。"2014美居奖"已经完美落幕，本届"美居奖"共设10大奖项，包括中国最美楼盘、中国最美别墅、中国最美风格楼盘、中国最美人居景观、中国最美商业地产、中国最美旅游度假区、中国最美酒店、中国最美文化建筑、中国最美样板间、中国最美空间。本书现将其推选出的中国地产界引领风尚的且具有宜居、宜商和具有设计艺术的中国美居项目集结出版。

本书作为介绍住宅楼盘与景观设计特色方面的第一品牌专业书籍，是在前五本基础上的延续与创新之作。本书将项目以"美居"为主线进行分类展示，在内容方面则以图文并茂的形式详解案例，从项目的获奖理由或入围理由、概况、建筑、规划、景观、室内等细节全方位介绍每一个案例。本书继续延续《100楼盘与景观Ⅴ》"理论+实例"的编排模式，理论部分重点阐述"美居"设计发展新趋势，从"美居"的角度探索未来住宅建筑发展的新趋势。本书整体释文简练、视角新颖、编排严谨，内容丰富而详实。同时，本书代表了当下住宅领域最新的建筑、景观设计理念和构思，是一本高水准的专业书籍，值得广大读者细细品鉴。

2014美居奖
各赛区和总评选活动介绍

总评选现场

2014年12月3日,2014年美居奖全国赛区总评选活动在广州顺利落幕,评选出10大美居奖奖项全国赛区总评选的前三名。

东赛区现场

2014年11月14日,2014年美居奖东赛区评选活动在上海顺利落幕,评定出10大美居奖奖项29个入围项目参加2014美居奖年度总评选活动。同期还举办了"2014美居奖系列高峰论坛之商业地产设计之变"活动。

评选项目：中国最美楼盘、中国最美别墅、中国最美人居景观、中国最美风格楼盘、中国最美文化建筑、中国最美商业综合体、中国最美办公建筑、中国最美旅游度假区、中国最美酒店、中国最美样板间。

分赛区评选标准：本着"公平、公正、公开"的评审原则，各奖项6个候选项目依次展示，专业评审团参照"美学性、宜居性、商业价值性、人文性、可持续性"等5个评分要素逐个评定，最终根据分数高低，每个奖项各推出三个项目进入2014美居奖年度总评选。

总评选评选标准：经过三大赛区的评选，最终共有90份作品进入12月3日全国赛区总评审。本着"公平、公正、公开"的评审原则，各奖项的候选项目依次展示，专业评审团参照"美学性、宜居性、商业价值性、人文性、可持续性"等5个评分要素逐个评定，最终根据分数高低，每个奖项选出全国赛区总评选的前三名。

北赛区现场

2014年11月21日，2014年美居奖北赛区评选活动在北京顺利落幕，评选出10大美居奖奖项29个入围项目参加2014美居奖年度总评选活动。同期还举办了"2014北京绿色建筑高峰研讨会之'合为'绿建"活动。

南赛区现场

2014年11月28日，2014年美居奖南赛区评选活动在深圳顺利落幕，评选出10大美居奖奖项32个入围项目参加2014美居奖年度总评选活动。同期还举办了"2014中国美居奖系列活动之最美楼盘巡礼品鉴及主题沙龙活动"，对获得2014美居奖南赛区"中国最美楼盘"第一名的红树别院项目进行了考察。

总评选评委阵容

 倪阳（评委组长）华南理工大学建筑设计院 副院长

 冼剑雄（评委副组长）广州市瀚华建筑设计有限公司 董事长

 曾剑（评委副组长）广东碧桂园集团（广东博意建筑设计院）规划设计院院长

 宋光辉 富力地产集团（广州市住宅建筑设计院有限公司）常务副院长

 张猛 万千国际 高级合伙人 / 首席建筑师

 李明 广州市万科房地产有限公司 总建筑师

 盛宇宏 广州汉森伯盛国际设计集团 董事长 / 总建筑师

 邓代明 棕榈设计有限公司（棕榈园林股份有限公司全资子公司）常务副总裁

 黄剑锋 SED 新西林园林景观有限公司 创始人 / 首席设计总监

 王志 《新楼盘》杂志社 总编

 邱春瑞 深圳大易国际室内设计有限公司 总设计师

 陈晓宇 加拿大 AIM 设计集团 总建筑师

 叶劲枫 广州普邦园林股份有限公司 园林规划设计院院长

东赛区评委阵容

 王煊（评委组长）水石国际 首席合伙人 / 总建筑师

 王欣（评委副组长）上海协信地产 研发总监

 潘允哲（评委副组长）UA 国际 合伙人

 卜域 恒盛地产 设计总监

 林钧 上海三益建筑设计有限公司 设计总裁 / 总建筑师

 夏莹 上海新外建工程设计与顾问有限公司 董事 / 总经理

 贾现军 上海匠人规划设计有限公司 副董事

 赵恺 上海霍普建筑设计事务所有限公司 董事 / 常务副总经理 / 首席设计总监

 秦戈今 上海秉仁建筑设师事务所 总经理

 陆臻 上海中房建筑设计有限公司 总建筑师

 沈雯 上海地尔景观设计有限公司 合伙人 / 项目总监

 蒋晓丹 万千国际 创始合伙人 / 总建筑师

 谢璇 鼎实国际 总经理

 王志 《新楼盘》杂志社 总编

 贺旭华 上海魏玛景观规划设计有限公司 总经理

北赛区评委阵容

刘克峰（评委组长）
万通地产立体城市研究院院长

陈音（评委副组长）
当代节能置业股份有限公司
研发中心总工

吕中元（评委副组长）
中央美术学院教授

郑锐
万达设计中心总工办
副总工程师

梁红
华夏孔雀城地产京南大区设
计总监

武春雨
万达设计中心景观副 总经理

全厚志
万通集团研发中心主任

唐艳红
易兰国际 副总裁

严涛
洲联集团 副总经理

王冬梅
中国建筑科学研究院中技集团
低碳研发中心主任

黄郁
CCDI 悉地国际 居住建筑产品
副总经理

顾斌
筑博设计股份有限公司
北京区域公司 总建筑师

南赛区评委阵容

胡树志（评委组长）
招商局地产控股股份有限公司
副总建筑师

李宝章（评委副组长）
奥雅设计集团
创始人 / 董事 / 设计总监

赵勇（评委副组长）
深圳花样年集团
总建筑师

朱晨
深圳市万漪环境艺术
设计有限公司
创始人 / 董事 / 首席设计师

王五平
深圳太合南方
建筑室内设计事务所
董事 / 设计总监

沈虹
GVL 国际怡境
景观设计有限公司
副总裁 / 总景观师

陈亮
深圳市陈世民建筑设计事
务所有限公司
董事 / 副总经理 / 设计总监

雷毅
澳大利亚·柏涛景观
副总经理

陈宏良
天萌国际设计集团
执行董事 / 总建筑师

彭光曦
深圳市立方建筑设
计顾问有限公司
合伙人 / 副总经理

吴华波
汉森伯盛国际设计集团
设计总监

金逸群
筑博设计股份有限公司
高级副总裁

牟中辉
深圳华汇设计有限公司
副总经理执行 总建筑师

Contents 目录

022 — New Development Trend of "Beautiful Housing" Design
"美居"设计发展新趋势

China's Most Beautiful Apartments
中国最美楼盘

"China's Most Beautiful Apartments" Award
荣获"中国最美楼盘"奖

Traditional Cultural Context, Modern Architectural Style
传统文化深厚,现代建筑风格

036

Shanghai Villa Clubhouse
The dignified and elegant architectural modelling emphasizes the overall proportion and detailed scale, inherits classical style and shows modern breath. Return to "Prairie Style" pursues the combination of beauty of technology and human interest, so that residents return to tranquility and nature.

上海院子
建筑造型厚重、典雅,强调整体的比例与细部的尺度,在继承古典风格的同时又赋予现代气息。"草原风"的回归重新追寻技术美与人情味的统一,使居住者情感回归于宁静与自然。

"China's Most Beautiful Apartments" Award
荣获"中国最美楼盘"奖

Ideal City with Varied Urban Skyline
城市天际线变化丰富的理想城

044

"Kangxin Jiayuan" Orientation Placement Housing, Zhongyi Village, Beijing
Monomer building design adopts linear rowlayout from south to north, to maximize the use of limited land and meet the need of relieve large population density of placement housing. High-rise and middle-rise is arranged alternatively to enliven spatial form of community and enrich the changes of urban skyline.

北京张仪村"康馨家园"定向安置房
设计单体采用南北向的行列式线性布局,最大限度利用有限土地,满足舒缓安置用房人口密度较大的需求,高层、中层穿插排布,活跃小区空间形态,丰富城市天际线的变化。

"China's Most Beautiful Apartments" Award
荣获"中国最美楼盘"奖

Modern Buildings Full Of Artistic Beauty
充满艺术美感的现代建筑

050

Mangrove Courtyard
Architectural design follows the principles of classical beauty of form. The single building was overall symmetric without obvious three-piece type whose symbolic meaning is achieved by set up aluminum alloy canopy at the top and dry-hanging fossil beige (marble) on the base.

红树别院
建筑设计遵循经典形式美原则,单体呈整体对称,虽无明显的三段式,但分别通过设置顶部铝合金雨棚及基座干挂化石米黄(大理石)来达到三段式的象征意义。

"China's Most Beautiful Apartments" Award 荣获《中国最美楼盘》奖	**Foreign-Style Hardbound House of Upgraded British Architectural Style** 英式建筑风格的升级版精装洋房	
056	**CIFI Arthur Shire** With full consideration of the banded terrain, the whole building adopts double extension type, trying to build the British-style architecture. The plan also takes account of household privacy, introducing the concept of private space into building interval, patio design, garden design and other aspects and giving full consideration to the feelings of households. **旭辉亚瑟郡** 项目充分考虑带状地形，建筑整体采用双排延展式，极力打造英伦风格建筑。规划还考虑了住户私密性问题、建筑间距、露台设计、庭院设计等方面都引入私密空间的概念，充分考虑住户感受。	
"China's Most Beautiful Apartments" Finalist 入围《中国最美楼盘》奖	**Open Neighborhood Space Creating New Living Value** 开放性邻里空间打造居住新价值	
062	**Changzhou Longfor Walking in Xiangti** The architectural form design that expresses the meaning of "home" pays attention to architectural humanization and the combination between design's modernity and affectivity, meanwhile highlights its differences with current popular style and reflects diversity of products, so that this residential form becomes advanced classical design in future's domestic residential products. **常州龙湖香醍漫步** 建筑形式设计汇中注重了建筑人性化，设计"家"的意义表达的建筑形式，侧重设计的现代性和情感性的结合，同时突出本设计与目前流行样式的差别，体现产品的差异性，使得该住宅形式成为未来国内住宅产品设计中的前沿经典之作。	
"China's Most Beautiful Apartments" Finalist 入围《中国最美楼盘》奖	**Multi-level group architectures with organic color collocation** 色彩有机搭配的多层组团建筑	
068	**North Garden** Architectural color is prominent by warm tone. Appropriately adding two-tone shutters, white window frame and framework highlights traditional, dignified, concise and neat style which building itself owns. **正北·名苑** 建筑色彩以暖色调作为主色调，适当增加重色的百叶窗、白色窗框及构架突出表现建筑本身既传统、稳重又简练、整洁的设计风格。	
"China's Most Beautiful Apartments" Finalist 入围《中国最美楼盘》奖	**Tuscan-style Interpretation of Xinjiang Residential Style** 托斯卡纳风格演绎新疆住宅风情	
076	**Jinke·Xinjiang Kingdom of Riverside** In generally, the building adopts exotic Spanish Tuscan style. Facade materials of commerce and villa employ yellow slate culture stone and pale yellow elastic coating and roof employs red terra-cotta. The color collocation as above brings bright color to architectural facade, which is moderately striking and feels manual and variegated. **金科新疆廊桥水乡** 建筑总体上采用具有异国风情的西班牙托斯卡纳风格，商业及别墅的外立面材料采用了黄砂岩板文化石与浅黄色弹性涂料，屋顶则采用陶红色陶瓦。这样的色彩搭配使建筑外立面色彩明快，既醒目又不过分张扬，给人一种手工、斑驳的感觉。	
"China's Most Beautiful Apartments" Finalist 入围《中国最美楼盘》奖	**Low-density Garden Apartments with Elevators** 低密度电梯花园洋房产品	
082	**Phase I of Chief Park City, Xi'an** The buildings are designed in modern British style with ternary structures. By the interweaving and interlocking of different blocks, it adds more vibrant elements to the originally serious facade. **西安建秦锦绣天下一期** 建筑采用三段式；但又通过一些体块的穿插、咬合关系打破了通常的三段切分方式，给稳重的立面增添了些许活跃的元素。	
"China's Most Beautiful Apartments" Finalist 入围《中国最美楼盘》奖	**New Residential Buildings of Oriental Flavor** 东方韵味的新式住宅建筑	
088	**Poly Leaves of the Forest, Shanghai** In modeling design, proportioning relationship in western architectures combines with detail elements in Oriental architectures. In architectural details of chapter, window frame and railing type, the application of simplified Chinese classical elements creates new residential architectural style of oriental flavor that is succinct and generous. **上海保利叶之林** 在造型设计中，将西式建筑的比例关系与东方建筑的细部元素相结合，在柱头、窗套、栏杆样式等建筑细部运用简化的中式古典元素，创造出简洁大气，且渗透着东方韵味的新颖的住宅建筑风格。	

"China's Most Beautiful Apartments" Finalist 入围"中国最美楼盘"奖	**Flexible Green Ecological Space** 形态灵活流转的绿色生态空间	
094	**Raycom·Tiancheng Phase IV** Five high-rise buildings in block A are designed ranging from 98 m to 83 m, forming rich undulating skyline and enriching urban space image. Commercial buildings in block B scattered high and low with flexible and dynamic form connect the parts into the dispersed but not isolated wholeness by the platform in 2nd floor. **武汉融科天城四期** A地块中的五栋高层公寓，分别设计为98m和83m不等，形成错落的天际线，丰富了城市空间形象。B地块商业高低错落，形态灵活流转，充满动感，又以二层的平台将几部分连接成一个整体，分散而不分离。	
"China's Most Beautiful Apartments" Finalist 入围"中国最美楼盘"奖	**Ecological and Pleasant Health Apartments** 生态宜人的养生住品	
100	**Suiyuan Jiashu** Architectural design of the project aims at creating a pleasant living environment for the elderly, using square, inner courtyard, corridor and other spatial design elements to create varied public spaces. **随园嘉树** 项目的建筑设计着眼于营造怡人的老年人居住氛围，采用广场、内院、廊道等多种空间造型元素，塑造富于变化的公共空间。	
"China's Most Beautiful Apartments" Finalist 入围"中国最美楼盘"奖	**Exquisite and Comfortable Waterfront Town** 精致、舒适、私享的水乡小镇	
106	**Eastern Mystery, Jinxi, Kunshan** The designers have organized the typical landscape elements of the Yangtze River Delta with modern skills, highlighting the gardens, streets, alleys, bridges and docks of the water town to create a living atmosphere of seclusion. **昆山首创青旅岛尚** 景观设计把江南诸景用现代的手法组织深化，景观元素突出"水乡的苑、水乡的街、水乡的巷、水乡的桥，水乡的码头"，营造出烟雨江南，漠漠水乡的隐居生活氛围。	
"China's Most Beautiful Apartments" Finalist 入围"中国最美楼盘"奖	**Well-organized Modernism Architecture** 错落有致的现代主义建筑	
112	**Ruize Jiayuan, Shenzhen** Project adheres to modernism architectural style. The scattered terraces and balconies design, color changes and contrast of material, create a tranquil and generous sense of scale. **深圳瑞泽佳园** 项目秉承现代主义建筑风格，通过错落的露台和阳台设计、色彩的变化、材质的对比，营造出一种宁静、大气的尺度感。	
"China's Most Beautiful Apartments" Finalist 入围"中国最美楼盘"奖	**Rich Spanish Style Architecture** 风情浓郁的西班牙建筑	
120	**Shahe·Century City Phase I** Multi-layer buildings in project adopt Spanish Style and high-rise buildings are modern style. The two styles are connected by similar material and colors. Spanish-style roofs, cornices, moldings and many other featured detail components, as well as undulating roof outlines, express the characteristics of Spanish architecture. **沙河·世纪城一期** 项目多层建筑均采用西班牙风格，高层建筑为现代风格，二者之间通过类似材质及颜色相互衔接，通过西班牙风格的屋顶、檐口、线脚等很多风格特征明显的细部构件，以及高低错落的屋顶轮廓线，来表达西班牙建筑的特性。	

"China's Most Beautiful Apartments" Finalist 入围"中国最美楼盘"奖	**Simple, Elegant and Modern High-rise Buildings of Chinese Style** 简约、清新、淡雅的现代高层中式建筑	
128	**Jiangnan Aristocratic Family** Designed mainly in modern New Chinese Style, the buildings have combined Chinese traditional architectural elements with modern architectural symbols by using black-white-gray color and point-line-surface composition, abandoning complicated traditional structures, simplifying the combination items, and choosing modern building materials. The exterior wall, bay window, balcony, louver and other decorative items are well combined to create a series of simple and elegant high-rise buildings of modern Chinese style. **湛江江南世家** 项目以现代新中式为建筑风格，引用传统中式元素，组合具有时代感的现代建筑标识，运用黑白灰、点线面的构图手法，简化构成素材，选用现代材料，把建筑的外墙、飘窗、阳台、百叶、装饰构件有机地协调统一起来，构筑成简约、清新、淡雅的现代高层中式建筑。	
"China's Most Beautiful Apartments" Finalist 入围"中国最美楼盘"奖	**Innovative, Unique and Large-scale Integrated Community** 新颖独特的超大规模综合社区	
134	**Times Property Eolia City (Phase IV, V, VI)** The buildings are designed in modern simple style to keep unified and present some rhythmic changes. Balconies, bay windows, roof trusses and the floating slabs of the exterior wall are combined to form innovative and unique building facade. In addition with the elegant colors, it will create a graceful and dignified atmosphere. **时代山湖海花园（四五六期）** 小区建筑采用现代简约主义的设计风格，设计手法强调统一中的韵律变化，通过阳台、凸窗、屋顶构架及外墙飘板等设计元素的组合，使建筑造型新颖独特，立体感强，凸显建筑的个性化，形成高尚、整齐、大气的建筑氛围。	
"Excellent Engineering" Award 荣获"优秀工程"奖	**Reconstructing the Memories of Alhambra Palace** 重构阿尔罕布拉宫的回忆	
140	**Hongbao Villas, Lohas Island, Suzhou** The project organizes each unit space by lanes and yards, from lanes into yards, then from the yards into doors, constructing spatial texture of Suzhou features. Hongbao creates the spatial mood of going along the lanes through yards and into doors. Molding style selects Islamic forms, taking Alhambra Palace as the motif design. **苏州中南·鸿堡** 苏州中南·鸿堡以巷、院来组织各个单元空间，由巷进院，由院入户，构建具有苏州特色的空间肌理。鸿堡营造出一种过巷入门庭、进院落的空间意境。造型风格选用伊斯兰形式，以阿尔罕布拉宫为母题来设计。	
"Excellent Engineering" Award 荣获"优秀工程"奖	**Living in Beautiful Landscape** 缔造在风景里生活的岁月	
148	**Hongju Mediterranean** With the theme of "leisure home of South European Style", the facade is designed in classic Mediterranean style with bright colors and lively lines. **宏聚·地中海** 项目以"南欧风情的休闲家园"为立意起点，建筑立面选择典型浓郁的地中海风格，使用鲜明跳跃的色彩和明快的线条作为建筑造型的基本语言。	
"Excellent Engineering" Award 荣获"优秀工程"奖	**High-end Residential Community of Exotic Style** 富于异域风情的高尚生活小区	
154	**CITIC Lake Forest Orchid Valley** Designed mainly in South Californian Spanish style, Orchard Valley has presented its own architectural languages in accordance to the local residential requirements. The design of the facade has focused on volume, western scale and colors, creating an exotic-style community for the residents. **中信森林湖兰溪谷** 兰溪谷整体以美国南加州西班牙建筑风格为主题，结合东莞本地居住者在建筑空间与平面布局上的要求，形成独具特色的建筑语言，建筑立面注重体量及西部尺度，色彩处理，为居住者精心打造一个富于异域风情的高尚生活小区。	

China's Most Beautiful Residential Landscapes

中国最美人居景观

"China's Most Beautiful Residential Landscapes" Award
荣获"中国最美人居景观"奖

Top Mansion in the Urban Aera
内环内品牌轨交豪宅

160

Poly Phili Mansion, Shanghai
Landscape designers integrate the connotation of Bayin, metal, stone, string, bamboo, gourd, clay, leather and wood, with the courtyard design ingeniously to play a beautiful symphony for the city.

上海保利翡丽云邸
项目将古代八音——金、石、土、革、丝、木、匏、竹的内涵巧妙融入进院落，使之与未来的居住场景有机融合，诚然为这座城市谱写了一首交响曲。

"China's Most Beautiful Residential Landscapes" Award
荣获"中国最美人居景观"奖

Revealing Gardens in Tang Dynasty Poem Through Interesting Landscape
趣味景观彰显园林唐朝诗韵

168

Poly Oriental Mansion
Adopting new Chinese garden as the design concept, designers are inspired by the popular poem Ti Po Shan Si Hou Chan Yuan, written by Chang Jian of Tang Dynasty. Following the tour path and conception of the poem to form different theme gardens, the project is eager to create an artistic garden but also with exquisite landscape.

北京保利东郡
从新中式园林的设计理念出发，项目追寻唐朝诗人常建脍炙人口的《题破山寺后禅院》诗句中游览的路径及意境的描述塑造出不同院落的意境主题，营造一幅具有景观趣味和意境的诗意花园。

"China's Most Beautiful Residential Landscapes" Award
荣获"中国最美人居景观"奖

First Impression Building of Capital
首都第一印象建筑

174

Wangjing SOHO
The unique surface design has made it possible for the buildings to show the beauty of dynamics and elegance from every angle. Aluminium plates and glass curtain walls are adopted for the facades, which have integrated with the blue sky revealing the artistic conception of mountains in the cloud and mist.

望京SOHO
项目建筑独特的曲面造型使建筑物在任何角度都呈现出动态、优雅的美感。塔楼外部被闪烁的铝板和玻璃覆盖，与蓝天融为一体，象征了山中云雾缭绕的意境。

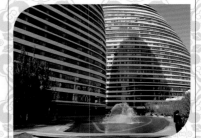

"China's Most Beautiful Residential Landscapes" Award
荣获"中国最美人居景观"奖

Landmark "Living Room" of Nanjing
金陵新城地标"会客厅"

182

Poly Joytown
Large scale and grand fountain, modern lamps, delicate theme sculptures and cloths, as well as landscape corridor with lush plants of display square are working together to reflect the beauty of Joytown against colorful lights. The contrast of quietness in the sunshine and dynamic at night represents brand new sensorial impact to visitors, which is highly interesting and attractive.

保利·堂悦
展示广场大气整齐的旱喷涌泉、线性现代的折线灯带、精致韵味的主题雕塑及布品、斑驳成影的绿道通廊在繁华灯光的烘托下映衬着堂悦的惊艳。阳光下的静雅与夜幕下的灵动为来客呈现着全新的感官冲击，令人流连忘返。

"China's Most Beautiful Residential Landscapes" Finalist
入围"中国最美人居景观"奖

New and Luxury Expression of Longfor Landscape
龙湖景观的华丽新阐释

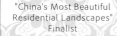

188

Longfor Chunjiang Central
Although with the same design language throughout Chunjiang Central, users will enjoy different psychological feelings in different spaces. The square on the Jiangnan Avenue shows the welcoming atmosphere, and the parking lot will rise up the users' expectation and curiosity about the development. While the linear spaces created in the sales center will offer the sense of dignity to users directly.

杭州龙湖春江郦城
春江郦城的设计语言高度统一，但传递出的心理感受在不同的空间却大不相同。江南大道上的广场给人以浓浓的邀请之感；停车场给人以满满的期待之感。而售楼处的轴线空间则给人以实在的尊贵之感。

| "China's Most Beautiful Residential Landscapes" Finalist 入围"中国最美人居景观"奖 | ## Adopting COSMO Concept to Reorganize the Urban life Elements
COSMO 理念重组都市生活要素 |
|---|---|
| 192 | ### COSMO, Hangzhou
According to the architectural styles, five building groups are divided for the residences, various elements including landscape walls, water features, sculptures and trees are working together to form changeable entrances and present different landscape.

杭州东润原筑壹号
住宅区域根据建筑多变的空间形态，将住宅区域划分出五个不同小组团，每个组团的入口通过景墙、水景、雕塑、树池等不同元素的组合形成多变的入口感觉，让每一个组团呈现不一样的景色。  |
| "China's Most Beautiful Residential Landscapes" Finalist 入围"中国最美人居景观"奖 | ## The Top Poetic Dwelling of Zhengzhou The Top Poetic Dwelling of Zhengzhou
郑州第一诗意栖居名盘 |
| 198 | ### Qinghua·Remember the Southern China
The lake running through the residential area is lying on the mountain and following the terrain of the slope, being sinuous, or a pond or running under the bridges, and feeding into the river outside the residential area to form a green gallery with changeable landscape but also collecting the functions of entertainment, recreation, fitness and communication.

郑州清华·忆江南
居住区以一条主要水系贯穿，这条主要水系依山就势在坡地中流淌、时而曲折蜿蜒、时而变换为清澈见底的水潭、时而有穿流过桥，最后汇入组团外的河道，形成一条自然绿廊，同时也形成一个富于变化的景观系列、一个集娱乐、休闲、健身和交谈的活力梯道。  |
| "China's Most Beautiful Residential Landscapes" Finalist 入围"中国最美人居景观"奖 | ## Green Transition on the Grey Site
灰色地块上的绿色蜕变 |
| 206 | ### Civic Cultural Square at Park Residence, Changchun
In order to make the site become "green" instead of "grey", designers adopt the most simple and effective way of creating multi-layer plant communities. Meanwhile, the plant communities have helped to divide the functional spaces, hide the messy spaces and provide shading to users.

长春柏翠园市民文化广场
为了实现场地从"灰色"到"绿色"的蜕变，项目直接营造层次丰富的植物群落并自我演替。同时，场地各功能空间需要浓密的植物群落对场地进行二次划分，消隐零乱的痕迹，提供良好的遮阴。  |
| "China's Most Beautiful Residential Landscapes" Finalist 入围"中国最美人居景观"奖 | ## Humanistic and Natural Landscape in French Style
人文、生态双重景观的法式演绎 |
| 214 | ### Eastern Provence
The water system in the villa area is the natural boundary to separate each villa. The designers strive to use humanistic art structures to create exquisite landscape details for the villa area.

东方普罗旺斯
别墅区的水体形成了分割其他别墅的自然界限，设计师力图利用有人文情节的艺术小品为别墅区的景观创造精巧的细节。  |
| "China's Most Beautiful Residential Landscapes" Finalist 入围"中国最美人居景观"奖 | ## World-class Resort Town
世界山居度假小镇 |
| 220 | ### Peacock City·Badaling
It has the design purpose of revitalizing and representing the essence of world-class resorts for targeted users. Every unit can enjoy beautiful landscape when sitting in their rooms, also large scale gardens are featured in the site that reflecting the pursuit for ecological lives and creating exquisite architectural and landscape design lying on the mountain. Fifth-layered landscape system is carefully designed for the site, landscape designers follow the terrain of mountain to build featured gardens.

孔雀城·八达岭
以"复兴世界山居精髓，开启山居庭院时光"为目标，园区规划布局内户户瞰景，并有超大立体花园，山景中承载着原生态的生活向往，塑造出半山上的精美建筑设计、繁花树影。项目的设计精心打造五重景观体系，依山势营造出特色山地园林。  |
| "China's Most Beautiful Residential Landscapes" Finalist 入围"中国最美人居景观"奖 | ## Natural Landscape Painting Nearby Yushan Lake
敔山湖畔的现代山水自然画卷 |
| 226 | ### Thai Hot Cathay Courtyard
After the analysis on district planning, architectural style and site condition, designers adopt the design philosophy of modern garden and combine with space design of irregular garden and gorgeous natural landscape to create a living model like a "retreat away from the world" that blended with nature, and to paint a natural painting leading users to step away from the hustle and bustle and walk into the natural, graceful and pleasant lives.

泰禾江阴院子
景观设计通过对项目的地域规划、建筑风格、场地条件的分析，运用现代园林的设计哲学，结合自然园林的空间手法、优美的自然景观，以与自然融为一体的居住形式，"桃花源"为模型，打造出敔山湖畔的现代山水自然画卷，引领人们远离烦嚣的城市，走向自然、优雅、舒适的山水生活。  |

Beauty of Elegance and Nature Created Through Neoclassicism
新古典主义缔造优雅与自然的精致美

"China's Most Beautiful Residential Landscapes" Finalist
入围"中国最美人居景观"奖

232

CIMC Zijin Wenchang
After careful study on the project, the landscape design of the community is in neoclassicism style, eager to create a brand new cultural landscape for commerce, office and residences that is elegant and exquisite.

中集紫金文昌
根据对项目的分析，项目的景观设计主要是以新古典主义风格为主题，极力打造一个全新的"古典优雅，精致从容"的商业办公及居住豪华、高档的人文景观环境。

Modern Simplified European Style Landscape Create the City Classic
现代简欧风情景观打造城市传世经典

"China's Most Beautiful Residential Landscapes" Finalist
入围"中国最美人居景观"奖

238

Zhengzhou Haima Park, Phase I
As a park-like garden with inspiration and beauty, its luxury Art Deco style integrates into the natural landscape; and the elegant, modern simplified European style landscape creates the Eco-environment of the community.

郑州海马公园（一期）
灵感加美感的公园式园林，奢华的ART DECO风格融入自然景观；优雅的现代简欧风情景观风格营造社区生态环境。

Exquisite Exotic Experience and Natural Sunshine Garden
别致的异域风情体验、自然清新的阳光花园

"China's Most Beautiful Residential Landscapes" Finalist
入围"中国最美人居景观"奖

248

Jinke Central Park City
Echoing with architectural style and theme, it makes full use of existing natural resources around to create Tuscany landscape style that blended into architecture to represent nature, harmony, art and interest to users.

金科大足中央公园
设计手法上与建筑风格及主题相呼应，利用周边的生态自然创建与建筑融为一体的托斯卡纳风格，同样雅致，体现"自然、和谐、艺术、风情"风格。

Modern and Fashion Working Environment
极具时尚现代气息的花园式办公环境

"China's Most Beautiful Residential Landscapes" Finalist
入围"中国最美人居景观"奖

254

Landscape of China Merchants Baoyao Area (Net Valley)
Modern and concise design style is adopted for the project, inserting humanistic concern into the environment and expressing through various sensory judgment including gustation, touch, hearing and visual, and small spaces such as elegant stop, tactual wall, time showcase, spray square, dreamland and rhythm garden, to create a modern and fashion work environment.

招商创业宝耀片区景观（招商网谷）
项目采用现代简洁的设计手法，在自然的环境中注入更多的人文关怀，通过味觉、触觉、听觉、视觉的不同感官体验，运用品位小站、触觉体验墙、时代橱窗、水雾广场、幻境花园和韵律之声等小空间，打造出一个多元化具有时尚现代气息的花园式办公环境。

The First Ecological and Low-carbon Office District Demonstration Area of Guangzhou-Foshan Economic Circle
广佛经济圈首个绿色低碳产业商务总部集群示范区

"China's Most Beautiful Residential Landscapes" Finalist
入围"中国最美人居景观"奖

260

Ecological Office District·Guangzhou-Foshan Basement
Inspired by Chinese spatial concept of "entering the hall before walking into the room", landscape design of the project is simple and concise, following the order of "introduction, elucidation, transition and summing up" expressing distinct concept and reflection.

中企绿色总部·广佛基地
景观设计以中式"登堂入室"空间概念作为景观创意蓝本，以起、承、转、合逻辑秩序进行设计，景观简约，反映出清晰的观念和思考。

A French Style and Vigorous Villa Area That is Both Ecological and Well-equipped
法式风情、充满活力且兼具生态修复功能的实用别墅区

"China's Most Beautiful Residential Landscapes" Finalist
入围"中国最美人居景观"奖

264

Master of The Mountain, Vanke Chengdu
Upon sufficient consideration of relationship between landscape and architecture, the designers create unique large-scale space of French garden, while assuring the landscape quality of the project. Plenty of French elements are adopted in landscape accessories, which highlight the nobility and elegance of French garden and create characteristic villa area.

成都万科五龙山公园
设计师充分考虑到景观与建筑的关系，在确保园区景观品质的同时，塑造出法式园林特有的大尺度空间。景观小品的设计中大量运用法式元素，突显出法式园林的尊贵与典雅，营造出别具风情的别墅区。

"China's Most Beautiful Residential Landscapes" Finalist 入围"中国最美人居景观"奖	### Model of Modern Landscape 曲岸观澜现代自然山水之典范	
270	**Showcase of Thai Hot High-rises** The water features around showcase are placed on the both sides of road and enclosed by plants to improve the visual effect and listening pleasure. **泰禾海沧高层展示区** 项目样板房附近的水系则通过绿化的围合，让水系若有若无的显现于道路两侧，增加了视觉的变化和悦耳的流水声。	

China's Most Beautiful Villas
中国最美别墅

"China's Most Beautiful Villas" Award 荣获"中国最美别墅"奖	### Eco Friendly Modern Quadrangle 生态环保的现代四合院	
278	**The Lake Dragon, Guangzhou (Zone F)** The architects have skillfully introduced surrounding views into the interior by using large room and large glass doors, and adjusted the height and orientation of each villa to get natural ventilation and unobstructed views. **广州玖珑湖小区发展项目（F区）** 在别墅设计上，建筑设计师巧妙地"引景入室"，大量采用大开间、大幅玻璃门元素，以此调整每一栋别墅的高度与方位，形成户户南北对流且不被遮挡的景观优势，力求令建筑与自然融为一体。	
"China's Most Beautiful Villas" Award 荣获"中国最美别墅"奖	### High-end and Low-density Livable Community 高档次低密度的宜居社区	
284	**Shanghai Yango Balmy Garden** The architectural design focuses on all details and shows respect to the natural texture and color of new materials. On the other hand, precise proportions and exquisite details enable people to experience the tradition, the history and the culture. **上海阳光城花满墅** 建筑从整体到局部，精雕细琢、镶花刻金都给人一丝不苟的印象，一方面尊重和保留新材质和色彩的自然风格，另一方面，通过准确的比例调整和精致的细节设计展现出浑厚的文化底蕴。	
"China's Most Beautiful Villas" Award 荣获"中国最美别墅"奖	### Multiple Levels, Flexible Spaces 视觉层次丰富，空间变化多样	
292	**Konka Moon River, Jiangsu** Villas are built near to water, and the boundary of the island features a density that's similar to a classic water town. A palette of vegetation sets a partition between the large water surface and the small water courses, which also provides great privacy for the villas. There are three different courtyard spaces far away from or near to water, namely, public courtyard, private gardens for the guestrooms and the landscape platforms along the water courses. **江苏康佳水月周庄** 别墅临水而设，在岛的边界上创造出与古典水城类似的密度。丰富的植被为大水体和小水道之间提供间隔，也为别墅创造隐私空间。别墅单元与水面或远或近，形成了公共庭院、私家花园和沿水景观大平台。	
"China's Most Beautiful Villas" Award 荣获"中国最美别墅"奖	### French-style Community of Nobility, Elegance and Romance 尊贵、典雅、浪漫的法式建筑居住社区	
298	**Yongding River Peacock Lake** Classic slope roof, together with multi-level lines, has skillfully integrated French-style colonnades, carvings and lines to highlight the elegance of the French-style residences. **永定河孔雀城大湖** 经典坡屋顶的设计加以层次丰富的线条，巧妙将法式柱廊、雕花、线条等经典元素一一归位美化，法式大宅的气度在层叠的工巧形式中得以尽显。	

"China's Most Beautiful Villas" Finalist 入围"中国最美别墅"奖	
304	## Top Single-family House Community 顶级纯独栋别墅社区 ### Kingdom Park The planning and design is inspired by nature and takes account of the comfortableness in modern western architecture, to avoid barrack-style layout and well arrange those villas among hills, forests and lakes. It reproduces a natural environment within the city and provides semi-open space that is independent from the private spaces. **金臣别墅** 在规划设计上，以自然为蓝本，萃取西方现代化建筑对于舒适度上的考量，突破了传统兵营式排布，巧妙地将山地、森林、湖泊这3大自然界元素融入别墅群落，在城市中再造自然，营造出独立于私属空间的半开放式空间布局。
310	## High-quality Buildings Combined with Golf Course 建筑布局与高尔夫球场有机结合的高品质建筑 ### Meilan Lake Silicon Valley Center, Shanghai The facade is designed in neoclassicism style, with special emphasis on classic proportion and details. Ternary form, elegant colonnade and the exquisitely designed details highlight the high quality of the buildings. **上海美兰湖硅谷中心** 建筑立面采用文艺复兴的新古典主义风格，注重古典的比例和细部，通过三段式的划分，大气的柱廊、细部的刻画突显建筑的高贵品质。
318	## Chinese Traditional Living Ideal 中国传统居住理想 ### CAREC SANCTUARY Based on Wright's prairie style which is characterized by its oriental complex, it introduces more Chinese cultural elements to interpret traditional Chinese living ideal from space, architecture and details. It makes the project significant to our times. **中航云玺大宅** 建筑风格在赖特颇具东方情结的草原风格基础上赋予更多中国文化的元素，在空间、建筑、细部几个层面诠释中国传统居住理想，并赋予一定的时代意义。
324	## Creating French-style Living Experience 创造法式生活体验的主题 ### Royal Mansion, Suzhou The facade is designed in French style with classic proportion, which looks elegant and dignified to enable people to experience the exotic style. **苏州中吴红玺** 建筑在立面处理上采用了法式建筑的手法，古典的比例创造了优雅尊贵的感受，为客户带来异国情调的体验感受。
330	## Family Mansion with Multi-style Facade Organically Integrated 多风格立面有机共融的家族大宅 ### Dragon Palace, Phase I It tries to create a symbol for high-end residences in China's market and a landmark of family mansion. The architectural form follows the example of European palace to present the luxury of Mediterranean Style, the pureness of British Style, the dignity of French Style and the elegance of Neoclassical Style. **博恩·御山水一期** 规划力求打造中国高端住宅符号，一个里程碑式的家族大宅，建筑形式效仿欧洲王庭：地中海建筑的宫廷华贵、英式建筑的原乡质朴、法式建筑的端庄宏伟、新古典主义建筑的高贵典雅在社区建筑上很好的得到了体现。
336	## Modern Residential Community Integrating Elegance, Rationality and Romance 融大气、雅致、理性与浪漫于一体的现代化居住区 ### Vanke Chefoo Island, Yantai The buildings are designed in new Chinese style and built with modern building materials. Design elements in traditional Chinese architecture have been abstracted and used to meet modern living requirements and show the respect to the history as well. **烟台万科海云台** 项目建筑为新中式风格，采用现代化建筑材料，在充分吸收传统中式建筑文化内涵的基础上，进行抽象总结与概括，在保证当代人的居住功能需求的同时也体现了对历史的尊重。

| "China's Most Beautiful Villas" Finalist 入围"中国最美别墅"奖 | **Multiple Skills to Realize Harmonious Coexistence Between Architecture and Nature** 多技法营造建筑与自然和谐共生的意境 |

| 342 | **Emerald Lemmon Lake, Jinan** The overall planning starts from creating urban interface and internal core landscape, making the inner and outer ring and the core water feature as the main theme. Low-rise residences within the inner ring are divided into three groups and organized in courtyard layout to form enclosed neighborhood spaces, which presents a spatial sequence by the the public spaces of the community, the public spaces of the residential groups, the public courtyard spaces and the private garden spaces.

济南翡翠莱蒙湖
整体规划以城市界面和内部景观核心的打造为主要出发点，以内外两环及核心水景为构思的主要中心。内环住宅分为三个组团，采用大合院的布置方式，形成围合式的邻里空间，完成小区公共空间、组团公共空间、合院公共空间、私家庭院的层层深入的空间序列。 |

| "China's Most Beautiful Villas" Finalist 入围"中国最美别墅"奖 | **Private Island Villas of Italian Style** 以意式风情为依托的私岛别墅 |

| 348 | **Purple Shore** Villas on this private island are noble and elegant, while the landscape is designed in Italian style, presenting the natural and cultural beauty of Tuscany. Buildings and landscape architectures echo each other, and natural landscape integrates with the local culture, creating a unique high-quality community for Handan.

紫岸
紫岸项目整体建筑突显私岛别墅的高贵特质，景观设计以意式风情为依托，还原托斯卡纳自然风光及历史人文之美，建筑与景观交相辉映，风景及文化锦上添花，成为邯郸独具特色的高品质社区。 |

| "China's Most Beautiful Villas" Finalist 入围"中国最美别墅"奖 | **High-quality Residential Courtyard Promoting Chinese Culture** 弘扬中式文化的高品质居住宅院 |

| 354 | **Cathay Courtyard** Buildings are designed mainly in new Chinese style with elegant appearance and exquisite details. The high quality and innovative form have distinguished the development itself in the low-density residential market which is dominated by European-style architectures. It has greatly highlighted the profound Chinese culture.

泰禾·北京院子
建筑风格以新中式为主导，整体设计简洁大气又不失细节，使整个居住区形象品质高端、形式新颖，在以西方欧式建筑为主体形态的低密度住宅市场中更加别具一格，将中式文化体现得淋漓尽致。 |

| "China's Most Beautiful Villas" Finalist 入围"中国最美别墅"奖 | **Small Town of European Style** 欧洲风情小镇 |

| 360 | **Sunny Melody, Yunfu** The project is positioned as a small town of European style. The building facade is dominated by Italian style with classical architectural elements of the Renaissance, the Gothic, the Roman times, the Tuscany, the Baroque, etc. Colors, materials and details are skillfully integrated to create an exotic style European town.

云浮远大美域
项目定位为欧洲风情小镇，立面采用以意大利为主的欧式风格，摒弃国内传统对意大利风格的狭隘认知，糅合了意大利文艺复兴、哥特、罗马、托斯卡纳、巴洛克等不同时期的建筑元素，在色彩、材料、细节等巧妙搭配，形成了有强烈异域风情的欧洲小镇。 |

| "China's Most Beautiful Villas" Finalist 入围"中国最美别墅"奖 | **Riverside Luxury Residence of Mediterranean Style** 地中海风格的江景豪宅 |

| 366 | **Riverside Houses** The residential facade is designed in Mediterranean style with red tiles and slope roofs to dialogue with the green mountain and forest. The application of bold colors, concave-convex forms, staggered structures as well as the changes of building heights, has presented a beautiful image.

江畔豪庭
住宅的立面设计为地中海风格，红色瓦坡屋顶的建筑与葱郁山林相映成趣。色彩的大胆应用和形体上的凹凸变化、错合、弧型拼接及建筑体量的高低错落，形成丰富的建筑景观。 |

| "China's Most Beautiful Villas" Finalist 入围"中国最美别墅"奖 | **Boutique Resort with Mountain and Lake Views** 山地湖景度假精品 |

| 372 | **Phoenix Valley, Changtai** These buildings, designed in natural and leisurely Southeast Asian style, feature hip roofs and colonnades as well as the details such as plaster, stone coping, wooden balcony and wood decorative cornice, to highlight the quality and uniqueness of the buildings.

长泰凤凰谷
建筑风格采取自然、休闲的东南亚设计风格，采用四坡屋顶、柱廊的形式，细部上通过局部的壁柱、石材压顶、木制阳台、木装饰檐板等做法，增加了建筑的品质感与独特性。 |

"China's Most Beautiful Villas" Finalist 入围"中国最美别墅"奖	**Boutique Holiday Residence of Modern Southeast Asian Style** 精益求精的现代东南式度假住宅	
380	**Vanke·Rancho Santa Fe, Phase II** Buildings of Phase II look elegant and dignified with simple colors. By using large-area glass windows and gray aluminum alloy frames, it creates great facade effect, provides open views and sufficient daylight, and strengthens the relationship between human beings and nature. **万科·兰乔圣菲二期东南式度假住宅** 万科·兰乔圣菲二期形体厚实大方、色彩朴素淡雅，立面上使用的大面积灰色铝合金玻璃窗及构架，既丰富了立面造型，也使得建筑室内空间采光良好、视野开阔，加强了人与自然的联系。	

China's Most Beautiful Stylish Apartments
中国最美风格楼盘

"China's Most Beautiful Stylish Apartments" Award 荣获"中国最美风格楼盘"奖	**Building Naturally Grown Residential Groups with Unique Geographical Advantages** 以得天独厚的地理优势塑造"自然生长"的居住组团	
390	**Vanke Anshan Whistler Town** The architectural form of this project uses the model of Canadian Whistler Town, and the planning layout complies with the tendency of the mountains, which is interspersed and staggered with rich and clear levels. The space expression uses the change of the spatial levels to break the rigidity of the traditional architecture and make the rhythm, proportion and dimension comply with the geometric aesthetic principles. **鞍山万科惠斯勒小镇** 项目建筑形式以加拿大惠斯勒小镇为模板，规划布局依就山势，穿插错落，层次丰富分明。空间表情通过空间层次的转变，打破传统建筑的呆板，其节奏、比例、尺度符合几何美学原则。	
"China's Most Beautiful Stylish Apartments" Award 荣获"中国最美风格楼盘"奖	**New Landmark of Fuzhou's New Humanistic Mansion** 福州新人文豪宅的新地标	
398	**Rongxin·Lan County** The design of this project is in modern minimalist style, and it brings the modern concept into the building. The wall uses glasses, which present the solemn atmosphere and also manifest the fashionable feelings incisively and vividly. The hundred-meter high-rise that stands straight from the ground with extraordinary magnificence embodies the constantly surpassing humanistic spirit and energy. **融信·澜郡** 项目设计为现代简约风格，把摩登理念引入建筑，外墙采用玻璃幕墙，气质稳重的同时又把时尚感展示得淋漓尽致。拔地而起的百米高层，傲然屹立的非凡气势，表达出不断超越的人文精神和力量。	
"China's Most Beautiful Stylish Apartments" Award 荣获"中国最美风格楼盘"奖	**Central Park in the Hustling and Bustling City** 繁华深处的中央公园	
404	**Vanke Le Bonheur** High-rise units of varied types are well organized with clear circulation and separate functions to meet different families' requirements. And the facade of the high-rises uses warm-tone coatings to create a simple and warm living atmosphere. **万科·柏悦湾** 高层住宅户型设计功能合理，流线清晰，动静分区明确，种类多样，满足各种家庭结构需要。高层立面设计采用明快的暖色基调涂料墙面，给人以简洁和如阳光般温暖的生活气息。	

"China's Most Beautiful Stylish Apartments" Finalist 入围"中国最美风格楼盘"奖	**Urban Space Focusing on Community Atmosphere and Neighborhood Relationship** 强烈社区感与良好人际关系并重的城市空间	
410	**Baoshan Greenland Linghai Phase II & III** The development features creative unit design with "double entry doors and double balconies" which is unique in Shanghai, meeting owners' requirements to divide a duplex space into two parts. Each part will have a width reaching 7 meters and a balcony to provide enough daylight and independent functions. **"绿地领海"二三期** 创意的户型设计，上海市仅有的"双入户门+双阳台"设计，可以满足业主将一套房复式空间分割成两套复式空间的需求，且每套面宽均可达到7 m左右，并各自带一个阳台，丝毫不影响采光及平常使用功能。	
"China's Most Beautiful Stylish Apartments" Finalist 入围"中国最美风格楼盘"奖	**Pleasant Community Integrating Space, Building, Landscape and Life** 空间、建筑、景观与生活充分交融的宜人社区	
416	**Fuzhou Poly Champagne International** The layout of the residences emphasizes space shaping, and the designers create a humane and close-to-nature living environment with ample spaces through the creation of single house, control of space and combination of outdoor greening environment design. **福州保利香槟国际** 住宅布局强调空间塑造，通过住宅单体的造型和空间限定，并结合户外绿化环境设计，创造空间丰富，亲近自然且具有人情味的居住环境。	
"China's Most Beautiful Stylish Apartments" Finalist 入围"中国最美风格楼盘"奖	**Brand Community with Comfortable, Modern & Noble Ambiance** 居住舒适、集聚现代高尚气息的品牌社区	
424	**Wanyuan Yu Jing** The facade design uses vertical lines to present the solemn, of high quality and rich cultural deposited Art Deco style. The wall of the residential building is mainly in stone, and the design uses the depiction of the material details to manifest the quality of this high-end residence. **万源御璟** 立面造型设计以纵向线条表现出庄重、富有品质感和文化底蕴的ART DECO风格。住宅楼外墙材料以石材为主，通过材质的细部刻画，表现出高端住宅的品质。	 
"China's Most Beautiful Stylish Apartments" Finalist 入围"中国最美风格楼盘"奖	**Apartment Design Approach Creates Urban Luxury Community** 公寓设计手法塑造强烈都市感的豪华社区	
432	**Cixi Hengyuan Condo Residence** The project breaks through the regular apartment design approach while designing this urban luxury community. It decorates both the exterior and interior with big opening and closing and gathers the high-end commerce and big residential units as a whole. The two blocks of hundred-meter luxury residences at the north and south stand face to face with unobstructed vision at the higher level, making the residents scan the prosperity of the city and enjoy the tranquil life as well. **慈溪恒元悦府公寓** 项目将常规的公寓设计手法突破至具有强烈都市感的豪华社区，以大开大合之姿兼修内外，集高端商业和高档大户型住宅为一体，南北两栋百米华宅相峙而立，无遮挡高层视野，既可纵览城市繁华，亦能尽享静谧生活。	
"China's Most Beautiful Stylish Apartments" Finalist 入围"中国最美风格楼盘"奖	**The Flagship Residential Property of Xiaoshan District, Hangzhou** 萧山区住宅地产的旗舰	
438	**Gemdale Tinyat Mansion** The delicate and elegant, charming and rich overall facade style fully shows the charm of the architectural style, tranquility of the residential community and the comfort atmosphere, and initiates the new fashion trends. The wall finishing materials is the combination of natural dry hanging granite stones and aluminum plates. **金地天逸** 整体立面风格，细致中见典雅，丰富中见韵味，充分体现建筑的风格魅力与住宅小区的宁静、舒适气氛，创导时尚潮流的新气势。外墙装饰材料采用干挂天然花岗岩石材与铝板相结合。	

"China's Most Beautiful Stylish Apartments" Finalist
入围"中国最美风格楼盘"奖

444 Living Territory for Future Young Man of Chaoqing Plot
朝青板块的未来年轻人居住领地

Zhonghong Pixel, Beijing
Inspired by the shape of honeycomb, the designers design the longitudinal facade with concave-convex walls and enrich the colors of the facade at the same time. The designers use the transition, volume increase and decrease of the building to form a split level of the building top so that to avoid the repression caused by the narrow space between the high-rise buildings.

中弘北京像素
以蜜蜂的家为创造灵感，将建筑的长向立面设计了众多凹凸墙面并同时丰富立面色彩，通过建筑的转折、体块的加减穿插，使建筑端部错落，从而避免高层建筑由于间距过近造成的压抑。

"China's Most Beautiful Stylish Apartments" Finalist
入围"中国最美风格楼盘"奖

452 Naturalistic Style Leads the Close-to-Nature Life
自然主义风格引领自然零距离生活

Gemdale River Town
The volume of the residence is diversified and without losing style and taste, making people touch the nature with their bodies for real and fully enjoy the sunshine, fresh air, rain and dew. The staggered volumes also form abundant facade texture.

金地朗悦
住宅建筑体量丰富却不失格调品位，让人真真实实地用身体触摸着自然，充分享受着阳光、空气和雨露，体块的错落也形成了丰富的立面肌理。

"China's Most Beautiful Stylish Apartments" Finalist
入围"中国最美风格楼盘"奖

458 High-end Model of Industrialized Prefabricated Technology
工业化预制技术的高端典范

Tianjin Vanke Jin Lu Garden
As the high-end residence that first uses the industrialized prefabricated technology in North China, the project has reduced the on-site construction time of the alpine region, and this is the initial realization of the dream of prefabricated house.

天津万科锦庐园
作为华北地区首个应用工业化预制技术的高端住宅产品，项目缩短高寒地区现场施工时间，是预制装配房屋梦想的初步实现。

"China's Most Beautiful Stylish Apartments" Finalist
入围"中国最美风格楼盘"奖

464 New Urbanism City with Ecological Landscape
新都市主义的生态山水之城

East Shore International Zone B
The project is organized in the way of "center and functional groups", and it develops in the way of residential neighborhood enclosure. The natural, smooth and curve road and greenbelt flexibly connect the buildings, and multilayer residence and high-rise apartment have enclosed a clear group space, forming the courtyard space and the natural open and close of various spaces.

东岸国际B区
项目采用"中心—功能组团"的结构组织方式，以住宅邻里围合的形式层层展开。自然流畅的曲线道路和绿带将建筑灵活串联，多层住宅和高层公寓围合成明确的组团空间，形成院落空间，各种空间收放自然。

"China's Most Beautiful Stylish Apartments" Finalist
入围"中国最美风格楼盘"奖

472 The Elegant Ambiance and Aesthetic Culture of the Upper Class Create the Living Standard of Yinchuan
上流阶层典雅气息和美学文化缔造银川人居标准

Yinchuan Hailiang International Community
The project is composed of small high-rise, high-rise, semi-detached house and townhouse. The ultralow density and the extremely large landscape floor distance manifest the extraordinary life manner. Large-scale villa groups are quite rare in Yinchuan, and the high-rises are symmetric in forms with hard and soft lines.

银川海亮国际社区
项目由小高层、高层、双拼别墅、联排别墅组成，超低密度及超大宽景楼间距，丈量非凡人生气度，大规模的别墅群在银川也堪称稀缺，高层楼型对称，线条刚柔并济。

人们物质生活水平的不断提高，已不再满足于一般意义上的居住概念。"美居奖"因应人们对居住审美的新要求，提出了美学性、宜居性、商业价值性、人文性以及可持续性五大要素。这五个要素不仅契合了当下住宅设计的重点，而且与住宅设计未来的新趋势息息相关。尤其是在当今新常态的中国经济大背景下，无论是住宅外观设计、景观设计还是住宅配套都呈现出了新的发展趋势。而这也进一步地塑造出人们未来心中的"美居"。

一、住宅结构与造型风格设计趋势

1. 建筑结构更趋高层化

尽管我国幅员辽阔，但人口众多却与之形成矛盾，因此，住宅开发宜提倡建设多层及高层建筑。鉴于我国人多地少的情况，城市建设用地尤其紧张，居住向空中发展势在必行。居民从小平房、大杂院到住高楼这种生活方式即居住文化的巨大变化，已逐步从过去的惧避心理转变为向往"一览众山小"、眼界开阔的高层住所。

2. 建筑风格的现状问题

1). 建筑外观普通，没有风格。在许多中小城市，房地产开发水平不高，对于建筑美观不太讲究，所建的楼盘没有明确的风格可言。政府大力推广经济适用房、两限房、福利微利房等，考虑到经济成本等因素，其建筑美观性堪忧。

2). 抄袭、跟风现象严重。回顾中国房地产市场的发展历程，欧陆风曾经刮遍全国。而一个楼盘用深色调立面好，结果很多楼盘相继采用，一个楼盘采用凸窗，其他楼盘也纷纷采用。这样的跟风、抄袭现象比较严重。

3). 简单机械处理。由于国内大部分建筑设计从业人员技术水平有限，对于风格理解比较简单，而开发商也常常因为成本的原因在建筑立面细节上打折扣，所以难免断章取义，或照搬照抄。

4). 在风格运用上较少考虑到本地的自然特征和文化特征。

3. 住宅建筑风格的未来趋势

1). 个性化。现代风格推崇个人主义，突出人的个性张扬。这其中尤其对楼盘外立面的新颖创新上有着更强烈的要求，形象鲜明突出的楼盘才能给人们留下深刻的印象。

2). 设计理念成熟理性。地道的古典主义与真正的现代主义都可以用一种尺度来衡量：秩序。将经济效益、环境效益、社会效益完美结合的成熟理性的设计理念，将是房地产市场的走向所在。

3). 多样化。异域风情从最早的欧陆风格，到现在各国风情的百花齐放，现代风格从简单机械的外立面而到现在的形式多变和个性化，加上中式风格的回归，随着经济的发展、建筑科技的进步以及消费需求的多元化，住宅风格也会日益多样化。

4). 精致化。随着品味的提高，建筑风格将更为完美、精致。粗制滥造，各不符实的风格是没有市场。

二、住宅立面设计趋势

1. 建筑立面更趋可变性

建筑立面是建筑接触外界的平台，也是反映建筑个性和风格的关键部分。人们对建筑物的感知一般都是从立面开始，因此想要建令人过目不忘的建筑，立面设计是首先需要重点考虑的环节。传统的立面设计都是力图用通过调整构图形式、处理色调、变化材质等方式来推陈出新，但无论怎么变化，这些建筑一经建成后就以固定的状态一成不变地呈现在人们面前，缺乏灵活性。随着近年来新鲜事物的不断涌出，人们思维的不断活跃，传统固定不变的建筑立面已经无

styles from various countries. The modern styles change from the simple and mechanical facades to the diversified and personalized forms. And along with the regression of Chinese style, economic development, advancement of architectural technology and the diversified consumption requirements, the residential styles will also be increasingly diversified.

d. Refined. Along with the improvement of taste, the architectural styles will be more perfect and refined, and there will be no market for the shoddy and gaudy styles.

The residential building styles should adjust measures to the local conditions and the important principle is to reach the harmonization of ecology, economy and culture and art. For example, to maintain the harmony and affinity with the surrounding environments, it can integrate the adventitious living culture with the local living culture and try hard to discover the essence of the living culture to create residences with Chinese characteristics.

2. Residential Facade Design Trend

2.1. The Building Facade Tends to Be Changeable

Building facade is the platform for the building to contact the outside world, and also is the key component to show the feature and style of the building. People's perception about the building usually begins with the facade, so to build an unforgettable building the facade design is the prior element that needs to be considered. The traditional facade design strived to innovate through the ways of adjusting the structures, dealing with the colors, changing the materials and so on, but no matter how it changed, the buildings invariably showed with a fixed state in front of people, lacking flexibility. With the continuous emerging of new things, people's thoughts are constantly changing; the traditional changeless building facade can no longer meet the demands of the times and the social desire for variable building facade becomes increasingly strong. In the meantime, with the rapid development of information technology and under the support of the increasingly strong architectural technology, many novel facade forms will emerge and come into being.

In the modern facade design, many state of the art designers use advanced structure techniques to create variable facades so that to bring people with novel feelings. Such as the currently well received intelligent sun shading system, it can adjust the building facade according to the changes of the surrounding environment, which not only makes the sun shading of the buildings more accurate and efficient, but also adds time dimension experience for the building, breaking the traditional facade knowledge and owning stronger recognition.

2.2. The Building Facade Tends to Be Concise

The architectural appearance has transformed from tedious to concise, and the facade gradually tends to be concise and bright, which adopts large areas of decorative colors and simple shaped decorative elements. The originally quite popular tedious and complicated European symbols, architrave and pilaster can barely be seen nowadays. Even if there are some European style residences, they only adopt the generalized and abstract decorative elements in terms of facade. The open window becomes increasingly large and is normally with bay window design; the sill has been lowered and the balcony is usually enclosed by glass. The design of large and low sill and entire glass balcony makes the relationship between human and nature more harmonious and full of breath of times. In the previous residential design, monotonous and stiff gables can be easily seen, which indeed destroys the whole landscape. Many designers solve this problem through increasing the length of the units at both ends, adding open windows on the gables and enlarging the units at both ends. After the enlargement of the units at both ends, they face one side of the urban street and are conducive to ensure the landscape view of the city.

2.3. The Building Facade Tends to Be Ecological

Green, energy-saving and ecological is the development trend of modern residential building. During the design and construction process of the residential building, the designers will try to properly use the natural resources, select the green and environment-friendly materials for the building facade, and reduce the burden that the building construction brings to the ecology and environment, achieving the balance between the obtaining and returning of natural resources, and taking the road of sustainable development of natural resources.

法满足时代的需求，社会对建筑立面可变性的渴望也愈发强烈。与此同时，信息技术飞速发展，在日益强大的建筑技术的支撑下，许多新颖的立面形式孕育而生。

在现代立面设计中，许多前卫设计师利用先进的结构技术来创造可变的立面，从而给人们带来新奇的感觉。例如目前大受欢迎的智能遮阳系统，它的使用可以使建筑立面随着周边环境的改变而作出相应调整，这不仅使建筑遮阳更加精确有效，还为建筑立面加入了时间维度的体验，突破了传统的立面认识，具有很强的辨识度。

2. 建筑立面更趋简洁性

在外观上，已完成从繁琐向简洁的转化，立面日趋简洁明快，多采用大面积的装饰色块及造型简洁的装饰构件。原来非常流行的繁琐复杂的欧式符号、线脚及壁柱已比较少见。即便现

在还有一些欧式风格的住宅，但从立面上来看，也只是采用了概括而且抽象化的装饰构件。开窗面积愈见增大，且多采用飘窗设计，窗台降低，阳台多采用玻璃封闭。大且低的窗台设计，全玻璃阳台，使人与自然的关系更加和谐，充满了时代气息。在以往的住宅设计中，单调、呆板的山墙林立，确实破坏整体景观。现在很多设计师采用通过加大端单元进深，在山墙上增加开窗，放大端单元的方法来加以解决这一问题。而且端单元放大后，对朝向城市街道的一侧，也有利保证城市景观。

3. 建筑立面设计更趋生态化

绿色、节能、生态是现代住宅建筑的发展趋势，在住宅建筑设计和建造过程中做到合理使用自然资源，尽量选择绿色环保的建筑立面材料，降低因建筑施工给生态环境造成的负担，实现自然资源索取与回报间的平衡，走自然资源的可持续发展的道路。

三、住宅户型设计趋势

一种户型决定了一种生活方式，户型设计实际上是生活方式的设计。而随着经济的快速发展，人们的生活方式较之之前已经有了很大的改变，因为，与人们生活方式息息相关的户型设计也在悄悄发生着变化。

户型设计的优化是住宅开发永远不变的趋势。常规性的户型设计主要考虑项目本身的地理环境及建筑规范，这种优化只能是按部就班；创新的户型设计需要结合社会文化，分析消费者特点从而设计出满足居者各方面需求的户型。

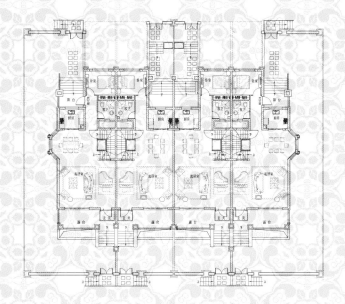

1. 户型模式更趋舒适性

舒适性已成为当前住宅室内设计的重要课题。怎样使住宅变得更舒适，具备什么样的条件才能舒适。在设计中，应以人们的日常生活轨迹为依据，现在的动静分离、洁污分离、方正实用的户型是人们所欢迎的。厅带阳台、前厅后卧、厨卧分离、厨房带生活阳台是适应日常家居所需要的。房间最好方正实用，避免使居住者感到不舒服。通风、采光条件得到普遍的重视，

3. Unit Design Trend

As a unit type decides a kind of life style, design for unit type is a design for life style indeed. Through the rapid development of economy, public life style have been changed a lot, hence, the unit design that closed related are being changed quietly. Optimization of unit design is an eternal tendency of residence development. Regular unit design always takes local environment and building code as the main consideration, the optimization just follows the prescribed orders; while for creative unit design, it should combine with social culture, and analyze characteristics of targeted users to make the unit meet various kinds of requirements.

3.1 Unit Type Tends to be Comfortable

Comfort has become a major topic of residential unit design, such as the ways to make unit more comfortable and what the conditions are needed for a comfortable unit. In the process of unit design, it should be based on the route of daily lives, square unit with separation between living and activity spaces, as well clean and kitchen spaces are the most welcoming unit type. With a balcony outside living room, living room and bedrooms set front and back respectively, division of kitchen and bedrooms, and a life balcony outside kitchen are all essential for daily lives. The unit should be square or rectangle to avoid any uncomfortable may be caused. Ventilation and daylighting are also very important, hence:

a. Plate-type building is a general welcoming product, while tower building is becoming less popular

b. Layout of neighborhood arrangement is taking place of residential community that with plenty of high-rises

c. Residential suburbanization is increasing distinctly

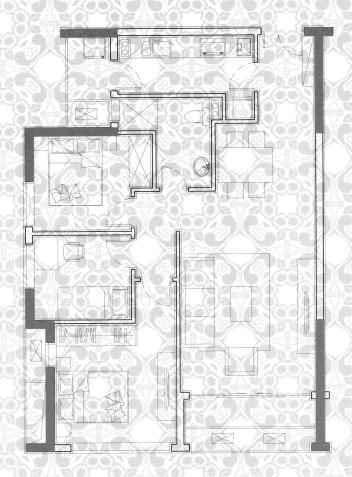

3.2 Unit Space Design Tends to be Diversified

Diversified unit type is meeting the increasing demands of personalized houses, also providing possibilities of resetting unit layout when there are some changes regarding family structure so that the unit has a longer lifetime, which is the tendency of unit type transformation. While designing the exterior and layout of the unit, it is acceptable to avoid simple and original unit layout, but adopt double deck and staggered floor structure, as well as other models such as large depth unit with ecological courtyard to make full use of spaces to form functional divisions naturally, furthermore, terrace and landscape outdoors are introduced into the rooms skillfully cooperating with French window, to mix up indoor and outdoor landscape and to create personalized unit space.

3.3 Unit Space Design Tends to be Delicate

a. Reasonable space allocation. Making the space of the unit allocated reasonably, and meeting the requirements of functions, it needs to decrease the interior route area or any areas without any defined functions. Each functional space is designed with enough size for household equipments, additional areas are set for assistance to meet any changes come with the improvement of living standards, such as an additional washroom, and a laundry house or sun room on the balcony.

b. Blear the functional zones and recombine and connect some zones. In conclusion, combination of master bedroom and washroom, master bedroom and walk in changing room, as well as master bedroom, washroom and changing room; connection of master bedroom and study room, living room and study room, kitchen and dining room; connection of kitchen and dining room, along with combination of operation and cooking spaces; multi-functional kitchen. It has various kinds of combination to connect each functional zone including decreasing number of fixed components, or adopting light weight components, light transmission materials and multi-functional furnishings.

c. Traditional components are developed to meet architectural aesthetics as well as demand of detailed functions. Traditional windows are developing to be bay-windows by lowering the windowsill and setting cabinets in the lower part, while seats, decorations or dressing table are all available in the upper part to increase interior space as well as view. Air conditioning board

因此：

1). 板式的单元组合得到普遍的欢迎，而点式（或塔式）即向短板式发展。

2). 街坊式布置也正在取代住宅小区的布局。

3). 居住郊区化发展速度明显加快。

2. 户型空间的设计更趋多元化

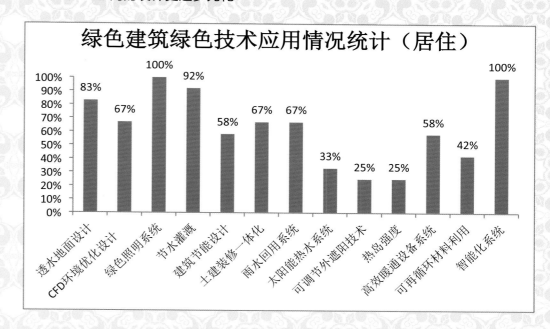

以多元化的户型来适应消费者日益增长的个性化住房需求，并能以灵活的户型结构适应因消费者家庭阶段性改变所导致的布局调整，使住房具有较长的使用期，是户型变化的趋势。在进行住宅外观和户型的设计时，应突破单一的户型空间格局，适当建设"楼中楼"、复式、错层结构，以及有生态天井的大进深户型等差异化的住宅模式，可以充分利用空间转换自然形成功能分区，同时巧妙地纳入户外阳台和整体绿化景观，辅以落地窗和露台等场域化的结构模式，使内外景观充分交融，创造足够个人化的户型空间氛围。

3. 户型空间的设计更趋精细化

1). 合理的面积分配。住宅设计中将总面积合理分配，在满足功能合理的前提下，减少没有明确用途部分的面积，减少户内交通面积，使各功能空间合理容纳各种家庭设施，将多余面积设置为住宅的辅助部分，满足人们生活水平提高带来的生活习惯的改变，如设置多个卫生间，满足分厕的要求，阳台上设置洗衣房、日光室等。

2). 模糊区域功能，将一些功能重新组合、连接、不做明确的界定。概括起来主要有：主卧室和卫生间的组合、主卧室和走入式更衣室的组合、主卧室和卫生间更衣室的组合；主卧房和书房的连接；客厅和书房的连接，客厅和餐厅的融合；厨房和餐厅连接、厨房的操作和烹饪空间的组合；复合型厨房的设计；卫生间、盥洗与洗浴的组合。其组合方法有，减少固定构件，或与轻质构件、透光材料、多用途家具分割连接不同的功能区域。

3). 住宅的传统构件发生演变，以满足建筑审美及人们对细部功能的要求。窗户由传统形式向凸窗和飘窗演变，窗台降低，下部设置储物柜，上部可以小坐、摆设饰物或调整为梳妆台，这样扩大内空间和景观视线。空调板作为立面构件不再独立存在，与窗台板、阳台板结合，在其表面设置金属多孔板将空调主机隐藏，净化建筑立面；出屋面楼梯间结合花架，将屋面作为公共活动场所。

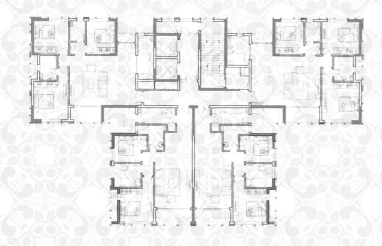

4). 室内布局注重实用。储藏室、步入式更衣室被普遍引进普通住宅；有的住宅已开始向立体分割方向发展，利用空间设计的不同高差隔出不同的功能区域，大大提高空间的利用率。

5). 横厅设计开始替代直厅。以往一般的住宅楼多为南北直厅布置。现在开始出现客厅和餐厅或书房均在南面的横厅设计。

6). 准一梯一户。清静、私密的一梯一户是居住者所向往的。如今设计已开始尽量向此目的靠拢。有的开发商在一梯二户的小高层中安装东、西两边都能开门的电梯，造成一梯一户的感觉。

7). 电梯大堂宾馆化。楼宇入口的电梯厅设计开始向酒店大堂靠拢，除了艺术挂画、吊灯、壁灯，还专设洽谈、休息区，让住户拥有酒店式的享受。

outside the unit as a facade component is combined with window board and balcony slab, being invisible by placing dotted mental plate beyond it to make the architecture facade clean and beautiful; the staircase that exceeds the facade is designed with pergola to become a public activity space.

d. Pay attention to practicability. Storerooms and walk in changing rooms are normally accepted by public. While some units are developing to divide the functional spaces vertically by making use of floor height difference in the unit, which have increased the use ratio of spaces.

e. Living room with larger width begins to take place of that with larger length. The residential buildings were designed with south-north living room with larger length, while it has become popular that living room and dining room or study room are both set in the south to form a larger width.

f. One unit per floor. A quiet and private unit type of one unit per floor is always what residences looking for. Nowadays, it has attracted more and more developers to get close to the pursuit, even elevators of some middle- and high-rise residential buildings that with two units per floor are designed to open at both east and west, to make residences feel like that they are living in a type of one unit per floor.

g. Lobby and elevator are designed close to the style of hotel. The design of elevator at the ground floor is getting close to the style of lobby of hotel, except artistic paintings, droplights and wall lamps, communication and resting areas are specially designed to provide the enjoyment of hotel for residences.

4. Residential Landscape Design Trend
With the improvement of landscape design level in the domestic real estate market, Chinese landscape design in the future will show five obvious trends.

4.1. Soft Landscape& Greening Configuration
All the time, residential real estate landscape has two different development directions. One is based on soft landscape greening, emphasizing on natural, quiet and warm effect of landscape; another is hard landscape, focusing on large squares, large axis and large structures. Real estate landscape is developed toward green configuration in the future trend, not only to increase the amount of green, but also to diversify green varieties and the layers. Late greening conservation has become increasingly important; plants grow better, house's effect will be better. Because the materials of hard structures will deform over time and aging, a lot of maintenance costs is required for renovations. So the design of hard structures should be minimized and be placed in essential position of main area.

4.2. Pleasing + Practical Value
To improve project quality, many developers spares no expense to purchase rare trees planted in the community from other parts of the country and even foreign countries. Rare species is the crowning touch to the houses, but still should not be planted too much. Monotonous tall trees lack of levels will reduce the landscape beauty. Community landscape should be created in accordance to the plant community, rather than blindly pile up. The appropriate design is an important condition for tree species selection. Surrounding buildings play a decisive role in trees planting, such as the classical style of real estate is suitable for banana trees, and modern building for geometric-shaped trees. Pleasing and practical tree species is the most valuable way of making landscape, while excessive pursuit of scarce varieties or foreign species is not healthy real estate landscape development trend.

4.3. Different Sceneries by Walking & Seasonal Changes
People tend to produce aesthetic exhaustion facing to landscape for months and years. Therefore, landscape planning has been developed from simple plant decoration to the appropriate design now. The landscape characteristics are mainly embodied in the plant, to create unlimited scenery in the limited space. Constantly changing landscape and different seasonal landscape could keep freshness and vitality for communities.

四、住宅景观设计趋势

随着国内房地产市场景观设计的水平正在不断提高,在我国景观设计的未来将呈现出五个明显的发展趋势。

1. 软质景观 + 绿化配置

一直以来,居住地产景观有两个不同的发展方向,一种是是以软质景观绿化为主,强调自然、静谧和温馨的景观效果;另一种以硬质景观为主,主推大广场、大轴线以及大型构筑物。地产景观在未来的趋势是向着以绿化配置为重点的方向发展,不仅绿量增大,而且绿化的品种、层次将会更加多样化。后期的绿化养护也越来越重要;植物长得越好,楼盘的效果也会越好。由于硬质构筑物的材料会随着时间推移而老化变形,需要投入大量成本进行翻修维护,因此硬质构筑物的设计量应尽量减少,主要放置于重点区域必不可少的位置。

2. 赏心悦目 + 实用价值

很多开发商为提升项目品质,不惜重金从外地甚至外国购来稀有成树,栽种在社区中。一个楼盘里植有珍稀树种固然有点睛之用,但不宜过多。千篇一律的高大树种与层次的缺乏会导致景观美感的缺失。景观社区景观打造应按照植物群落来布局,而非盲目堆砌。应景也是树种选择的重要条件。周围的建筑体对树木的栽种有很大的决定作用,如古典风格的楼盘适合芭蕉树,现代建筑体则适合几何形状的乔木等。让人赏心悦目又能给业主创造实用价值的树种才是最可

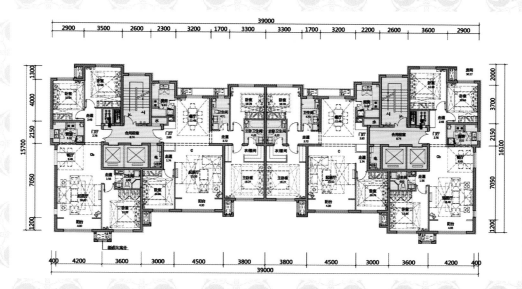

贵的造景方式,过分追求稀缺品种或者洋品种并不是健康的楼盘景观发展趋势。

3. 移步异景 + 季节变化

对于长年累月所面对的景观,人容易产生审美疲劳。因此,景观规划已经从简单的植物铺陈发展到现在的应景设计。这种景观特色主要体现在植物上,在有限的空间营造出无限的风光,而不断变换的景观、各季不同的景观能让社区保持更多的新鲜感与活力。

应景能带给业主更多的审美情趣。在小区里闲庭信步,不会感觉乏味,每走到一个地方可以通过不同角度领略小区的景色,这就是移步异景。在植物的栽种上,现在的四季植物代替了过去的常绿植物,让一年四季的景观在景观社区里得到更好呈现,如春天的绿萝、夏天的香樟、秋天的银杏、冬天的落叶松。不同阶段都有美的表现手法,让业主感受到季节的自然变化。

4. 植物布局 + 水景点缀

曾经在相当长的一段时间内,水景被看作是最能提升楼盘品质的景观设计元素。但是由于大型水景存在耗水成本大、渗漏损失大、水质难以持续保障以及运行安全等诸多问题,越来越多的消费者在选择水景大盘时更为理性,不再把水景的多少作为判断景观好坏的标准。同时,开发商也意识到水景设计之初缺乏完备的水量保障、水质保障方案是引发诸多问题的根源。因此,近几年的项目中以大型水景为主题的景观设计越来越少,水景现在多以点缀形式出现。这不仅是适应消费者和市场的需求,而且响应了国家所提倡的"节水、节能、节地"、建设环境友好型和节约型社会的可持续发展战略。"植物布局 + 水晶点缀"并举,则成为现代楼盘造景的新风尚。

Appropriate design can bring owners more aesthetic appeal. Strolling in the community with not feeling boring, each step brings different views through different angles. On planting, seasons plants take place of former evergreen plants to create better landscape in the community throughout the year, such as scindapsus aureus in spring, camphor in summer, ginkgo in autumn, and larch in winter. Different expression techniques of beauty in different stages let owners experience natural changes in the season.

4.4. The Plant Layout & Waterscape Embellishment
For quite a long time, waterscape is considered to be the best landscape design elements that can improve the quality of house. However, due to the issues of large water feature, such as large water consumption costs, large leakage losses, water quality protection problem and operational safety, more and more consumers in the choice of water features are more rational and no longer judge landscape quality standards by the number of water features. At the same time, developers are beginning to realize that the lack of completed water volume security and water quality protection program at the beginning of waterscape design is at the root of many problems caused. Therefore, in recent years, large-scale landscape projects with the theme of waterscape design become less and less, but mainly play a role of embellishment. This is not only to adapt to the needs of consumers and the market, but also responses to the country promotion of "saving water, saving energy, and saving land" and the sustainable development strategy of constructing environment-friendly and conservation-oriented society. "Plant layout + crystal embellishment" simultaneously has become a new trend in modern real estate landscape.

4.5. Regress to the Native & Regional Features
With the development of history and culture, developers and designers in the landscape design no longer mechanically copy western-style landscape mode, and advocate "regress to the native". Nowadays landscape begins to be planned and constructed in the context of Chinese culture. These landscape features according to the regional characteristics and natural features to show the continuity of history and culture and regional characteristics. "Blue sky, white walls and red tile" in Qingdao embody the coastal features; "Coconut tree wind sea rhyme" in Haikou is the Cantonese custom; "Small bridges" in Suzhou is the landscape in Jiangnan charm. These cases return to the local landscape in order to develop pure Chinese style works with a sense of history and geographical characteristics. As the original landscape's homogeneity phenomenon will be broken, "regress to the native and regional features" trend will be increasingly obvious.

5. Residential Facilities' Design Trend
When the real estate industry goes from its Golden Age into the Silver Age, the developers begin to look for new profitable possibilities. Vanke first proposed the concept of "service provider of urban facilities" to still focus on real estate industry but to find more opportunities related to urban facilities. Therefore, the previously sales-oriented business has become more diversified in addition with operation, management and service. This integrated business model becomes more and more popular, driving the new development of China's real estate market.

Responding to the improvement of urban facilities, the residential facilities should also be upgraded. Therefore, in China's future residential design, special attention will be paid to improving the residential facilities. To build a highly comfortable living environment, it should not only focus on the organization and decoration of the interior spaces, but also fully consider the the lighting environment, acoustic environment, thermal environment, air quality as well as the supporting facilities around the residence. Specifically, there are mainly five factors that are considerably important:

5.1 Sunlight and Natural Light
The sunlight and natural light are highly valuable and have a great influence on people's physiological and psychological states. Maximizing the natural light and making full use of it is very important for a residence. Thus the future residential design will be based on the sunlight analysis and fully use the CAD devices to get the best lighting effect within the specific site. The nature lighting effect is also influenced by the location and structure of the windows as well as the building materials, which will certainly encourage the innovation in design.

5. 回归本土 + 地域特色

随着历史文化的不断升温，开发商和设计师在景观设计上不再机械地搬用西方式园林的模式，主张"回归本土"。时下的景观开始在中华文化的大背景下进行规划和建设。这些景观有根据各地方区域特征和基地的自然特色来表现历史文化的延续性与地域特色。如青岛的"碧水蓝天白墙红瓦"体现了滨海特色；海口的"椰风海韵"则是一派南国风情；苏州的"小桥流水"式景观则是江南水乡的韵致。这些景观案例无一不是回归本土中进行深入挖掘，从而开发出兼备历史感和地域特色的纯正中国风格作品。而随着原有的景观同质化现象将被打破，"回归本土 + 地域特色"的趋势将日益明显。

五、住宅配套趋势

随着房地产行业由黄金时代进入白银时代，地产开发商都寻找新的利润点。万科地产带头提出"城市配套服务商"的概念，即仍专注于房地产行业，但要围绕城市配套寻找相关业务的发展机会，从单纯向销售要业绩转变为向销售、经营、管理和服务要业绩。这种城市综合性运营的趋势越发明显，中国正迎来一场新的地产变革。

与这种"城市配套服务"相对应的是住宅内部配套的升级。我国未来住宅的设计，将重点转向提高住宅设备配置方面。要获得一个高舒适度的居住环境，不仅要注重套型内部平面空间关系的组合和硬件设施的改善，还要全面考虑住宅的光环境、声环境、热环境和空气质量环境的综合条件及其设备的配置。主要包括以下五个方面的内容。

1. 日照及自然光

日照及自然光的生理卫生价值极高，对人的生理、心理状态影响较大。在住宅中，最大限度地利用并合理开发自然光资源有着重要意义。未来住宅的设计将充分利用日影分析原理和计算机辅助设备，来改善日照和用地之间的矛盾，有效地发挥土地的综合利用效率。自然采光还涉及窗口布置、窗的构造及所使用的材料等方面，这些也必将引起设计上的创新。

2. 隔声问题

隔声在我国的住宅设计中始终是一个薄弱环节。轻质材料在未来住宅中的广泛使用，又会加重隔声的难度。隔声技术包括空气隔声和固体隔声两部分。住宅内，人们可忍受噪音约 40～45 分贝。为达到这一指标，必须增加门窗的密闭性，并改善墙体构造。多年来，由于对住宅楼地面的固体传声问题重视不够，住宅内隔声效果较差。在未来的住宅中，加强对楼板的隔声叠层构造和面层处理，将会受到更多的关注。

3. 热环境

热环境是直接影响居住舒适度的重要因素。对采暖和空调而言，又涉及节能、造价、维修、管理等诸多方面。目前，对各种采暖方式的研究和采暖系统的开发各具特色，可适用于不同的使用条件。我们提倡相对集中的采暖方式，但应当改为双管系统、分户计量的方式进行。这种

5.2 Sound Insulation

In China, sound insulation is always a "week link" in residential design. The widely application of lightweight materials in future residences will make sound insulation more difficult. The acoustic insulation techniques include airborne sound insulation and insulation against solid-borne sound. Generally, people can live under a sound pressure of about 40~45 dB at home. So, to meet the requirements, the doors and windows must be more soundproof and the wall structure should be improved. For many years, no enough attention has been paid to the insulation of solid-borne sound, which resulted in poor sound insulation in the residences. Therefore, for the future residences, importance will be attached to the sound insulation design of the building floors.

5.3 Thermal Environment

The thermal environment will influence the comfort level directly. Heating and air conditioning usually touch upon many problems such as energy conservation, construction cost, maintenance and management. Now heating systems are varied and distinctive to meet different requirements. What we advocate is the central heating system which can be used flexibly for a whole residential group, or for an individual residential building, or just for several residential units.

In recent years, advanced European energy-saving technologies have been constantly introduced. With more than 20 years of research and practice, the Europe has explored many effective ways in residential energy saving, especially in technology for the residences of low energy consumption and high comfort level. This kind of technology has applied low-power heating and cooling devices to adjust the indoor temperature and save the construction cost. It is expected to be promoted and widely applied in China's future residential buildings.

5.4 Emission of Harmful Gases

The emission of harmful gases from house is one of the biggest concerns for the residents. However, the emission of exhaust gases from kitchen and bathroom is less than satisfactory. The vertical air-exhausting pipe for high-rise residences is almost useless, and the failure of gas emission will cause many problems. Actually, the exhaust system and the sewerage system are interconnected and will have a influence on each other.

The future ventilation system for the residences will improve the technologies for air exhausting and use new discharging devices. The horizontal linear smoke piping is highly effective which is widely used in many countries and regions, and in the future it is expected to be upgraded and widely applied in China.

To ensure timely replacement of fresh air in winter or summer, in the future residences, new type of heating and cooling system as well as air exchanger will be used and receive more and more attention.

5.5 Arrangement of Piping

The unsatisfactory piping arrangement has always restricted the improvement of the housing quality. Under the planned economic system, the pipes for water, heat, electricity and gas are fiercely independent and are arranged disorderedly in the houses, which greatly affects the integration of the kitchen and bathroom. Therefore, the future residential design will solve these problems and arrange these pipes in order.

Varied pipes running through the floor is the main cause of leakage. It is also a controversial problem in commodity housing. To solve this problem, these pipelines will be laid horizontally between the under layer and the surface of the floor, or be installed straightly in the pipe walls.

To establish a concentrated tube well is a mature method in modern residential design. The tube well will help to clean the indoor air and reduce the interference from the pressure piping. It is also easy to install, manage and repair the relative devices (In Japan, the tub well is usually installed with varied meters and water heaters). Though the well tube will occupy some spaces, it will greatly improve the indoor environment.

相对集中采暖方式的热源供应，既可以组团为单位，也可以单幢楼或居住单元为单位来组织。

近年来，欧洲先进的节能技术不断地被引进。经历了20多年的大量研究和实践，欧洲在住宅节能方面探索出了很有成效的方法。其中，以低能耗、高舒适度的住宅节能技术尤为突出。该项技术通过使用功率非常低的辅助采暖和制冷设备，既可调整室内温度，保持舒适的状态，又节约了住宅的综合造价。此项技术在我国未来的住宅建设中，有望得到应用和推广。

4. 有害气体的排除

住宅室内污气及有害气体的排除，是居住者最为关心的问题之一。但迄今为止，在有效排除厨房、卫生间的污气和有害气体方面仍不尽人意，高层住宅竖向烟风道形同虚设。串烟、串气、串声现象十分严重，且直排热水器屡屡出现事故。居住套内排气、排污装置实际上是一个大系统，尽管装置很好，但由于排风管道或烟道不畅，其设备与设施同样可能达不到功能目标。

未来住宅的通风系统设计，将重点解决烟风道技术问题，研究开发新型专用的烟风道系统和接口配件，形成完整的竖向排烟气的成套技术产品。水平直连排出烟气的作法，具有简便、直接、高效的特点，在很多国家和地区得到普遍采用。预计此项技术完善后，将在我国未来的住宅建设中广泛应用。

为了保证采暖和空调房能够及时补充和更换新鲜空气，预计未来住宅中，将采用补新风式冷热交换空气补充装置。此项技术也会在未来的住宅设计中越来越受到重视。

5. 管道的布置

住宅中各种管道布置的不尽人意，一直制约着我国住宅品质的提高。在计划经济体制下，水、暖、电、气各自为政，造成住宅内各种厨卫管道无序空行。在商品住宅中，这种影响也始终没有消除，各种管道尽管经过精心包藏，仍问题丛生，极大地影响了我国住宅厨卫整体功能品质的提高，影响了厨卫整体化、集成化技术的进步。未来住宅设计将以"自家管道不到邻居家去"和"压力管道出户"为原则。

各种管道穿楼板是造成住宅跑、冒、滴、漏的主要根源，在商品住宅中也是一种产权不清的表现，这种现象必须改变。解决的办法是，在下沉楼面和楼面垫层中铺设水、暖气各种水平管道。解决水平管道铺设的办法是设置管道墙，将所有的设备沿着管道墙进行布置，并将各种水平管道放置在管道墙内。

建设集中管井是现代住宅设计中的一种成熟的做法。为了净化室内空间，减少由于压力管道给住户带来的干扰，明晰产权，属公共使用的设备管道应当设置在套型以外，以便于设备的安装、维修、抄表及设置其他一些家用设备（在日本，管井内设置各种表具及热水装置）。管井可以放大尺寸，扩大进深。尽管面积加大了，却将极大地改善住宅内部的空间环境。

China's Most Beautiful Apartments
中国最美楼盘

Aesthetics	美学性
Livability	宜居性
Commercial Value	商业价值性
Humanity	人文性
Sustainability	可持续性

Traditional Cultural Context, Modern Architectural Style
传统文化深厚，现代建筑风格

Shanghai Villa Clubhouse   上海院子

Location/ 项目地点：Jiading, Shanghai China/ 中国上海市嘉定区
Developer/ 开发单位：RK Properties/ 路劲地产集团有限公司
Architectural Design/ 建筑单位：Lacime Architectural Design Co., Ltd. / 上海日清建筑设计有限公司
Design Team/ 设计团队：Song Zhaoqing, Pang Weihua, Zhao Xubin/ 宋照青 庞伟华 赵旭彬
Floor Area/ 建筑面积：179 063 m²
Clubhouse Floor Area/ 会所建筑面积：560 m²
Photographer/ 摄影：Xisan Image/ 禧山映像

Keywords   关键词
Coordinating Proportion; Artistic Beauty; Different Sceneries by Walking
比例协调　艺术美感　步移景异

"China's Most Beautiful Apartments" Award
荣获"中国最美楼盘"奖

The dignified and elegant architectural modelling emphasizes the overall proportion and detailed scale, inherits classical style and shows modern breath. Return of "Prairie Style" pursues the combination of beauty of technology and human interest, so that residents can find back tranquility and nature.

建筑造型厚重、典雅，强调整体的比例与细部的尺度，在继承古典风格的同时又赋予现代气息。"草原风"的回归重新追寻技术美与人情味的统一，使居住者情感回归于宁静与自然。

Overview   项目概况

Located in Waigang Town, Jiading District of Shanghai City, the development is to the south of Jinghe River, to the west of Bai'an Road and to the east of the city park, enjoying convenient transportation and beautiful environment.

　　项目位于上海市嘉定区外冈镇，北临泾河，东临百安公路，西临城市公园。交通便利，环境优美。

Project Positioning 项目定位

Project design, under the basis of national laws and regulations, uses an appropriate, humane design concept to create an unique, warm and natural urban living community. Strong cultural atmosphere in architectural style and environmental design meets the high-grade housing needs.

项目设计在贯彻国家相关法律法规前提下,以一个贴切的、人性化的设计理念营造出一个别具一格的温馨、自然的城市生活社区,在建筑风格及环境设计方面营造浓厚的人文气氛,满足高档次高品位的居住需求。

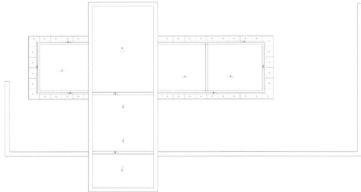

Roof plan 屋顶平面图

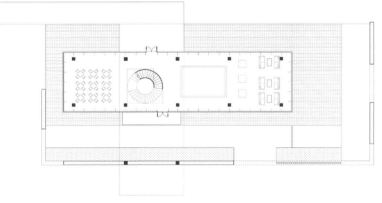

First Floor Plan 一层平面图

Planning Concept 规划理念

"A lake per kilometer, a forest per hectometer" features Jiading Newtown's landscape ecology. The program reflects the respect and attention to urban environment in many aspects, such as overall layout, facade forms and landscape design, and actively integrates this plot into the overall atmosphere of Waigang district by reasonable layout strategies and unique form languages and into an organic part of the urban form.

嘉定新城有着"千米一湖,百米一林"的景观生态特征,本方案在总体布局、立面形式、景观设计等诸多方面体现出对城市环境的尊重和关注,并通过合理的布局策略和独特形式语言,使这一区块积极地融入外冈区域的整体氛围中,成为城市形态的一个有机组成部分。

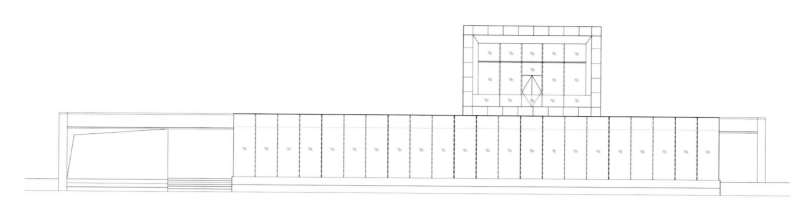

North Elevation 北立面图

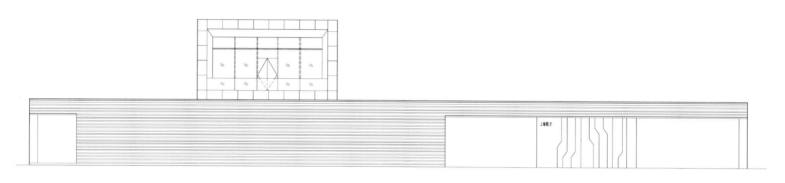

South Elevation 南立面图

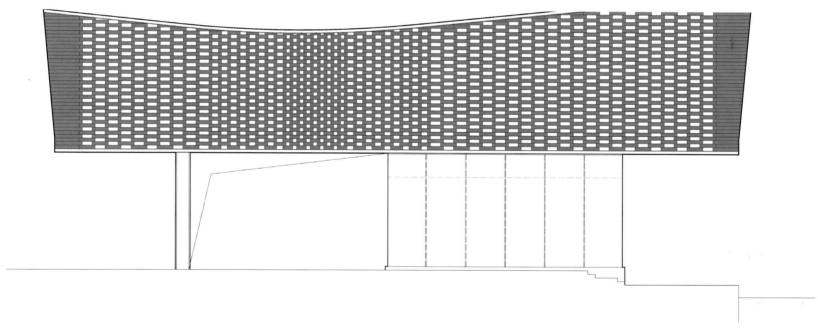

End Elevation 侧立面图

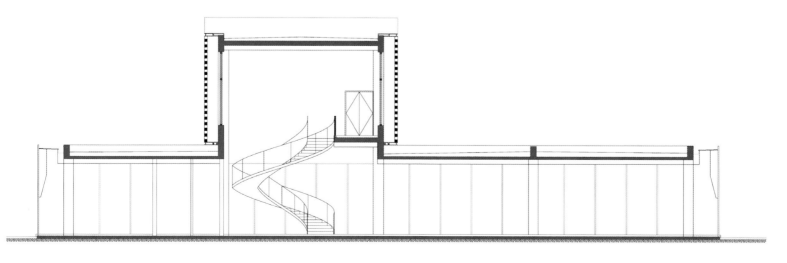

Sectional Drawing 剖面图

041

Landscape Design 景观设计

Two rivers run through each other, which is the main clue of the whole system of residential landscape. In addition, landscape corridor, sight-viewing walkway, riverside landscape roads along the Wutang River and Miaojing River constitute the backbone and ridge of the whole large landscape system. Along this clue, a meaningful and beautiful picture scroll slowly expanded. Terrain waterscape outlines and covers rich layers of trees and water. Twists and turns of roads and open space make everywhere scenery and form unique visual feeling of different sceneries caused by walking.

两条河流相互贯通，是整个小区景观体系的主要线索。另外，景观长廊和步行景观道，沿吴塘河和庙泾河的滨河景观道，共同构成整个大景观体系的骨干和脊梁，沿着这个线索，一幅隽美的画卷缓缓展开，地形水景的勾衬掩映展示出绿水相依的丰富层次，道路的曲折变化和空间的大开大阖带来处处皆景，步移景异的独特视觉感受。

Clubhouse Design 会所设计

Buildings adopt the modern architectural style, use method of volume and block inserted each other and form strong visual impact by virtual-actual comparison. The waterscape around main building and dark-colored walls brings a moment of tranquility, which interprets courtyard culture.

The interpretation of the language of building materials: ground floor adopts curtain walls purely made by transparent glasses, which reflect on the water surface to increase the feeling of serenity. Second storey adopts hollow brick walls, whose heavy facade form takes on dignified sense and also brings fantastic lighting effects. The two forms bring the building to life by strong virtual-actual contrast.

建筑采用现代建筑风格，并运用体块穿插手法，虚实对比处理从而达到强烈的视觉冲击力。主体周边的水景及深色院墙则给建筑带来片刻宁静，院子文化在这里得到诠释。

建筑材料语言的的诠释：建筑首层采用纯玻璃肋幕墙形式，通透整洁的墙面映在水面上，增加了宁静的感觉，二层采用镂空砖墙的形式，厚重的立面形式给建筑带来稳重感，也给室内带来奇幻的光影变化效果。二者的结合，强烈的虚实关系对比则给建筑带来了活力。

Ideal City with Varied Urban Skyline
城市天际线变化丰富的理想城

"Kangxin Jiayuan" Orientation Placement Housing, Zhongyi Village, Beijing
北京张仪村"康馨家园"定向安置房

Location/ 项目地点：Xuanwu, Beijing, China/ 中国北京市宣武区
Architectural Design/ 设计单位：WSP ARCHITECTS / 维思平建筑设计
Chief Designer/ 主设计师：Chenling/ 陈凌
Design Team/ 设计团队：Sui Lubo, Sun Ying, Chenzhen/ 隋鲁波 孙莹 陈真
Total Land Area/ 用地面积：163 000 m²
Construction Land Area/ 建设用地面积：123 000 m²
Plot Ratio/ 容积率：3.4
Green Coverage Rate/ 绿化率：30%
Type/ 项目类型：Residence/ 住宅
Status/ 项目状态：Completion/ 已建成

Keywords 关键词
Livable House; Ecological Technology; Private World
宜居住宅 生态科技 私属天地

"China's Most Beautiful Apartments" Award
荣获"中国最美楼盘"奖

Monomer building design adopts linear rowlayout from south to north to maximize the use of limited land and meet the need of relieving large population density of placement housing. High-rise and middle-rise buildings are arranged alternatively to enliven spatial form of community and enrich the changes of urban skyline.

设计单体采用南北向的行列式线性布局，最大限度利用有限土地，满足舒缓安置用房人口密度较大的需求，高层、中层穿插排布，活跃小区空间形态，丰富城市天际线的变化。

Overview 项目概况

As the first self-planning and self-constructing residential community in Xuanwu District, the project is located out of the West 4th Ring and within the West 5th Ring, 2 km to Beijing-Shijiazhuang Expressway and 1 km to Zhangyicun Station of the planned Metro Line 11. Occupying a total land area of 163,000 m², of which the land for construction is about 123,000 m², it is expected to provide 5,000 units for 14,000 residents. The government of this district has invited the famous design studio to improve the living quality and ensure a better living condition, showing great care to people's livelihood.

北京张仪村保障房项目是宣武区第一个自主规划、自主建设的住宅小区。地块位于北京市丰台区西四环外，西五环内，京石高速公路以北2 km，规划地铁11号张仪村站以北1 km。项目总用地面积163 000 m²，其中建设用地面积123 000 m²，预计安置居民5 000户，安置人口1.4万人。区政府为提升住宅质量，保障居民更为舒适的生活条件，力邀著名设计事务所主持设计，体现了对民生的深切关注。

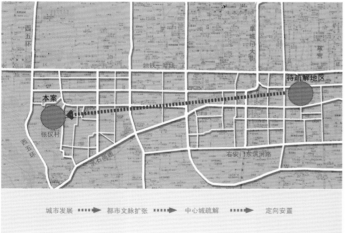

location analysis diagram　区位分析图

Site Plan　总平面图

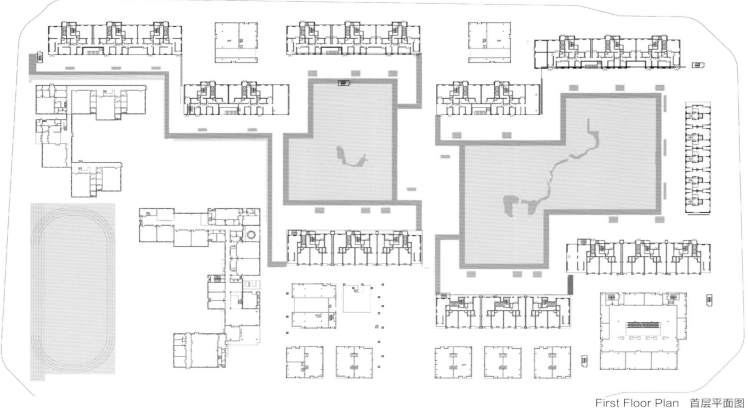

First Floor Plan 首层平面图

046

Design Challenge and Design Objective 设计难题与设计目标

The design faces many challenges, namely, the transferring of the urban functions, the low-income or aging residents, and the lacking of supporting facilities and public landscape resources. To solve these problems, the designers have built an ideal city that is convenient, high-efficient and full of vitality to offer job opportunities and encourage neighborhood communication; that is green and makes people feel relaxed; that has realized the sustainable development of the traditional culture; that is low-carbon and easy to manage.

当设计师面对原有居民商业等城市机能被疏解到城市外围、特征群体大多是低收入缺少创业手段并且老龄化严重、配套设施和公共景观资源缺乏的问题，将如何解决呢？他们建设了一个理想城。一个充满都市活力、方便、高效、提供就业、促进邻里和谐的理想城；一个最大限度绿色、令人放松精神的理想城；一个传统城市文化得以持续发展的理想城；一个低技低碳的易于管理运营的理想城。

Design Concept 设计理念

The design concept of "green, science and technology, humanities, efficiency and ecology" in this project creates iconic, demonstrative and high-quality orientation placement housing. Public buildings area enclosing completely the site forms two "urban power rings" which open to the public around the clock. The site is a pure central park, belongs to private land, combines with flowing lines and viewing points and lines and brings into oddments, water and tall trees to keep human, environment and architectures in harmony.

项目以"绿色、科技、人文、高效、生态"的设计理念，打造标志性、示范性、高品质的外迁安置用房。完整围绕用地的公建带，形成两个"都市动力环"，全天候向所有公众开放。被"都市环"围合的是没有地下建筑的纯中央花园，是属于居民的私属领地，与人流线、视点视线结合，引入小品、水体与高大乔木，使人与环境、建筑和谐共生。

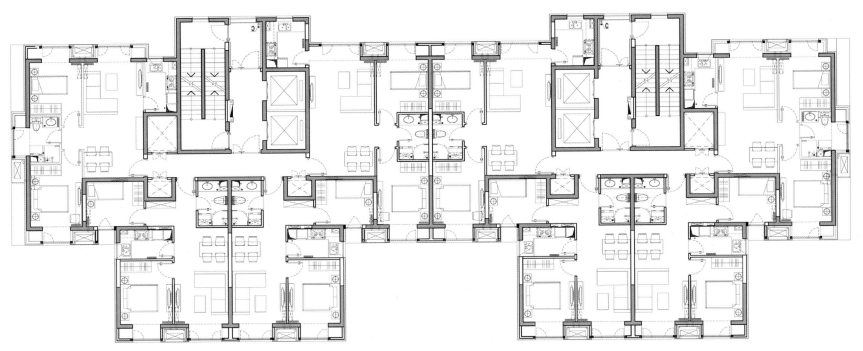

Plan for Standard Floor　标准层平面图

Architectural Design 建筑设计

Monomer building design adopts linear rowlayout from south to north to maximize the use of limited land and meet the need of relieving large population density of placement housing. High-rise and middle-rise buildings are arranged alternatively to enliven spatial form of community and enrich the changes of urban skyline.

Mixing function and high volume bring about job opportunities and harmonious community development; high-rise residences have the best sunlight, ventilation and lighting condition; straight architectural form reduces itself shape coefficient meanwhile obtains the largest greening space; "urban ring" looks like grilles in quadrangle courtyard from the air and frames the artificial and natural landscapes in central park. Be personally on the scene, the "urban ring" turns out to be the old alley's walls from which the branches extend out like castles in the air, thus each courtyard is the one and only.

设计单体采用南北向的行列式线性布局，最大限度利用有限土地，满足纾解安置用房人口密度较大的需求，高层、中层穿插排布，活跃小区空间形态，丰富城市天际线的变化。

功能混合和高容积带来了就业机会和社区的和谐发展；居住在空中的高层住宅有着最好的日照、通风、采光条件；平直的建筑形态在降低自身体型系数的同时获得了最大的绿化空间；"都市环"从空中看像四合院的窝花格，框住的是中央花园的风景：人工与自然在这里对话。身临其境时，都市环变身老胡同的院墙，高高在上的空中楼阁是伸出院墙的大树，每个院落都独一无二。

Meiju Encyclopaedia 美居小百科

The Zhangyicun project won the Beijing Residential Planning Award in 2011, which epitomizes the development concept of "internationalization, modernization and refinement" in Xuanwu District. Its implementation will help to ease the pressure on old city of Xuanwu District and be of profound significance to the transformation of the old city.

张仪村保障房项目于2011年荣获北京市住宅规划奖，是宣武区"国际化、现代化、精品化"发展理念的集中体现，它的实施将有利于缓解宣武区旧城压力，对旧城改造就有深远意义。

Modern Buildings Full of Artistic Beauty
充满艺术美感的现代建筑

Mangrove Courtyard　红树别院

Location/ 项目地点：Shenzhen, Guangdong, China / 中国广东省深圳市
Architectural Design/ 建筑设计：Shenzhen CUBE Architecture Design Consulting Co., Ltd. / 深圳市立方建筑设计顾问有限公司　深圳市库博建筑设计事务所有限公司
Land Area/ 用地面积：10 384.2 m²
Plot Ratio/ 容积率：3.0
Status/ 项目状态：Completed/ 已建成

Keywords　关键词

Simple and Lively; Artistic Beauty; Exquisite Material
简洁明快　艺术美感　材料考究

 "China's Most Beautiful Apartments" Award
荣获"中国最美楼盘"奖

Architectural design follows the principles of classical beauty of form. The single building is overall symmetric without obvious three-piece type whose symbolic meaning is achieved by setting up aluminum alloy canopy at the top and dry-hanging fossil beige (marble) on the base.

建筑设计遵循经典形式美原则，单体呈整体对称，虽无明显的三段式，但分别通过设置顶部铝合金雨棚及基座干挂化石米黄（大理石）来达到三段式的象征意义。

Overview　项目概况

The project is located in the core area of Huarun Dachong Reconstruction Center, adjacent to the MIXC on the south and overlooking Shahe Golf Course on the east. Designed to be surrounded by the planned Nanjing Foreign Language School on the east, south and west, it is far away from the hustle and bustle of the city to have birds singing everywhere.

项目位于华润大冲旧改核心区域，南侧与华润万象新城咫尺之隔，东侧可远观沙河高尔夫，近被规划待建的南外学校东南西三面相围，远离城市喧嚣，处处可闻鸟啼声。

Site Plan 总平面图

Planning Concept 规划理念

The overall planning maximizes the use of peripheral landscape resources, while fully exploits the internal geographical environmental resources, and strives for each house having two-way visual experience. Thus designers put three high-rises of 1#, 2#, 3# outside, which are arranged to face south or southeast to have a great view of the outdoor landscapes. 11-layer 4# building stands alone in the corner of northwest. Not to appear closed, courtyard landscape is cleverly maximized to create a different kind of quiet and elegant courtyard

整体规划最大化借外围景观资源，同时要充分挖掘内部地理环境资源，力求每栋住宅都有双向的视觉体验。因此设计师将 1#2#3# 三栋高层布置在外侧，呈南向或东南向布置，充分借助外景，11 层的 4# 独自设置于西北角，让庭院不至于显得过于闭塞，巧妙地营造出最大化庭院景观，为创造静谧雅致的别样庭院埋下伏笔。

Architectural Design 建筑设计

Architectural design follows the principles of classical beauty of form. The single building is overall symmetric without obvious three-piece type whose symbolic meaning is achieved by set up aluminum alloy canopy at the top and dry-hanging fossil beige (marble) on the base. Simple and lively facade, modern horizontal balconies and partial white balconies form a sense of rhythm, combined with glass curtain wall of dry hanging ceramic plate to strengthen the wholeness of building and the erective sense; neutral colors and a large area of beige, gray are used to weaken volume, dotted with dark rust coppery ceramic plate.

建筑设计遵循一些经典形式美原则，单体呈整体对称，虽无明显的三段式，但分别通过设置顶部铝合金雨棚及基座干挂化石米黄（大理石）来达到象征意义的三段式。立面简洁明快，极具现代感的横向阳台、局部跳跃一些白色阳台，形成韵律感，与干挂陶板的玻璃幕墙穿插组合，增强建筑的整体与竖向挺拔感；中性的色调，大面积为米白色，弱化体量设置灰色，个别点缀了暗红偏锈铜色陶板。

Architectural Detail Design and Material Selection 建筑细部设计与材料选择

To reflect the exquisite and delicate modern industry, detail design uses a lot of aluminum alloy, glass, ceramic plate, face brick and stone. Different with dark aluminum alloy used in the past, this design chooses relatively shallow lead gray with strong metallic feeling which matches with transparent glass to achieve fresh and high-end visual experience; ceramic plate choses two colors of dark red like rust copper color and lead gray with unified module size of 300 × 900; large face brick uses gray and beige with size of 600 × 900; fossil beige marble is close to large beige brick.

The balconies of the tower buildings are designed with dry-hanging ceilings made of aluminum alloy. The delicate lines of balcony uniform with glass walls curtains. At the same time railings adopt very modern stainless steel components and transparent glass. The entire balcony becomes grand and magnificent, experiences the unparalleled sophistication, and shows the delicacy and luxury of modern technology to the limit. The main wall uses face brick similar to the texture of stone, side wall enhances the erective sense of building by vertical staggered joints. Partial corner position even adopts integrated L-shaped face brick to ensure the quality of construction.

为体现现代工业的精致与细腻，细部设计大量采用铝合金、玻璃、陶板、面砖、石材。同以往采用深色铝合金不同，该设计选择了比较浅的铅灰色，金属感极强，且与高透玻璃搭配达到比较清爽、高档的视觉感受；陶板选择了两种颜色，一种暗红偏锈铜板色，一种偏铅灰色，都统一模数300×900尺寸；大面积的面砖采用灰色与米白，划分尺寸控制在600×900；化石米黄的大理石与大面积的米白色面砖接近。

建筑塔楼通过设置铝合金吊顶干挂的阳台，阳台线条精致细腻，与玻璃幕墙线条整齐划一，同时栏杆采用了极现代的不锈钢构件与透明玻璃，整个阳台给人大气磅礴的同时，又能体验到无与伦比的精致，把现代技术的精美与奢华发挥到极致。主墙面采用类石材质感的面砖，通过分缝处理达到与石材浑然一体的感觉，侧面山墙通过竖向错缝处理，加强建筑的挺拔感，在局部转角位置甚至采用一体成型的L面砖，更加保证建造的品质。

Landscape Planning Concept 园林规划理念

Garden design introduces the hotel resort concept of Southeast Asia, at the same time learns from the space techniques of Chinese classical garden, to create infinite pool dominated landscape and multi-level, exquisite small landscapes which are interconnected. Vegetation complements the landscape pieces to create rich space and the visual experience like large space for small garden.

园林设计引入东南亚酒店式度假概念，同时汲取中国古典园林以小见大、欲扬先抑的空间手法，创造出以无边际泳池为主的大景观，多层次、异样精致的小景观，相互贯通，又通过植被与景观小品互映互掩，把小的花园营造出丰富的空间，从而把小庭院做出如同大空间的视觉体验。

Foreign-Style Hardbound House of Upgraded British Architectural Style
英式建筑风格的升级版精装洋房

CIFI Arthur Shire　　**旭辉亚瑟郡**

Location/ 项目地点：Pudong, Shanghai, China/ 中国上海市浦东区
Architectural/ Landscape Design/ 建筑 / 景观设计：W&R Group / 水石国际
Land Area/ 用地面积：30 208.3 m²
Floor Area/ 建筑面积：43 711.73 m²
Plot ratio/ 容积率：1.19

Keywords　关键词

Adjusting Measures to Local Conditions; Unique Taste; Natural Garden
因地制宜　品味独特　自然花园

"China's Most Beautiful Apartments" Award
荣获"中国最美楼盘"奖

With full consideration of the banded terrain, the whole building adopts double extension type, trying to build the British-style architecture. The plan also takes account of household privacy, introducing the concept of private space into building interval, terrace design, garden design and other aspects and giving full consideration to the feelings of residents.

项目充分考虑带状地形，建筑整体采用双排延展式，极力打造英伦风格建筑。规划还考虑了住户私密性问题、建筑间距、露台设计、庭院设计等方面都引入私密空间的概念，充分考虑住户感受。

Overview　项目概况

Construction and development of the project are based on the experience of building high- quality residential communities for years, the designer constantly makes innovation and improvement, and strives to create the perfect combination of architectural aesthetics and quality of life. CIFI Arthur Shire adopts the British-style architecture, which as the symbols for noble life is not only pure architectural form, but also blends into more spiritual things. These buildings are therefore showing strong adaptability and excellent integration, and producing spiritual communication and emotional sympathy on the basis of meeting oriental residence.

　　项目建设开发基于多年来打造高品质住宅社区的经验，并不断做出创新与改善，力求打造出建筑美学与生活品质完美结合的作品。旭辉亚瑟郡项目采用了英式风格，英式建筑作为一种尊贵生活的代表符号，代表的已不仅仅是纯建筑形态，而是融入了更多精神层面的东西。这些建筑也因此显示了极强适应性和极好融合性，在满足东方人的居住基础上，更产生了精神上的沟通和情感上的共鸣。

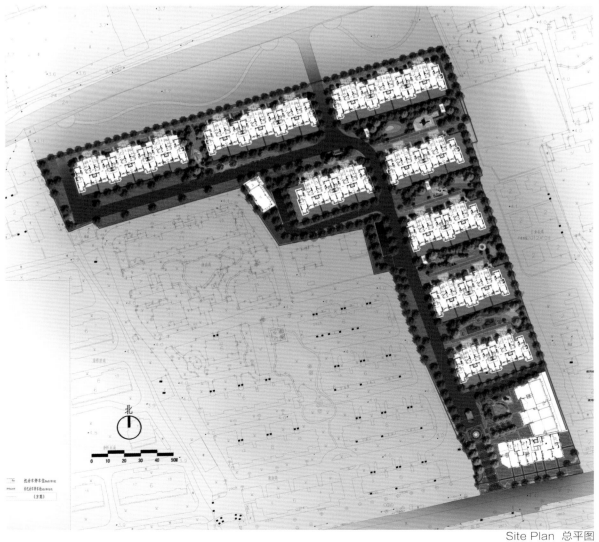

Site Plan 总平图

Project Positioning 项目定位

The project aims to create high-end Chinese community, creating and leading advanced lifestyle and concept. Meanwhile as the benchmarking project of CIFI Group in Kangqiao District of Shanghai, it is composed of nine 8-storey British-style buildings to offer upgraded and well-decorated garden houses. This project consists of most first-time houses and first-time upgrades, hardbound houses, part of affordable houses above the commercial area.

项目意在打造中国高端社区，创造和引导超前的生活方式和理念。同时，作为旭辉集团在上海康桥地区的标杆性项目，主要以九栋八层的英伦建筑主打升级版精装洋房，本项目大多以首置首改为主，精装房，有局部经济适用房，位于商业上方。

Architectural Design 建筑设计

British architectural style is introduced into Shanghai that is deeply influenced by western culture. Thus western elements are well retained and integrated with charming Old Shanghai, created carefully by modern construction methods to present vivid and unique British style for new generation of middle-class white-collar workers. Walls adopt cultural bricks which have brown stone culture for a century. The cultivation is contained in every brick, stone and ironwork. There are solemn gable roofs with two slopes on each side. Architectures in garden and mountain flowers in roofs serve to adorn life. Exposed wood frames and dormer windows feature British architectures. Hand-made iron railings and manual window parquet patterns are permeated with natural atmosphere.

在上海这块深受西方文化浸润的土壤上，引入选用英式建筑风格，在保留浓郁的西方风情元素基础上融入老上海的神韵，并以现代建造手段精心打造，为新一代中产阶级白领呈上形神兼备、品味独特的英伦风。建筑墙身全采用文化砖，百年沉淀的褐石文化，历经岁月洗礼。涵养与修为凝结于褐石的每块砖石与铁艺光影中。人字型双坡陡屋面庄重而严肃。在园林里种植建筑，在屋顶设计山花结构点缀生活人。外露的木构架、老虎窗诠释着英伦建筑的特点。手工打制铁艺栏杆、手工窗饰拼花图案，更是渗透着自然的气息。

Planning and Layout 规划布局

With full consideration of the banded terrain, the whole building adopts double extension type by ways of adjusting measures to local conditions. The plan also takes account of household privacy, introducing the concept of private space into building interval, patio design, garden design and other aspects and giving full consideration to the feelings of households.

考虑到项目所在区域的带状地形，利用因地制宜的方式，建筑整体采用双排延展式。规划还考虑了住户私密性问题，建筑间距、露台设计、庭院设计等方面都充分考虑住户感受。

North Elevation 北立面图

South Elevation 南立面图

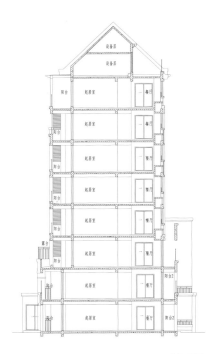

Sectional Drawing / 剖面图

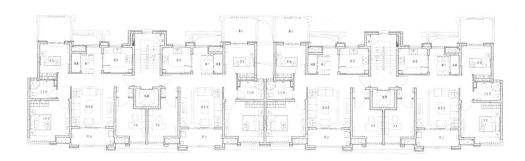

Standard Floor Plan 标准层平面图

Landscape Design 景观设计

The project intends to create high-quality residential landscape with British-style natural garden. There are "five gardens and one road", A distinguished post road runs through the Royal Knights Garden, Rose Garden, Hyde Park, Chatsworth Park, Four Seasons Garden, forming rich spatial sequence, to create a garden life experience space. Ground floor private garden design enhances the exclusive sense of home garden to improve the quality of floors meanwhile create a landscape surface from terrace sight and natural infiltration.

旭辉亚瑟郡意打造高品质英式自然花园化住宅景观。在亚瑟郡中打造了"5园1道",一条尊贵庄严的皇家驿道贯穿了骑士花园、玫瑰花园、海德园、查兹沃斯园、四季花园,形成了丰富的空间序列,营造了花园化生活体验空间。首层私家花园化设计增强专属感,专属花园入户体验,楼层营造入户花园提高楼层品质感,同时营造了露台视线景观面与自然渗透。

House Type Design 户型设计

The units within this plot consist of 90-130 m^2 house types, and most are 90 m^2 ones, which emphasize on the sense of dignity, strive to meet daily needs in function and make the best use of landscape resources. Living rooms and bedrooms adopt large sight-viewing windows or French doors that can let full natural light into the interior. The biggest highlights are private garden on ground floor, attic on top and large bay window and large terrace on the middle. House type's layout focuses on high-quality living needs, manages to keep activity and rest place apart, builds humanized living space with reasonable flow lines.

本地块内产品为90～130 m^2户型,以90 m^2户型为主。强调尊贵感力求在功能上满足生活需要和善用景观资源。起居室、卧室皆采用大面积景观窗或落地玻璃门,充分把自然光引入室内,最大亮点在于首层赠送私家庭院,顶层赠送阁楼,中间段可享受大飘窗及大露台的双重享受,房型布置注重高品质的生活需求,做到动静分离,流线紧凑合理,构筑人性化的居住空间。

Open Neighborhood Space Creating New Living Value
开放性邻里空间打造居住新价值

Longfor Chianti Walk, Changzhou　　常州龙湖香醍漫步

Location / 项目地点：Changzhou, Jiangsu, China / 中国江苏省常州市
Developer / 开发单位：Longfor Property / 龙湖地产
Architectural Design / 建筑设计：Shanghai Hyp-arch Architectural Design Consultant Inc. / 上海霍普建筑设计事务所有限公司
Land Area / 占地面积：308 820 m²
Floor Area / 建筑面积：735 400 m²
Year of Completion / 建成时间：2014 年

Keywords　关键词
Distinct Layers; Flexible Connection; Water Landscape Belt
层次分明 灵动贯穿 亲水景观带

 "China's Most Beautiful Apartments" Finalist
入围"中国最美楼盘"奖

The architectural form design that expresses the meaning of "home" pays attention to architectural humanization and the combination between design's modernity and affectivity, meanwhile highlights its differences with current popular style and reflects diversity of products, so that this residential form becomes advanced classical design in future's domestic residential products.

建筑形式设计中注重了建筑人性化，设计"家"的意义表达的建筑形式，侧重设计的现代性和情感性的结合，同时突出本设计与目前流行样式的差别，体现产品的差异性，使得该住宅形式成为未来国内住宅产品设计中的前沿经典之作。

Overview　项目概况

Project is located in northwest of downtown in Changzhou City, extending west to Qingfeng Road, north to North Tanghe Road, east to Sanxin Road and south to Qingye Road, with Hejiatang River and Yongning Road running through. The site is adjacent to the East Longitude 120 Park, 1,500 m away from Dinosaur Park, 4,000m from New North District Center, and 2,500 m from new Changzhou administrative center.

项目位居常州市中心城区西北部，西至青丰路，北至北塘河路，东至三新路，南邻青业路。贺家塘河及永宁路贯穿其中。基地毗邻规划中的东经 120 公园，距恐龙园 1500 m，新北区中心 4 km，常州市新的行政中心 2.5 km，常州市中心城区（核心商圈）5 km，区位条件优越、交通方便。

Site Plan 总平面

Planning and Layout 规划布局

The entire project planning is developed in three phases and all are low-rise and high-rise residential products. Phase I has finished the buildings of low-rise residences while the high-rises are under construction. Thus partly view can be presented.

The project planning starts with fluency, sharing and harmony. The overall layout puts high-rise residences to surround the villas. Different with traditional approach of complete separation of high-rise area and villa area, this project will extend the high-rise area into villa area by open neighborhood space, link the axis of villa area and water landscape belt, maximize the value of high-rise dwellings and enhance the privacy through small nodes in villa area. Meanwhile the high-end residences developing experience of Longfor Property is incorporated to make delicate and smooth spaces.

整个项目工程规划分三期开发。均为低层及高层住宅产品。目前在建一期开发项目，其中低层住宅均已交付，高层在建。部分实景已呈现。

项目规划将流畅、共享及和谐作为设计出发点。整体布局将高层呈围合状布置于别墅外围，有别于传统将高层区与别墅社区完全分隔的手法，本项目将开放性邻里空间作为纽带，将高层的活动范围延伸至别墅区，连接别墅区主轴与亲水景观带，最大化高层居住价值，并通过设置别墅小尺度节点加强私密性。同时融入龙湖地产多年高端住宅开发经验，整体氛围一气呵成，空间精致流畅。

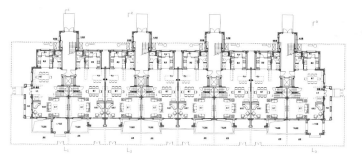
First Floor Plan 一层平面图

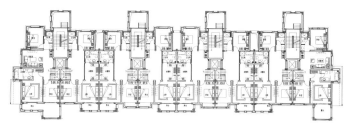
Second Floor Plan 二层平面图

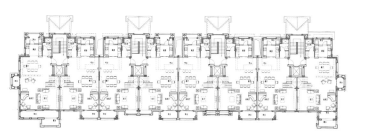
Third Floor Plan 三层平面图

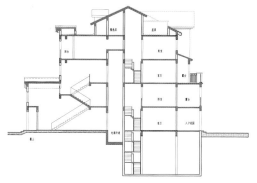
Sectional Drawing 剖面图

Sectional Drawing 剖面图

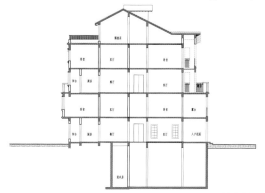
Sectional Drawing 剖面图

Housing Design 住宅设计

The townhouse mainly is 220 m² and its different architectural styles provide rich choice for owners' different needs. The housing design with reasonable function division satisfies owners' living requirements. House's inside centers on living rooms, with clear separation of privacy and public, living and sleeping area, cleanness and dirt. Compact interior layout and short passage improve the area's usage rate and comfort degree. High-rise residential unit design on the basis of the overall layout is flexible and changeable and creates more interior spaces with natural ventilation and lighting to save energy. The plane division is clear and fluent, with economic and reasonable layout.

The floor plan emphasizes the design principles of natural lighting and ventilation for living room, kitchen and bathroom. The living room is square and practical, uniformly equipped with outdoor air conditioner bit, condensate pipes, kitchen flue, water heaters air outlet etc. Each unit has work balcony, and the main balcony's depth is no less than 1.5 m, and with clothesline facilities. Water, electricity and gas meters are out of the house. Residential area has underground garage. Single design emphasizes building's accessible scale, thus spatial variation and form treatment of detail design and end unit make full use of landscape resources. In the architectural modelling design, the architectural form design focuses on architecture's humanization, expresses the meaning of "home", emphasizes on combination of modern and affection, when highlights the difference between this design and current popular form to reflect product's diversity and makes this residential form become the advanced classic project among future domestic residential design. Design use concise forms, delicate details as well as modern and vibrant colors to embody future trends in residential architecture style and create unique image.

联排别墅以 220 m² 为主，不同的建筑风格给有不同需求的业主提供丰富的选择。功能分区合理的房型设计满足了业主的生活需要，住宅套内设计以起居室为中心，内部空间公私分离、居寝分离、洁污分离、室内布置紧凑，走道短捷，提高面积的使用率和使用的舒适程度。高层住宅套型设计依据总体布局灵活多变，创造更多的自然通风、采光的室内空间，节约能源。平面分区明确，流线顺畅，布局经济合理。

住宅平面强调明厅、明厨、明卫及自然采光与通风的设计原则，厅房方正实用，统一设计室外空调机位、冷凝水管、厨房烟道、热水器排气口等，每户皆有工作阳台，主阳台进深不少于 1.5 m，并有晒衣设施。水、电、煤均出户。住宅设地下车库。单体设计强调建筑近人尺度，细部设计以及尽端单元的空间变化和形式处理，充分利用景观资源。建筑造型设计中，在建筑形式设计汇中注重了建筑人性化，设计"家"的意义表达的建筑形式，侧重设计的现代性和情感性的结合，同时突出本设计与目前流行样式的差别，体现产品的差异性，使得该住宅形式成为未来国内住宅产品设计中的前沿经典之作。设计中以简洁的形式、精致的细部，以及现代而鲜明的色彩来体现未来住宅建筑风格的趋势，塑造独特的个性特征。

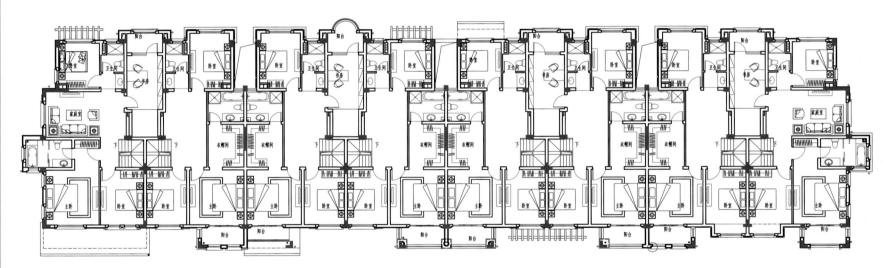

Fourth Floor Plan 四层平面图

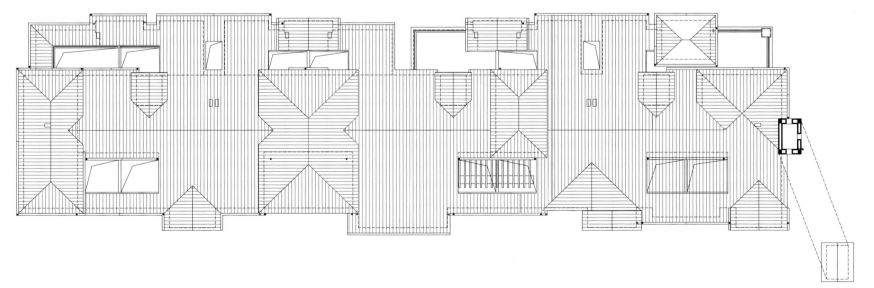

Roof plan 屋顶平面图

Landscape Design 景观设计

Design mainly presents Longfor landscape features, and integrates villa's landscape concept into high-rise space. Special attention is paid to create river way and the main axis of community which enables other landscape nodes to form the landscape layout of point, line and surface and combine with Longfor landscape concept of multi levels and many plants to visually expand landscape's depth. High-rise zone is connected flexibly and continuously. And villa zone is well arranged, and each movement of step will bring the different sceneries.

设计主要体现龙湖的景观特色，将别墅的景观理念融入高层空间中。将河道及社区主轴线进行重点打造并带动其他景观节点形成点、线、面的景观布局，并结合龙湖多种植多层次的景观理念，对景观的进深感进行视觉拓展，高层区灵动贯穿、一气呵成，别墅区层次分明、步移景异。

Meiju Encyclopedia 美居小百科

End unit, which means the unit at the end of a floor, plays an important role in forming the facade effect. The treatment of end unit is usually achieved by increasing unit numbers or changing the floor plan (eg. expand the spaces and design according to the building form). Thus the floor plan of the end unit should well consider the windows' form in the gable wall, and shape an elegant outline for the building by setting corner window, corner terrace or top terrace.

尽端单元，居住区中立面造景的重要元素；尽端单元户型处理常见的套型处理手法有改变户型（如扩大面积、套型结合形体设计）和增加户数。因此在进行套型设计时，不仅要照顾到山墙面的开窗形式，还可以运用设转角窗、转角处露台、顶部退台等方法使楼栋边缘轮廓通透轻盈、体形丰富。

Multi-level group architectures with organic color collocation

色彩有机搭配的多层组团建筑

North Garden　正北·名苑

Location/ 项目地点：Jiaozhou, Shandong, China/ 山东省胶州市
Architectural Design/ 建筑设计：Southeast University Architectural Design and Research Institute Shenzhen Branch/ 东南大学建筑设计研究院深圳分院
Total Land Area/ 总用地面积：120 640 m²
Total Floor Area/ 总建筑面积：301 819.5 m²
Building Density/ 建筑密度：27.1%
Greening Coverage Rate/ 绿化率：37.2%

Keywords　关键词

Organic Collocation; Multilevel Group; Well Arranged
有机搭配　多层组团　层次分明

 "China's Most Beautiful Apartments" Finalist
入围"中国最美楼盘"奖

Architectural color is prominent by warm tone. Appropriately adding deep-colored shutters, white window frame and framework highlights traditional, dignified, concise and neat style which building itself owns.

建筑色彩以暖色调作为主色调，适当增加重色的百叶窗、白色窗框及构架突出表现建筑本身既传统、稳重又简练、整洁的设计风格。

Overview　项目概况

Project is located in the junction of old town and new town, with convenient traffic. The plot is to the west of Changzhou Road, to the north of Beijing Road, to the east of Guangzhou Road and to the south of Yangzhou Road, reaching 204 national highway only across a street.

项目地块位于胶州市旧城区和新城区交界处，为新旧城区交通的必经之地，交通便利。

地块东临常州路，南挨北京路，西接广州路，北依扬州路，距离204国道也仅隔一条街道。

Site Plan 总平面图

规划结构分析

通过高层与多层间的景观轴线和规划道路自然将小区分成五个小组团，每个组团之间即互有呼应又相对独立。

交通系统分析

在道路的设计上，充分依据便捷、高效的原则，将车行出入口合理地与市政道路相连接，同时内部车行流线的设计也尽可能做到快捷流畅，并有效地结合地上停车，形成高效的车行系统；
步行道路的设计则充分与景观有机结合，达到"移步换景"的效果，使人们陶醉于景观转换的同时轻松达到目的地；
地下车库出入口与车行出入口紧密联系，也体现出便捷高效的设计理念。

分期建设建议

根据现有地形条件及周边用地情况，考虑市政规划道路和学校的影响，将地块分为一、两期分别建设。
一期主要为多层房；
二期结合规划道路组团建议，主要为景观洋房、高层及部分多层，一期与二期用地之间由小区规划道路自然分隔。

景观系统分析

景观设计的原则为"一横三纵"的景观网，东西方向景观主轴将两个主要景观节点串连在一起，与三条南北方向景观次轴相结合成四个景观次节点，三条景观次轴又将建筑组团内部第三级景观节点联系在一起，形成完整的景观系统，使小区内部景观层次分明，变化丰富。

产品布局分析

多元化的产品设计，包含商铺、景观洋房、多层景观房、小高层、高层以及多层经济适用房，同时加入会所的设计，使小区的配套设施更加完善和便利。

Design Philosophy　设计原则

1. Making the influence of the municipal roads and the adjacent school into consideration, the design has ensured the integrating of the community and made different building groups relatively independent for easy construction.

2. To avoid the interference on the south and north, the buildings are arranged against the borders to creating a favorable internal environment.

3. Taking advantages of the characteristic internal landscapes, it forms a clear spatial system with multiple levels.

4. The stores are well arranged along the street according to the functional requirements.

　　1. 结合市政道路和学校的影响，设计既考虑小区的完整性又保证各个组团相对独立，便于建设；

　　2. 避免南、北两侧的影响——背向围合方式，将不利影响阻挡于外围；

　　3. 充分利用内部景观规划形态特点，形成由低到高、由疏到密、层次分明的空间体系；

　　4. 按照功能要求和流线特点设置沿街商铺。

Planning Structure 规划结构

The landscape center between high-rise building groups and the multi-storey one is extended to form the spatial axis of the whole community, which combines with the planning road to naturally divide the development into five groups: two high-rise building groups, two multi-storey groups in the east and west, and one garden housing group in the center.

设计单位将高层组团与多层组团之间的景观中心沿展成整个项目的规划空间轴线，结合规划道路自然将整个小区划分为五个组团：两个高层围合的组团、东边的多层组团、中部的花园洋房的中心组团以及西边的多层组团。

Building Facade 建筑立面

The facade design has well combined tradition with modernity to keep the buildings in harmony with the surrounding environment. It not only meets the requirements of the urban planning, but also creates a good city image and highlights the characteristics of the building complex. The buildings are dominated by warm colors and complemented by dark shutters, white window frames and framework, highlighting the traditional, dignified, simple and elegant style.

住宅立面考虑传统与现代相结合，与其所在城市中的环境相符合，满足城市规划的要求，争取创造出良好的城市形象，突出建筑群体本身的个性。建筑色彩以暖色调作为主色调，适当增加重色的百叶窗、白色窗框及构架突出表现建筑本身既既传统稳重，又简练整洁的设计风格。

Commercial Design   商业设计

Commercial design handles the relationship between street stalls and surrounding sites, forming extremely special commercial streets. Commercial facade materials mainly consist of stone, and vary kinds of organic color collocation enriches commercial atmosphere.

商业设计处理好街铺与周边场地之间的关系，形成极其特别的沿街商业街。商业立面材料以石材为主，各种色彩有机搭配，丰富商业气氛。

Landscape Design 景观设计

The grid landscape system is composed of "one horizontal line and three longitudinal lines". East-west landscape axis that connects two main landscape nodes combines with three south-north landscape secondary axes that connect the third-degree landscape nodes within each group to form four landscape secondary nodes, which forms a complete landscape system that is well arranged and changeable.

　　景观设计的划分为"一横三纵"的景观网格系统，东西方向景观主轴将两个主要景观节点串联在一起，与三条南北方向的景观次轴结合形成四个景观次节点，三条景观次轴又将各组团内部第三级景观节点联系在一起，从而形成完整的景观系统，使整个小区景观层次分明，变化丰富。

Tuscan-style Interpretation of Xinjiang Residential Style
托斯卡纳风格演绎新疆住宅风情

Jinke·Xinjiang Kingdom of Riverside　金科新疆廊桥水乡

Location / 项目地点：Wujiaqu, Xinjiang Uygur Autonomous Region, China/ 新疆维吾尔自治区五家渠市
Developer / 开发单位：Jinke Real Estate / 金科地产
Architectural Design / 建筑设计：W&R GROUP / 水石国际
Land Area / 用地面积：138 434 m²
Floor Area / 建筑面积：165 544 m²
Green Coverage Rate / 绿化率：32.9%
Plot ratio / 容积率：1.2
Category/ 项目类型：Semi-detached Villas and Townhouses/ 双拼及联排别墅
Status/ 项目状态：Phase I completed;Phase II under construction/ 一期竣工，二期在建

Keywords　关键词
Natural Environment; Humanistic Connotation; Bright Color
自然环境　人文内涵　色彩明快

"China's Most Beautiful Apartments" Finalist
入围"中国最美楼盘"奖

Designed mainly in exotic Spanish Tuscan style, the building facades of the commerce and the villas employ yellow slate culture stone and light yellow elastic coating, and the roofs are covered by terracotta tiles. The color collocation as above brings bright color to architectural facade, which is moderately striking and feels manual and variegated.

建筑总体上采用具有异国风情的西班牙托斯卡纳风格，商业及别墅的外立面材料采用了黄砂岩板文化石与浅黄色弹性涂料，屋顶则采用陶红色陶瓦。这样的色彩搭配使建筑外立面色彩明快，既醒目又不过分张扬，给人一种手工、斑驳的感觉。

Overview　项目概况

Located in the south of Wujiaqu District of Xinjiang, the residential plot is to the east of 102 provincial road, to the north of the planned Horizontal 1st Road, the west of the planned Longitudinal 2nd Road and to the south of the planned Horizontal 2nd Road. Occupying a total area of about 138,000 m², the site is flat and near to an urban green land on the south, with many planned roads and supporting facilities under construction.

　　项目位于新疆五家渠市区南部。地块西临102省道、南临规划横一路、东至规划纵二路、北临规划横二路；地块性质为居住用地。项目市政道路及生活配套规模正在完善当中。项目用地面积约13.8万 m²，地块南侧为城市绿地，地块内地势较为平整。

Site Plan 总平面图

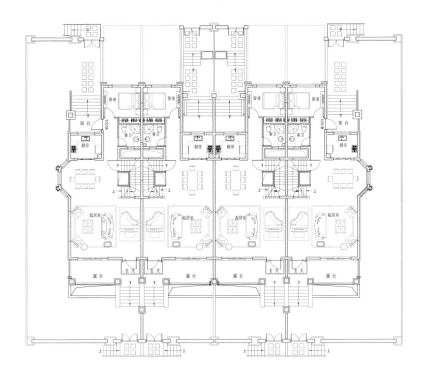

First Floor Plan　一层平面图

Plot Value　板块价值

Strategically located in Qinghu Economic and Technological Development Zone of Urumqi, to the west of Diwang Headquarter Base and Eurasian International Automobile City, to the east of Wuwu Road, to the north of the city park and to the south of MPERIAL Garden, the project enjoys great potentials.

项目地处乌鲁木齐青湖经济技术开发区，东临地王总部基地、亚欧国际汽车城，西邻乌五公路，南邻市政公园，北邻君豪御园小区。位置得天独厚，升值潜力无限。

Positioning Strategy 定位策略

The project relies on government's new district planning, employs the product type group of high-rise residences, low-rise residences, garden houses and low-rise commercial buildings along the street, aims to create a high-grade living demonstration area with first-class natural environment and humanistic connotation and provides a noble, elegant, harmonious and healthy lifestyle for residents.

项目依托政府新区规划,采用高层住宅、低层住宅、花园洋房以及低层沿街商业的产品类型组合,旨在营造一个具有一流自然环境和人文内涵的高档生活示范区,为住户提供尊贵、典雅、和谐、健康的生活方式。

Planning and Layout 规划布局

Land planning is arranged as higher in the north and lower in the south, higher in the west and lower in the east. Garden houses are set up in the north, high-rise residences in the west next to provincial road, low-rise residences in the southeast and commercial buildings along the street in the south and east. The whole land is divided into three large areas by shady trunk roads and public greening waterscape spaces and several residential groups are integrated organically and richly through the environment.

Low-rise residential area takes central cross-shaped waterscape as its feature, further to form the central landscape area of the community. Garden houses in the north and high-rise residences in the west enclose into central low-rise buildings. In order to achieve the sharing of community, the planning pays attention to the spatial organization of base's corners, trying to create an environment to strengthen the visual and spatial contacts between corner's group space and district center even community center and to influence districts' layout, thus affecting people's life.

地块规划以北高南低,西高东低布置,北侧地块布置花园洋房,西侧邻近省道布置高层住宅,东南边布置低层住宅,地块南边及东边布置沿街商业。通过社区中林荫主干道路和公共绿化水景空间将其划分形成3大区域,并通过环境将其整合使几个住宅组团成为一个有机而丰富的整体。

低层住宅片区以地块中央十字水景为低层区主要景观,进而形成整个小区的中心景观区,北边洋房和西边的高层围合了中心区的低层建筑。为达到社区的均好性,规划重视对基地的边角空间组织,试图通过环境的营造,强化边角组团空间与片区中心乃至社区中心在视觉与空间上的联系,同时用环境来影响片区的布局形态,进而影响人们的生活。

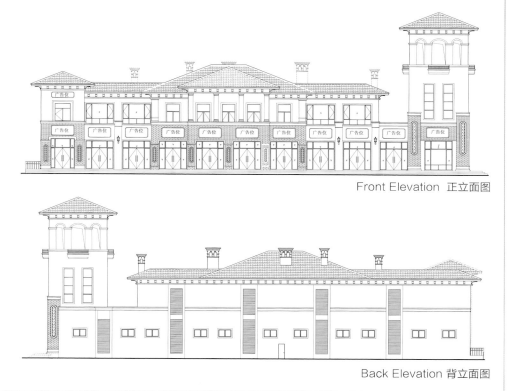

Front Elevation 正立面图

Back Elevation 背立面图

Design Concept 设计理念

To build the first-class humanistic mansion in Wujiaqu City, the planning and design of this project follows the following concepts:

1. Project combines the residences and environment by making full use of the natural landscape in southern urban green land to present environment's significance for residential life.
2. Interior and exterior landscape spaces should be expanded as far as possible to integrate organically with surroundings.
3. Floor planning focuses on forming the difference between building layout and creates rich community space through the changes of density and morphology.
4. Project bases on maximizing landscape. High-rise buildings enjoy both urban green landscape in the south and low-rise area's landscape which highlights the neighborhood and occupation of landscape are arranged around the central landscape.
5. Complete, reasonable and practical public facilities are set up to provide comprehensive and convenient community life services.

为将项目打造成为五家渠市一流的人文大宅，在规划设计上遵循以下理念：

1. 项目充分利用南侧城市绿地自然景观，把住宅融于园区环境之中，体现环境对于居住生活的意义。
2. 尽可能地拓展住区内部和外部的景观环境空间，使之与周边环境有机融合，相得益彰。
3. 平面规划注重形成建筑布局上的差异性，通过密度和形态的变化创造丰富的社区空间。
4. 项目以景观最大化为原则，高层建筑既享有南部城市绿地景观，又享有低层区景观，而低层区则强调景观的邻近及占用，围绕中心景观布置。
5. 设置完备、合理、实用的公建设施，提供完善、便利的社区生活服务。

Architectural Design 建筑设计

Designed mainly in exotic Spanish Tuscan style, the building facades of the commerce and the villas employ yellow slate culture stone and light yellow elastic coating, and the roofs are covered by terra-cotta tiles. The color collocation as above brings bright color to architectural facade, which is moderately striking and feels manual and variegated. In facade design, the designers pay attentions to both details such as architectural architrave, towers of club, bay window of villas; and to commercial interface facilities, such as combining "signage" with the main building, which ensures the entire quality of community's facade so that the whole building complex is more harmonious and unified.

建筑总体上采用具有异国风情的西班牙托斯卡纳风格，商业及别墅的外立面材料采用了黄砂岩板文化石与浅黄色弹性涂料，屋顶则采用陶红色陶瓦。这样的色彩搭配使建筑外立面色彩明快，既醒目又不过分张扬，给人一种手工、斑驳的感觉。设计师在外立面设计中既注重细节的勾勒，如建筑的线脚、会所的塔楼、别墅的八角窗；又注重商业界面的设施，如"店招"与建筑主体的结合设计，以保证小区整体的外立面品质，使建筑整体更加和谐统一。

Low-density Garden Apartments with Elevators
低密度电梯花园洋房产品

Phase I of Chief Park City, Xi'an 西安建秦锦绣天下一期

Location/ 项目地点：Xi'an, Shaanxi, China/ 中国陕西省西安市
Architectural Design/ 建筑设计：BDCL(Blewett Dodd Ching Lee Partners)/ 博德西奥国际建筑设计有限公司
Total Land Area/ 总用地面积：129 800 m²
Total Floor Area/ 总建筑面积：157 100 m²
Plot Ratio/ 容积率：1.21

Keywords 关键词

British Style; Elegant and Dignified; Denotative Image
英伦风格 沉稳大气 外延形象

"China's Most Beautiful Apartments" Finalist
入围"中国最美楼盘"奖

The buildings are designed in modern British style with ternary structures. By the interweaving and interlocking of different blocks, it adds more vibrant elements to the originally serious facade.

建筑采用三段式；但又通过一些体块的穿插、咬合关系打破了通常的三段切分方式，给稳重的立面增添了些许活跃的元素。

Overview 项目概况

Located in Weiyang District of the north Xi'an City, which belongs to Chanba Ecological Base, the project is near to Weiyang Lake University Town on the north and Chanba River on the east, overlooking Xi'an's new administrative center in the west. It is envisioned to be a benchmarking community with low-density, elevator-equipped garden apartments, superior to the others in this area for its attractive architectural form and high quality.

项目位于西安城北未央区内，属浐霸生态基地规划范围。北临未央湖大学城，东依浐霸，西望西安新的行政中心。目标定位打造低密度电梯花园洋房产品，营造高品质标杆产品，以产品形态及品质建立突破区域竞争的核心条件，奠定高尚大盘住区形象。

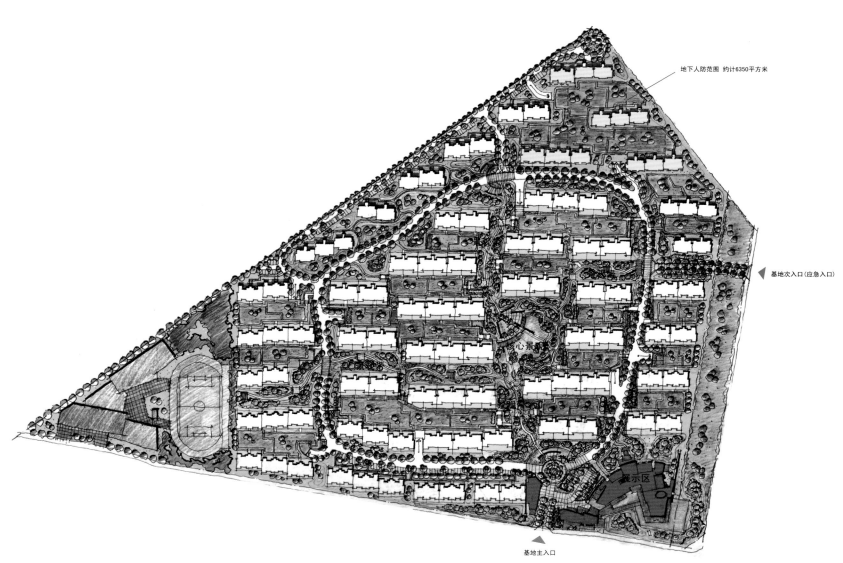

Site Plan 总平面图

Overall Plan 总体布局

The plan is characterized by "one ring", "two belts", "one center" and "one gateway". Curved road brings dynamic space experience, and two landscape axes provide changeful public green spaces. At the crossing where these two axes meet, the landscape center of the community the decorated by different elements, to be a comfortable space for neighborhood communication. At the main entrance, the sales center, commercial facilities, as well as the gate and peripheral wall combine together to form a kind of high-end and comfortable atmosphere. The entrance space has served as the gateway of the community and the starting point for the south-north landscape axis.

规划布局以"一环"、"两带"、"一心"、"一门户"为主要特点。曲线的道路为小区营造了动态丰富的空间体验；两条主要景观轴线提供了宜人而富有变化的公共绿色空间；两轴交汇处则是小区的景观中心，通过景观元素营造良好的视觉中心，成为宜人的公共交往空间；主入口结合销售中心、商业、大门及围墙营造了高端宜人的入口氛围，成为小区的门户和南北景观轴的起点。

Facade Style 立面风格

The buildings are designed in modern British style with ternary structures. By the interweaving and interlocking of different blocks, it adds more vibrant elements to the originally serious facade.

建筑采用简约现代版的英伦风格。建筑采用三段式；但又通过一些体块的穿插、咬合关系打破了通常的三段切分方式，给稳重的立面增添了些许活跃的元素。

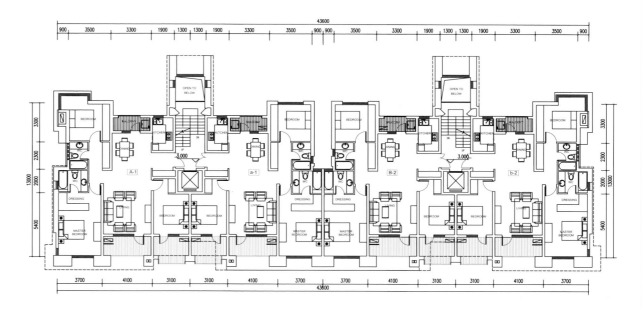

Second Floor Plan 二层平面图

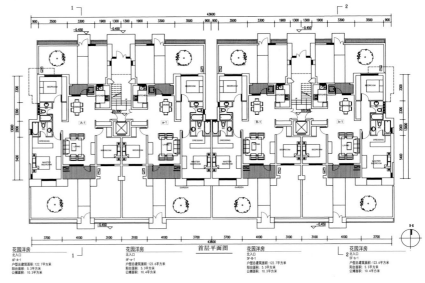

First Floor Plan 首层平面图

North Elevation 北立面图

Facade Material 立面材料

The buildings look elegant and dignified with tiles and stones as the main materials. In addition with some metal and glass components, it adds vitality to the buildings. Red split tiles define the tone of surface and tiles of other colors are interspersed to highlight the texture of the buildings. The foundation part is suggested to use gray granite to emphasize its approachable scale and high quality. Part of the building top is painted in light color to make the building breathable.

建筑的主要材料为面砖与石材，沉稳大气；局部穿插的金属与玻璃又为建筑带来一串清脆的音符。以暖红色温暖的面砖为主，为劈开砖。同时局部可穿插一点同色系不同色相的劈开砖，以提升建筑质感的感知度。基座建议采用灰色花岗岩，以提升近人尺度的品质感。顶部局部采用浅色涂料，给沉稳的建筑以局部的透气感。

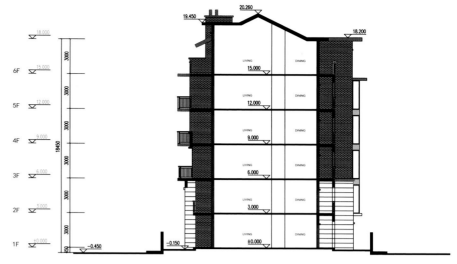

Sectional Drawing 剖面图

Unit Types Design 户型设计

The apartments are designed with different floor plans. 80% of them are units with two bedrooms or three bedrooms, together with one living room and one dinning room. The apartments on the ground floor are accessible through the gardens, and the duplexes on the top are designed with double-height living room and terrace, enjoying the space experience of the villas. The apartments above the second floor employ home gardens, terraces or balconies to create a garden-like environment. All these apartments are designed with independent vestibule, flowing living room and dinning room to extend the space experience.

住宅产品丰富，主力户型为二房二厅和三房二厅，占开发面积比例和总套数比例均应在 80% 左右，其它户型占 20% 左右。首层户型从花园直接入户，顶层复式户型客厅挑空，设露台，为首层、顶层户型营造别墅感，提升附加值。二层以上户型尽可能营造花园感（如利用入户花园、露台、阳台等方式）；入户皆设独立玄关；起居厅与餐厅空间流动，延伸空间体验。

Clubhouse Design 会所设计

The sales center, the commercial facilities, the gate and the surrounding landscapes are under integrated design to save cost and achieve a shocking effect: the coherent sales area will impress people a lot.
Based on the landmark tower axis, the buildings are well organized to be opposite, staggered or twisted, with strong connection between them. The tower has become the focus and key element for this area.
It employs courtyard, cornice and landmark tower to highlight the denotative image of the buildings. Though the land for building is limited, to meet the sales requirements, it needs to create impressive selling atmosphere. Thus the designers employ courtyard, cornice and landmark tower to expand building structures and achieve the best exhibition effect with lower cost.

将销售中心、商业、大门及周围景观当做一个整体来设计，便于早期销售阶段整体造势，用低成本的代价取得震撼的效果，避免销售中心、商业、大门出现支离破碎的格局；令销售动线的整体印象非常连贯、一气呵成。

通过标志塔轴线组织建筑形体关系。为形成有张力的组织关系，几栋建筑出现了一些对位、错动、扭转等动感的构图元素；但又通过强有力的轴线关系将它们组织在一起。标志塔成为整组建筑的构图焦点和关键元素。

项目运用院墙、挑檐、标志塔等构筑物，扩大建筑的外延形象。由于建筑面积比较有限，但为满足销售造势的需求，需要营造气势恢宏的卖场形象。因此设计师借用院墙、挑檐、标志塔等构筑物，延展建筑的形象范围，从而达到用较低的成本获得最大的展示效果。

New Residential Buildings of Oriental Flavor
东方韵味的新式住宅建筑

Poly Leaves of the Forest, Shanghai 上海保利叶之林

Location/ 项目地点：Baoshan, Shanghai China/ 中国上海市宝山区
Developer/ 开发商：Poly Real Estate / 保利地产
Architectural Design/ 建筑设计：Urban Architecture Design Co., Ltd/ UA 国际
Land Area/ 用地面积：98 805.1 m²
Floor Area/ 建筑面积：197 610.2 m²
Type/ 项目类型：High-Rise Residences, Public Supporting Facilities/ 高层住宅、公建配套
Status/ 项目状态：Completed/ 已建成

Keywords 关键词

Succinct and Generous; Naturalism; Human-Oriented Spirit
简洁大气　自然主义　人本精神

 "China's Most Beautiful Apartments" Finalist
入围"中国最美楼盘"奖

In modeling design, proportioning relationship in western architectures combines with detail elements in Oriental architectures. In architectural details of chapter, window frame and railing type, the application of simplified Chinese classical elements creates new residential architectural style of oriental flavor that is succinct and generous.

在造型设计中，将西式建筑的比例关系与东方建筑的细部元素相结合，在柱头、窗套、栏杆样式等建筑细部运用简化的中式古典元素，创造出简洁大气，且渗透着东方韵味的新颖的住宅建筑风格。

Overview 项目概况

Located in Dachang Town, Baoshan District of Shanghai, with the Planned 2nd Road on the east, Qihua Road (Fengrun Huating) on the south, Xinkaihe River (Edan River Shore) on the west, and Tangqi Road (recently proposed) on the north, it is a key and high-quality project developed by Poly Group. It is highly accessible and well-served by Metro Line 7 and Metro Line 16 as well as Hutai Road and the Outer Ring Road. A high school, a food market and a kindergarten will be built on the south of the site, and a 500 mu open park is planned to locate on the north.

项目位于上海市宝山区大场镇，东至规划二路、南至祁华路（葑润华庭）、西至新开河（鹅蛋浦）、北至塘祁路（近期拟建），是保利集团在该区域重点打造的高品质楼盘。周边有 M7 号和 M16 号轨道交通，沪太路和外环线交通十分便捷。地块南侧规划有一所高中和菜场及幼儿园。地块北侧塘祁路以北规划有占地 500 亩的开放公园。

Design Objective 设计目标

After completion, the project becomes high-grade boutique residential community near to Gucun Park on the edge of the Outer Ring of Shanghai. Design adheres to naturalism of modern architecture and human-oriented spirit of comfortable living and explains the supreme state of "harmonious life, natural comfort"; based on the rule of harmony, pursues the harmony between buildings, people and society to create owners' happy life; fully considers the various needs of people in modern living and conforms to the concept of sustainable development, close to natural warm home.

项目建成后成为上海外环边邻近顾村公园的高档精品住宅小区。设计秉承现代建筑的自然主义与舒适性居住的人本精神，诠释"和谐生活，自然舒适"的至高境界；以和为道，追求建筑、人与社会的和谐共生，营造业主的和美生活；既充分考虑到人们对现代家居生活的各种需求，又符合可持续发展理念，贴近自然的温馨家园。

Planning Layout 规划布局

The whole plot is divided into three parts of B1-04, B2-08 and B2-13 by two pieces of waters flowing from south to north, forming the composition relation of "five groups, three water systems and one green axis" and constituting clear and powerful urban surface. Three groups near the 500 mu planning park make full use of park's landscape. Some small units are arranged in the southern group next to vegetable market and Fengrun Huating. Natural landscape of three water systems and green axis running through block B2-08 and B2-13 form a strip of landscape belt provides good landscape environment for households and enhances the overall quality of community.

整个地块被两条南北向的水域划分成 B1-04、B2-08、B2-13 三个地块，形成五组团三水系一绿轴的构图关系，同时也构成清晰而有力度的城市界面。分布在临近 500 亩规划公园的 3 个组团充分利用了公园的景观。分布在南侧的组团因为临近菜市场和葑润华庭，也布置了一些小户型。通过三条水系的天然景观以及贯通 B2-08 地块和 B2-13 地块的绿轴，形成一条景观带，为更多的住户提供良好的景观环境，进而提升小区整体品质。

Elevation 立面图

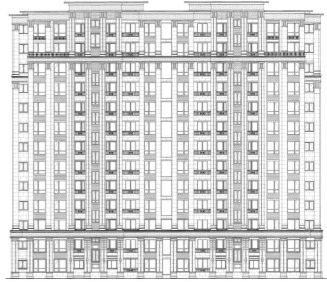

Elevation 立面图

Facade Design 立面设计

In modeling design, proportioning relationship in western architectures combines with detail elements in Oriental architectures. In architectural details of chapter, window frame and railing type, the application of simplified Chinese classical elements creates new residential architectural style of oriental flavor that is succinct and generous. At the same time, facade has no complicated decoration, thus reducing the construction cost, simplifying construction and showing restrained and elegant temperament.

在造型设计中，将西式建筑的比例关系与东方建筑的细部元素相结合，在柱头、窗套、栏杆样式等建筑细部运用简化的中式古典元素，创造出简洁大气且渗透着东方韵味的新颖的住宅建筑风格。同时，由于立面没有繁复的装饰，不但降低了建筑造价，而且施工简便，又具备内敛高雅的建筑气质。

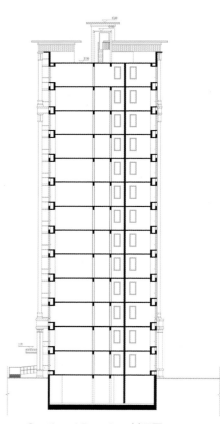

Sectional Drawing 剖面图

Landscape Design 景观设计

Landscaping on the periphery keeps noise and adverse impact from surroundings and ensures a good internal environment in community. Each group has a large garden enclosed in the center to give every household a good viewing sight and scenery.

To display the good landscape features of residential area and to enhance the spatial contact between three groups, the plan sets up a shared landscape axis integrating with pedestrian system to form a larger and richer landscape system. The main landscape axis runs from east to west closely links three blocks, forming a unique grid landscape system.

Landscape integrates with architectures to achieve relatively perfect overall environment. Thus favorable privacy and recognizable neighbors and common sense of boundary and belonging form a healthy community suitable for cultivating body and mind and a pure land in bustling city.

基地通过大外围绿化，屏蔽周围的噪音和不良影响，保证小区良好的内部环境。各组团中心均有超大面积的集中围合花园，使每家每户都有良好的景观视线，做到户户有景。

为体现居住区良好的景观特色，同时为加强三个组团的空间联系，规划中设置了共享绿化景观轴，共享绿化主轴与步行系统相融合，从而形成更大更丰富的绿化景观系统。横贯东西的主景观绿轴将三块地紧密相连，形成独有的网格化景观体系。

景观与建筑空间的结合设置，使建筑不是脱离在自然环境之外，而是与自然融为一体。使整体环境达到相对完美的境界。从而创造良好的私密性、易识别的邻里、共同的界域感和归属感，使得整个小区成为健康生活、修养身心的住区，成为闹市中难得的净土。

Elevation 立面图　　　　Elevation 立面图

Flexible Green Ecological Space

形态灵活流转的绿色生态空间

Raycom·Tiancheng Phase IV 武汉融科天城四期

Location/ 项目地点：Wuhan, Hubei, China/ 中国湖北省武汉市
Developer/ 开发商：Sinochem Franshion Properties (Beijing) Limited/ 中化方兴置业（北京）有限公司
Architectural Design/ 设计单位：WSP ARCHITECTS / 维思平建筑设计
Land Area/ 用地面积：18 565 m²
Ground Floor Area/ 地上建筑面积：80 765 m²
Plot Ratio/ 容积率：4.35
Building Density/ 建筑密度：30%
Greening Ratio/ 绿地率：35%

Keywords 关键词

Flexible Layout; Main Image; Green Ecology
布局灵活　主体形象　绿色生态

"China's Most Beautiful Apartments" Finalist
入围"中国最美楼盘"奖

Five high-rise buildings in block A are designed ranging from 98 m to 83 m, forming rich undulating skyline and enriching urban space image. Commercial buildings in block B scattered high and low with flexible and dynamic form connect the parts into the dispersed but not isolated wholeness by the platform on 2nd floor.

A地块中的五栋高层公寓，分别设计为98m和83m不等，形成错落的天际线，丰富了城市空间形象。B地块商业高低错落，形态灵活流转，充满动感，又以二层的平台将几部分连接成一个整体，分散而不分离。

Project location 项目区位

Raycom·Tiancheng Phase IV is located in core area of Hankou, next to two main urban roads of Jiefang Road and Beijing-Hankou Road, with convenient traffic and important geographic position.

　　融科·天城四期地块工程项目位于汉口核心区，临城市最主要的两条主干道：解放大道和京汉大道，交通便利。地理位置十分重要。

Spatial Structure 空间结构

Main image of Block A contains five high-rise residences in north-south layout, ensuring good ventilation and lighting. 4-layer commerce and studio office building between the residences together enclose into a diamond-shaped polygon community. Block B mainly is 2-4 layer commercial buildings flexibly distributed which form dispersed but not isolated building group connected by aerial platform on the second floor.

A 地块以 5 栋高层住宅为主体形象，采用正南北向布局，保证了住宅良好的通风与采光。住宅之间以 4 层的商业和 studio 办公楼相连，共同围合成一个钻石状多边形社区。B 地块以 2~4 层商业为主，布局灵活、分散，并在二层以一个空中平台相连，形成分散但不分离的建筑组群。

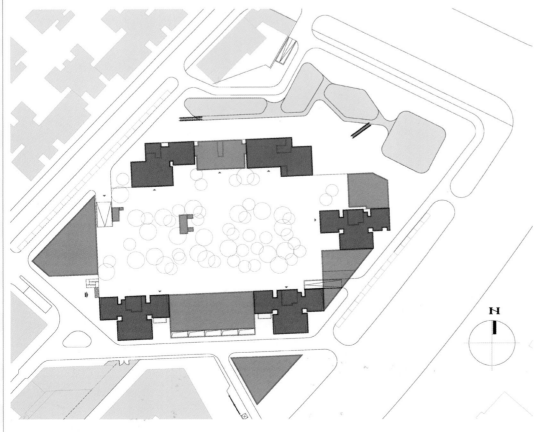

Site Plan 总平面图

Function Division 功能区划

Block A is mainly for residential use, which is complemented by ground floor retails and low-floor studio offices along Beijing-Hankou Road. Block B is mainly for commercial use. And the basement is designed with a double-floor parking and some equipment rooms to serve the whole community.

A 地块主要以住宅为主，混合加入底层商业，沿京汉路一侧商业上部还设有 studio 办公。B 地块以商业为主。地下室为双层停车场、设备用房等设施，服务于社区，提供必要的生活配套。

Landscape System 景观系统

The central green land is the landscape center of the whole residential area, enabling all residents to enjoy great views and have the experience of living in the central park. In the large central garden, exquisitely designed and well-proportioned landscape items create a people-oriented environment for neighborhood activities and communications.

中心绿地是整个住宅区的景观中心，使每座楼的居民都能享有最大化的景观感受，提供生活在中央公园的居住体验。在大规模的中心花园里，精心设计的充满人的尺度的小品景观匠心独具，塑造出以人为本的环境空间，为社区居民的活动与交往营造出良好的意境。

Architectural Form 建筑形态

Architectural form is an external expression of the inherent meaning, emphasizing on the symbolic and open architecture in volume and shape. Five high-rise buildings in block A are designed ranging from 98 m to 83 m, forming rich undulating skyline and enriching urban space image. Commercial buildings in block B scattered high and low with flexible and dynamic form connect the parts into the dispersed but not isolated wholeness by the platform on 2nd floor.

建筑形态是内在意义的外部表达，在体量和形态上强调了建筑的标志性和开放性。A 地块中的五栋高层公寓，分别设计为 98m 和 83m 不等，形成错落的天际线，丰富了城市空间形象。B 地块商业高低错落，形态灵活流转，充满动感，又以二层的平台将几部分连接成一个整体，分散而不分离。

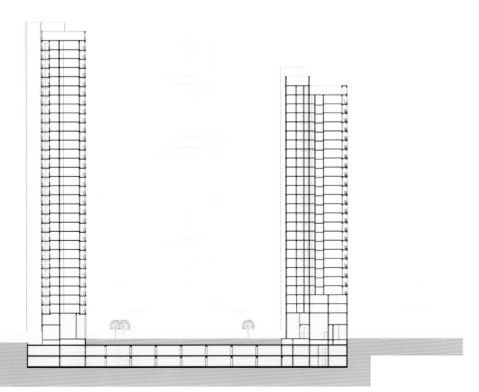

Sectional Drawing 剖面图

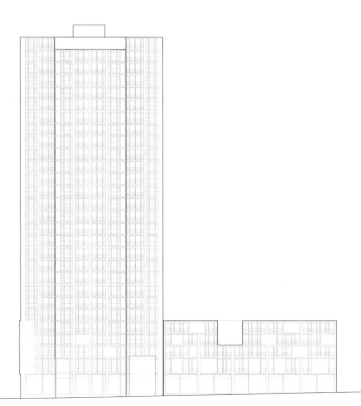

Standard Floor Plan of Building 22　22 号楼立面图

Plan for Standard Floor　标准层平面图

Green Systems　绿色环保系统

Under LEED, the evaluating standard of international ecology and energy design organization, the designers produce many environment-friendly designs to create a benchmark for environment-friendly buildings. As the current design concept changes gradually, the concern to details will help to transfer the high energy-consuming buildings into energy-efficient ones, and then establish the sustainable architectural mode.

In term of choosing materials, green building materials is prior to that of doing harm to environment. Trying to use local materials can effectively reduce the pollution to the environment caused by freight transportation. It is needed to choose durable and low-energy materials. Attention to use thermal insulation materials improves architectural thermal environmental performance.

In term of recycling materials, choosing recyclable architectural materials and recording the types and quantities of all wastes during construction will achieve the recycling of building materials.

Rainwater recycling system is used in the collection and treatment of rainwater. In the rainy season, the overflowing water will be collected by the concave vertical system; in the other seasons, the collected water will be used for landscape irrigation, which establishes a sustainable landscape system within the community.

In term of architectural equipment, efficient equipment of heating, air conditioning, hot water, gas and lighting is adopted. If possible, solar energy is used to provide electricity and lighting. To strengthen the efficient use of water resource, recycled water supply system, water-saving flush toilet and no leakage faucet are promoted.

Rational use of roof can increase greening rate, such as roof garden to reduce roof runoff. Part roof can use highly reflective material to reduce roofing heat island effect.

In term of garbage disposal, classifying and recycling living waste, reusing plastics and glasses, dumping abandoned building materials and making organic matter into compost so as to maximize reuse the garbage and reduce environmental pollution.

LEED是国际生态和能源设计组织的评价标准。在这个标准下,设计师进行了许多基于环保理念的设计。使得该区域同时也能成为环保建筑的标杆。现行的设计观念正在逐渐改变,一些对细节的关注能使建筑从高能耗向低能耗转化,创造可持续发展的建筑模式。

在材料选择上,选择绿色建材,少用或不用对环境有害材料。尽量使用地方材料,可以有效地减少交通货运给环境造成的污染。选择耐用材料、低能耗材料。注重使用保温材料,提高建筑热环境性能。

在材料再利用上,侧重选择含可回收成分的建材,施工期间对所有废物垃圾类别和数量进行记录,实现建材的再利用。

在地表径流处理上,采用雨水回收系统,将地表径流进行收集和处理。在雨季,溢出水量通过内凹型的竖向规划对其进行集中回收;其他季节的地表径流四分之三通过集中回收后对景观进行灌溉,确保园区景观系统的可持续发展。

建筑设备的使用上采用高效能的采暖、空调、热水、燃气、照明设备。如果有可能,采用太阳能提供照明用电。为加强水资源的有效利用,提倡采用中水供水系统,和节水型抽水马桶、无渗漏水龙头。

合理利用屋面能提高绿化率,考虑采用屋顶花园,减少屋面径流。部分屋面可采用高反射材料,减少屋面热岛效应。

在垃圾处理上,对生活垃圾进行分类处理和回收,对塑料、玻璃再生利用,废弃建材进行填埋,有机物用作堆肥,最大限度化废为宝,减少环境污染。

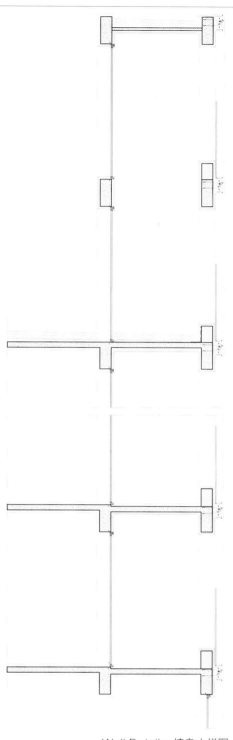

Wall Details 墙身大样图

Ecological and Pleasant Health Apartments
生态宜人的养生住品

Suiyuan Jiashu 　**随园嘉树**

Location/ 项目地点：Hangzhou, Zhejiang, China / 中国浙江省杭州市
Developer/ 开发商：Vanke Real Estate Group/ 万科地产集团
Architectural Design/ 建筑设计：Shanghai ZF Architectural Design Co., Ltd / 上海中房建筑设计有限公司
Land Area/ 占地面积：63 853 m²
Floor Area/ 建筑面积：70 000 m²
Plot Ratio/ 容积率：1.0
Type/ 项目类型：Apartments for the Elderly / 养老公寓

Keywords　关键词

Health Apartments; Friendly and Pleasant; Elegant Modeling
养生住品　亲切宜人　造型简洁

 "China's Most Beautiful Apartments" Finalist
入围"中国最美楼盘"奖

Architectural design of the project aims at creating a pleasant living environment for the elderly, and using square, inner courtyard, corridor and other spatial design elements to create varied public spaces.

项目的建筑设计着眼于营造怡人的老年人居住氛围，采用广场、内院、廊道等多种空间造型元素，塑造富于变化的公共空间。

Overview　项目概况

Occupying an area of 63,852 m², the project is located in the core area of Liangzhu Culture Village, including a 4,500 m² central health and recreation zone which is nicknamed "Gold Cross". It is customized especially for the elderly of modern China to meet their physical and psychological requirements.

随园嘉树位于良渚文化村核心区，占地面积 63 853 m²，其中包含 4 500 m² 中央"金十字"养生休闲区。随园嘉树是基于对中国当代长者生理、心理特征和行为特点所进行的深入研究，为他们特别定制的养生住品。

Site Plan 总平面图

Design Planning Concept 设计规划理念

1. Space design is prior to behavior design. According to the characteristics of an aging society, as well as the behavior characteristics of the elder preferring living in a compact community, designers introduce the concept of "group", relying on Liangzhu's beautiful natural environment and maturing facilities and strive to create a suitable inhabited place for the elderly within Liangzhu Culture Village.
2. Conforming to the terrain lower from west to east, the whole site is dealt into several platforms whose elevation difference is used to form layers of rich skyline meanwhile set up supporting public buildings and underground garage. Which on the one hand reduces the amount of earthwork excavation, on the other hand maximizes the use of ground resources into green land and activity places, to unify the use function and form.
3. "Cross garden" design forms two axis from east to west and from south to north. The reception hall, fountains, small theatre, Central Plaza and a series of ordered space design, naturally lead sight and behavior of people to the outdoor, organically integrating with indoor space.

1. 设计空间先设计行为。根据社会老年化的特点，以及老年人喜爱聚居的行为特点，设计师引入一个"群"的概念，依托大良渚优美的自然环境和日臻成熟的配套设施，力求在良渚文化村内打造一个适宜老年人聚居的场所。

2. 顺应基地西高东低的地势，将整个基地处理成为几个台地，利用台地高差，形成沿山层层叠叠的丰富天际线。同时顺应地形利用高差设置配套公建和地下车库，一方面减少开挖土方量，另一方面让地面资源最大限度让给绿化和活动场所，达到使用功能和形式的统一。

3. "十字庭院"的设计，形成了东西和南北两条纵深轴线。接待大厅、叠水、小剧场、中心广场一系列有序的空间设计，将人们的视线与行为自然引至室外，使室内外空间有机融合、浑然一体。

Planning Structure 规划结构

The project phase III health center's planning structure generally is "one center, two axes", which takes public buildings as the center and forms the east-west axis and north-south axis. North-south axis is connected by reception hall in the ground floor, lounge, central courtyard, and senior activity classroom, while the east-west axis is connected by curved landscape trails, mini theater, a series of enclosed outdoor courtyards and the green centers within the groups. All groups could share landscape resources along these two crossing axes. The elderly living here can appreciate the scenes in different steps and natural organic pleasant scenery.

随园嘉树三期工程养生中心规划结构，概括而言就是"一个中心，两条轴线"；即以沿途的配套公建为中心，形成东西和南北两条轴线。南北轴线由一层的接待大厅、休闲吧、中心庭院、老年人活动教室等串联起来，东西轴线将曲形景观步道、迷你剧场、一系列围合的室外庭院、组团中心绿化联系起来。各个组团均能共享这两条十字轴线串联的景观资源。身处其中的老年人，均能体会到步移景异、自然有机的宜人风景。

Architectural Design 建筑设计

Architectural design of the project aims at creating a pleasant living environment for the elderly, using square, inner courtyard, corridor and other spatial design elements to create varied public spaces. Design takes consideration of the continuity between blocks and interface without losing changes, eliminates the sense of monotony brought by vertical and horizontal style, and enhances the visual contingency and identifiability of landscape.

项目的建筑设计着眼于营造怡人的老年人居住氛围，采用广场、内院、廊道等多种空间造型元素，塑造富于变化的公共空间。设计中重点考虑了街廊和界面的连续性，同时使其在连续中又不失变化，消除行列式带来的单调感，增强视觉对景的偶然性和可识别性。

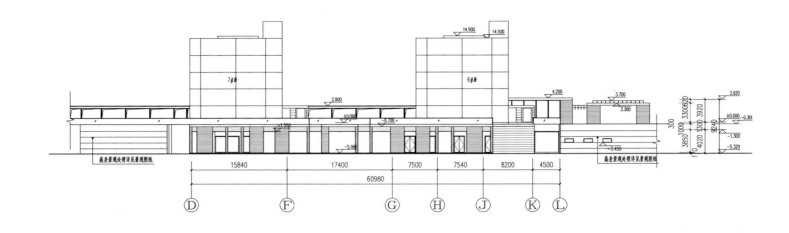
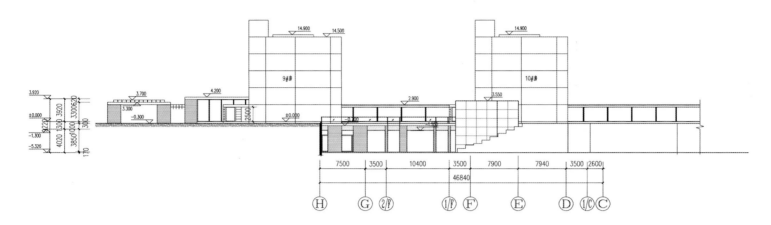

Elevation 立面图

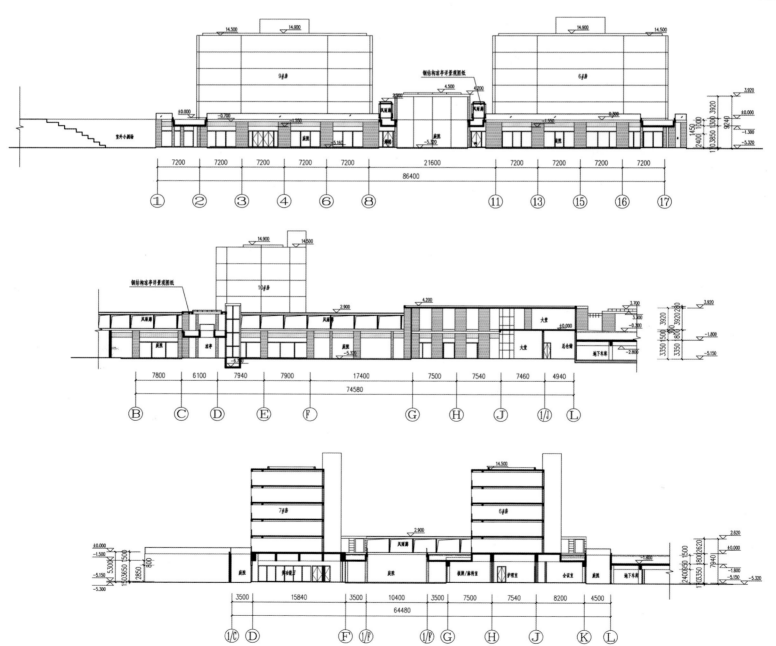

Sectional Drawing 剖面图

Single Design 单体设计

Taking into account the traditional living ideas of the Chinese elderly, every single building is designed with aging-in-place provisions which mainly come in two types: one bedroom + one living room + one bathroom; two bedrooms + two living room + one bathroom. Four units of these two types are combined flexibly on each floor to share an elevator. Each unit is designed with two rooms facing south, ensuring the elderly to enjoy plenty of sunlight.

考虑到中国老人的传统观念和传统的居家养老方式，项目单体设计为居家式的养老套型单元，套型由一房一厅一卫与两房两厅一卫两种主要套型组成。两种套型以一梯四户的形式组合，每种套型均有 2 个开间朝南，保证老年人居住空间拥有充足的阳光。

Exquisite and Comfortable Waterfront Town

精致、舒适、私享的水乡小镇

Eastern Mystery, Jinxi, Kunshan 昆山首创青旅岛尚

Location/ 项目地点：Kunshan, Jiangsu, China / 中国江苏省昆山市
Developer/ 开发商：Beijing Capital Co., Ltd. & CYTS / 北京首创及中青旅
Landscape Design/ 景观设计：L&A Design Group/ 奥雅设计集团
Landscape Area/ 景观面积：99 800 m²

Keywords 关键词

Natural Ecology; Succinct Line; Comfortable and Pleasing
自然生态 线条简洁 舒适宜人

 "China's Most Beautiful Apartments" Finalist
入围"中国最美楼盘"奖

The designers have organized the typical landscape elements of the Yangtze River Delta with modern skills, highlighting the gardens, streets, alleys, bridges and docks of the water town to create a living atmosphere of seclusion.

景观设计把江南诸景用现代的手法组织深化，景观元素突出"水乡的苑、水乡的街、水乡的巷、水乡的桥、水乡的码头"，营造出烟雨江南，漠漠水乡的隐居生活氛围。

Overview 项目概况

The project is located in the southwest corner of Kunshan City, and is adjacent to Zhouzhuang Town, with Dianshan Lake to the east, Chenghu Lake on the west, Wubao Lake on the south as well as Fanqing Lake and Bailian Lake on the north. Therefore, Jinxi has always been known as "golden and jade waves", being richly endowed by nature culture.

项目位于昆山市西南隅，与古镇周庄相邻。东临淀山湖，西依澄湖，南靠五保湖，北有矾清湖、白莲湖。"东迎薛淀金波远，西接陈湖玉浪平"，故锦溪历来有"金波玉浪"之称，拥有得天独厚的文化依托。

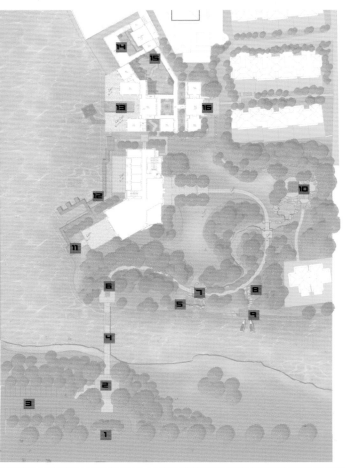

1. 入口特色雕塑
2. 入口节点广场
3. 公共停车场
4. 景观桥
5. 叠流水景
6. 特色大门
7. 木桥
8. 特色雕塑
9. 亲水平台
10. 假山观景亭
11. 游客返回码头
12. 亲水木栈道
13. 特色泳池
14. 休憩庭院
15. 水景庭院
16. 特色水景

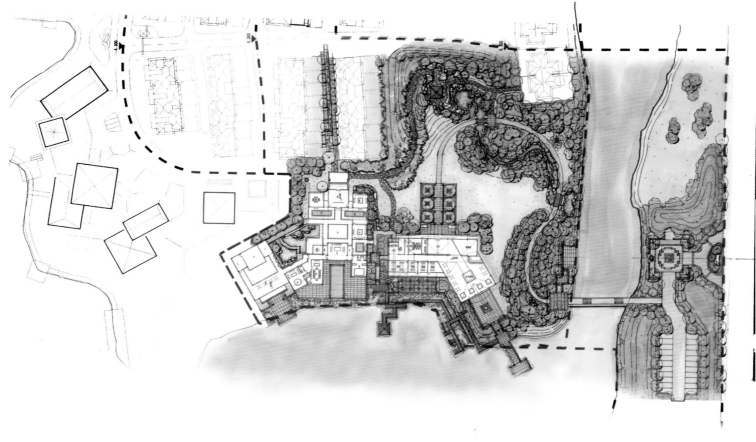

Landscape Design 景观设计

The project has offered modern Chinese-style villas built by water, with Chinese-style wide eaves, white walls and black tiles as well as succinct lines. The landscape design has started from the regional characteristics and been inspired by the water environment to create an exquisite, comfortable and private water town. Considering that the project in Jiangnan area where various water landscapes forms different kinds of landscape of "bridges people" in the long-term historical development process, is also surrounded by gurgling water and the vision of landscape design is defined as "water village impression": river flows around the ancient town; small and exquisite waterside pavilion and wooden house lies along the river; narrow and winding lanes between high walls stretch forward under the weeping willow, disappearing in the misty rain. The designers have organized the typical landscape elements of the Yangtze River Delta with modern skills, highlighting the gardens, streets, alleys, bridges and docks of the water town to create a living atmosphere of seclusion.

项目的建筑为临水而筑的现代中式纯别墅，中式阔檐，白墙黛瓦，线条简洁。景观设计从地域出发，依托于江南水乡这个大环境，打造精致、舒适、私享的水乡小镇。考虑到项目地处江南地区，流水景观多样，在历史长期发展过程中，形成了"小桥流水人家"的别样景致，而项目基地亦是流水环绕，流水潺潺，景观设计的愿景定义为"水乡印象"："绕镇而过的河流，水阁木楼，临河而起，绰影幢幢，小巧轻卧，杨柳依依，高高的垣墙夹着狭窄曲折的街巷，延伸向前，消失在烟雨中"。景观设计把江南诸景用现代的手法组织深化，景观元素突出"水乡的苑、水乡的街、水乡的巷、水乡的桥、水乡的码头"，营造出烟雨江南，漠漠水乡的隐居生活氛围。

Plant Design 植物设计

Plant design based on Chinese style, nature, ecology and other aspects, pays attention to the combination of constructing and gathering space and to the comparison between texture and color of various plants, to create a natural, comfortable and unique landscape effect by use of plant's seasonal changes. In addition, the limited lane spaces between the houses, combined with home spaces and planted with traditional Chinese garden plant types, such as bamboo, camellia, thunbergii, magnolia and crabapple, give unlimited humanistic conception of space.

植物设计围绕着中式、自然、生态等几个方面展开，注重空间营造收放相结合，注重各种植物的质感和色彩的对比，利用植物四季变化来营造自然、舒适且不一样的景观效果。此外在有限的宅间巷道空间，结合入户空间，配植中国园林中颇具典故的传统植物类型，如竹子、山茶、金桂、玉兰和海棠等，赋予无限的人文意境空间。

Well-organized Modernism Architecture
错落有致的现代主义建筑

Ruize Jiayuan, Shenzhen 深圳瑞泽佳园

Location/ 项目地点：Shenzhen, Guangdong, China / 中国广东省深圳市
Architectural Design/ 建筑设计：Architects & Engineers Co., Ltd. of Southeast University / 东南大学建筑设计研究院
Total Land Area/ 总用地面积：46 397.13 m²
Total Floor Area/ 总建筑面积：130 612.79 m²
Plot Ratio/ 容积率
Green Coverage Ratio/ 绿化率

Keywords 关键词
Free Form; Orderly; Simple Modern
自由形态 井然有序 简约现代

"China's Most Beautiful Apartments" Finalist
入围"中国最美楼盘"奖

Project adheres to modernism architectural style. The scattered terraces and balconies design, color changes, contrast of materials, create a tranquil and generous sense of scale.

项目秉承现代主义建筑风格，通过错落的露台和阳台设计、色彩的变化、材质的对比，营造出一种宁静、大气的尺度感。

Overview 项目概况

The project is located in Pusha, Liuyue Village, Henggang Street, Longgang District, Shenzhen City, with Pusha Road and the built plants on the east, Hongmian Road on the south, the future development land of Zhenye Town on the west, and the land for affordable housing on the north. .

瑞泽佳园位于深圳市龙岗区横岗街道六约村埔厦。该宗地东面为埔厦路和已建成的厂房，南面为红棉路，西面为振业城的后期发展用地，北面为经济适用房地。

Design Concept 设计理念

Residential planning was designed in layout form of "point and line" to form ethereal and rich urban skyline, coordinated with the entire urban planning.

The high-rise residences along the west side of central courtyard are arranged in order in form of point. The east side of the 8-layer garden houses adopt linear layout, using the traditional plate-type layout to form a semi-enclosed courtyard space. The two kinds of buildings co-produce a central landscaped courtyard. The free-form layout of the landscape meanwhile is used to form more transparent and richer personal space and public space, which greatly enriches the visual impression and life experience.

规划住宅呈"点、线"布局方式，形成空灵的、层次丰富的城市天际线，与整个城市规划相协调。

沿中心庭院西侧高层住宅用点式排列，井然有序。东侧8层洋房采用线性布局方式，沿用传统的板式布局，形成一个半围合的庭院空间。高层住宅与8层洋房共同产生一个中心景观庭院，利用自由形态的景观布局方式，同时形成更通透、更多层次的个性空间和公共空间，大大丰富了视觉印象和生活体验。

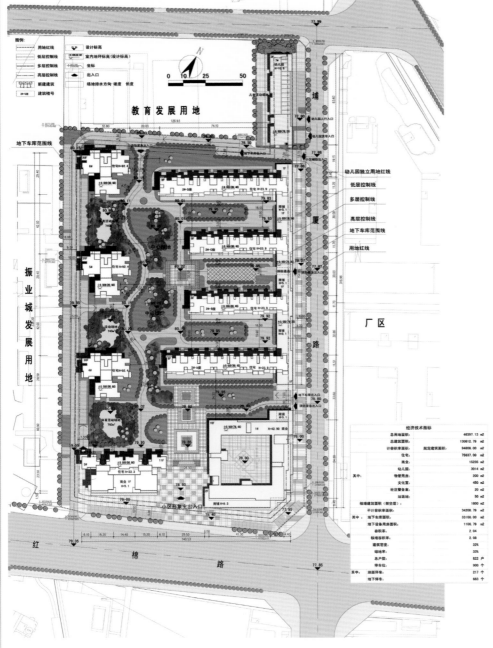

Site Plan 总平面图

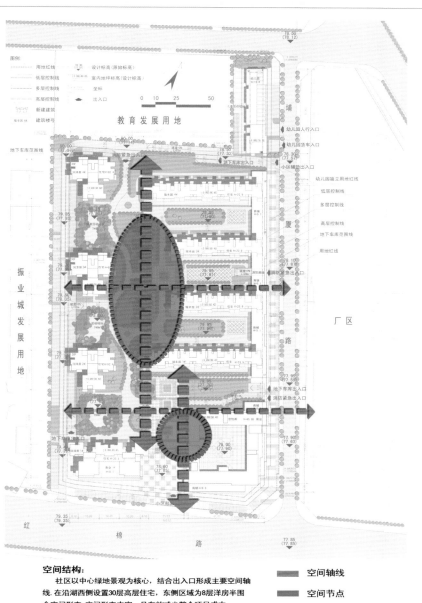

空间结构:
社区以中心绿地景观为核心,结合出入口形成主要空间轴线。在沿湖西侧设置30层高层住宅,东侧区域为8层洋房半围合空间形态,空间形态丰富。且有效减少整个项目成本。

■ 空间轴线
● 空间节点

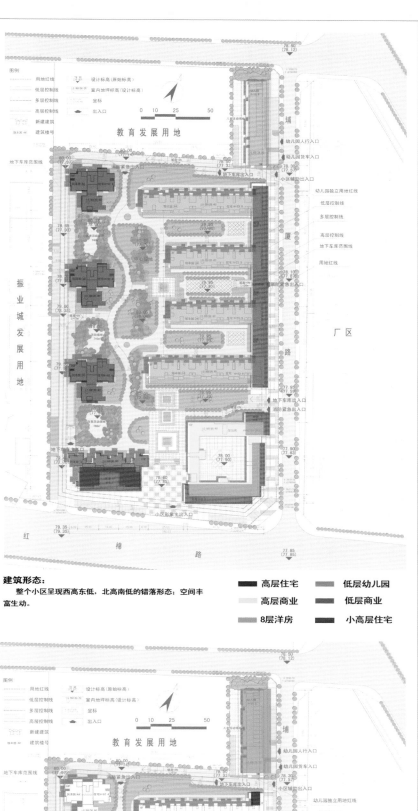

建筑形态:
整个小区呈现西高东低,北高南低的错落形态;空间丰富生动。

■ 高层住宅　　■ 低层幼儿园
■ 高层商业　　■ 低层商业
■ 8层洋房　　■ 小高层住宅

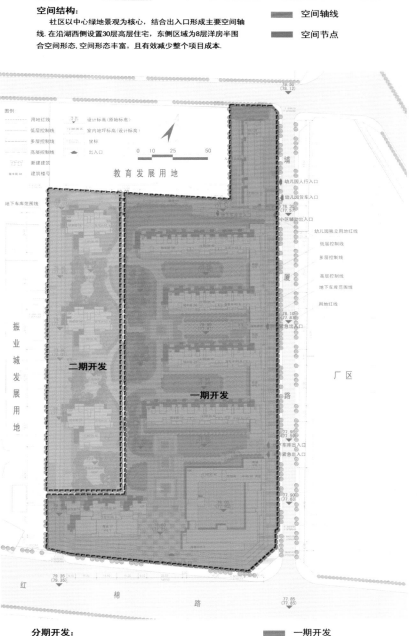

分期开发:
社区以中心湖面为核心,且以此为分区的界线,分为二期开发。

■ 一期开发
■ 二期开发

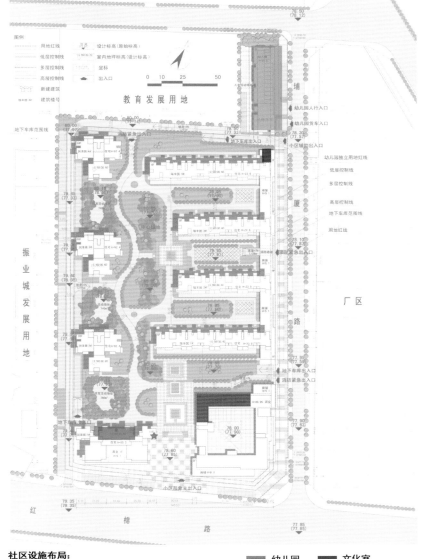

社区设施布局:
公共配套设施以中心绿地为依托。有序的布置在公共小区主入口花园内底层,方便管理。9班幼儿园设于地块最北面独立占地2500M2。

■ 幼儿园　　■ 文化室
■ 垃圾站　　■ 物管用房
★ 社区警务室

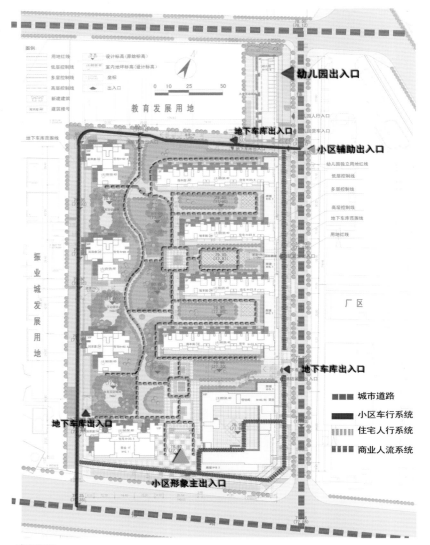

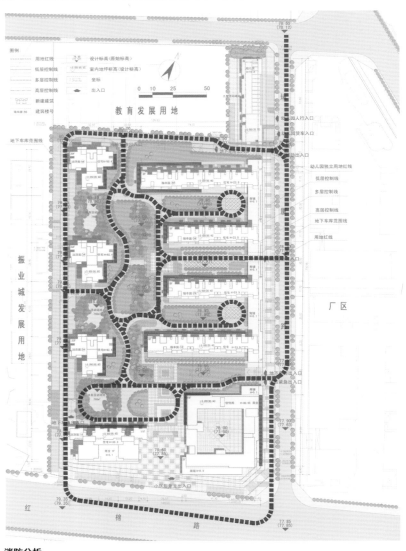

交通组织：
　　社区以红棉路设一个主形象出入口，于埔厦路设一个辅助出入口。主要人行流线沿中心绿地展开。重视景观对景与公共设施的结合，促成景观与功能的合理衔接。同时以此分期开发，方便管理。半地下车库出入口分别结合主出入口设置，方便管理，在主要车行道路两侧设置露天停车，实现停车地上地下的结合，节省成本。

消防分析：
　　小区消防组织以小区主要出入口及车行道路为基础，结合紧急消防出入口及紧急消防车道。消防车能有效的到达小区各栋建筑，登高扑救。

■■■ 消防环道
■■■ 消防登高面

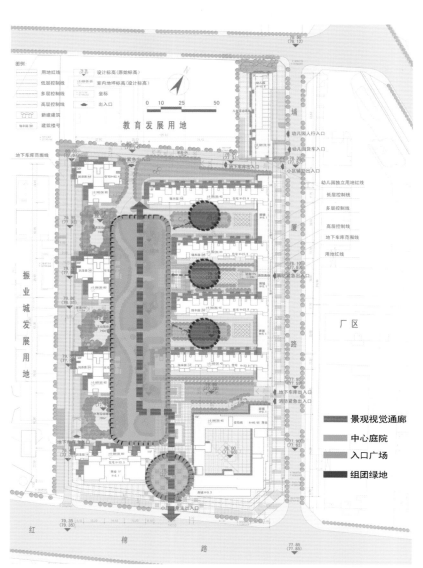

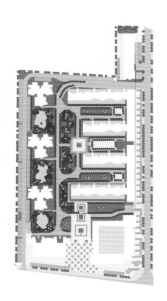

景观设计：
　　本方案以中心绿地形成核心景观区域及主要公共活动空间，入口广场、中心庭院、组团绿地形成有序的景观视觉通廊。有效地区分公共区域及私密区域。在景观设计上以绿地为纽带，与亭、树、小桥，等有机结合突出小区特色。

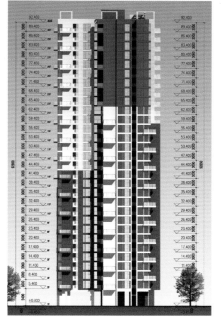
East Elevation 东立面图

West Elevation 西立面图

South Elevation 南立面图

North Elevation 北立面图

North Elevation 北立面图

South Elevation 南立面图

Spatial Structure 空间结构

Central green landscape combines with the entrance to form main space axis. The community sets up 30-story high-rise residences along the lake on the west side. The east side of the area is eight-layer houses in semi-enclosed space form, creating rich spatial form and effectively reducing the overall cost of the project.

社区以中心绿地景观为核心，结合出入口形成主要空间轴线，在沿湖西侧设置 30 层高层住宅，东侧区域为 8 层洋房半围合空间形态，空间形态丰富，且有效减少了整个项目成本。

Architectural Form 建筑形态

The buildings within the community are well organized to be lower in the east and south, and higher in the west and north, creating rich and vivid spaces.

整个小区呈现西高东低、北高南低的错落形态，空间丰富生动。

Architectural Modeling Design 建筑造型设计

Project adheres to modernism architectural style. The scattered terraces and balconies design, color changes, contrast of materials, create a tranquil and generous sense of scale. Modernism buildings and facade changes bring restrained, classic and solemn culture and temperament, making it exude a unique sense of comfort.

This design, focused on the future and based on the developer's plan, has a distinctive image feature and architectural style, inheriting classical tradition for long time. The use of bay windows, overhanging balconies and terraces and other design techniques creates communication spaces between human beings, human and nature, enhancing the sense of people living in the field.

Simple modern architectural appearance echoes with the lush nature by mainly using sedate ocher brown and ethereal white. The well-arranged large glass and high-grade coating, modern and impressionistic homes, powerful building facades and unique free-form garden together form a strong and powerful visual perception.

项目秉承现代主义建筑风格，通过错落的露台和阳台设计、色彩的变化、材质的对比，营造出一种宁静、大气的尺度感。现代主义风格建筑及外冷内热的立面变化带来的文化和气质上的内敛、经典和冷峻，使它散发出独有的舒适感。

这种设计着眼于未来，立足于开发商的规划，具有鲜明的形象特征及建筑风格，历久弥新，传承经典。利用飘窗、出挑阳台、露台等设计手法，创造出人与人、人与自然进行交流的空间，增强居住人群的领域感。

简约现代的建筑外观呼应着自然的浓郁葱绿，立面表情将以沉着的赭褐、灵动的白色为主，大面积玻璃和高档涂料的错落有致、写意居所的现代格调、颇具张力的建筑立面效果与自由形态园林的独特风格共同形成强而有力的视觉观感。

Landscape Design 景观设计

The plan forms major public space of core landscape based on the central green land. Entrance plaza, central courtyard, and green land groups form orderly landscape visual corridor and effectively distinguish public areas from private areas. In landscape design, green land organically links with pavilions and bridges to highlight characteristics of community.

本方案以中心绿地形成核心景观区域主要公共活动空间，入口广场、中心庭院、组团绿地形成有序的景观视觉通廊，有效地区分公共区域及私密区域。在景观设计上以绿地为纽带，与亭、榭、小桥等有机结合突出小区特色。

Rich Spanish Style Architecture
风情浓郁的西班牙建筑

Shahe·Century City Phase I　　沙河·世纪城一期

Location/ 项目地点: Changsha, Hunan, China / 中国湖南省长沙市
Developer / 开发商: Changsha Shahe Investment Real Estate Co., LTD/ 长沙沙河水利投资置业有限公司
Architectural Design/ 建筑设计: Peddle Thorp Pty Ltd./ 澳大利亚柏涛（墨尔本）建筑设计有限公司
Total Land Area/ 总用地面积: 718 700 m²
Total Floor Area/ 总建筑面积: 820 000 m²
Plot Ratio/ 容积率: 1.11
Building Density/ 建筑密度: 20.63%

Keywords　　关键词

Rich Flavor; Three-dimensional Landscape; Diverse Style
风情浓郁　立体景观　多样风格

"China's Most Beautiful Apartments" Finalist
入围"中国最美楼盘"奖

Multi-layer buildings in project adopt Spanish Style and high-rise buildings are modern style. The two styles are connected by similar material and colors. Spanish-style roofs, cornices, moldings and many other featured detail components, as well as undulating roof outlines, express the characteristics of Spanish architecture.

项目多层建筑均采用西班牙风格，高层建筑为现代风格，二者之间通过类似材质及颜色相互衔接，通过西班牙风格的屋顶、檐口、线脚等很多风格特征明显的细部构件，以及高低错落的屋顶轮廓线，来表达西班牙建筑的特性。

Overview　　项目概况

The project is located in Gaoling Village, north of downtown in Changsha and south of North Central Ring Road. Phase I, in the north of the whole project, sets up five entrances in the east, north and west sides, of which the pedestrian entrance is in the center of North Ring Road and business street of the main entrance extends to the lake in south. There is sports center, business center and sales offices here. Phase I contains 10 buildings of low-rise residences, located in the center of the community along the western business center. North of the low-rise residences sets up townhouses, folding and garden villas according to gradually rising terrain. townhouses and 9 buildings of 11-layer residences are set up along the lake on the east of business center. The 24-layer apartment at the main entrance are the landmark of community. West land of phase I has 9 kindergartens. All types of residential and public buildings are set up to provide a complete support for phase I development.

沙河·世纪城位于长沙市中心区以北中环线以南高岭村。一期建设位于整个项目的北部。一期共设五个出入口，分别位于东、北、西侧，人行主入口位于北环路中部，主入口步行商业街一直向南延伸到湖面。体育中心及商业中心售楼处都集中在此设置。一期共10栋低层住宅，位于商业中心西侧沿岸的小区最中心位置，低层住宅以北地势逐渐升高，依次设有TOWNHOUSE、叠拼及花园洋房。商业中心的东侧沿湖岸设有TOWNHOUSE及9栋11层住宅，主入口的24层公寓为小区的标志性建筑，一期用地西侧设有9班幼儿园。各种类型住宅及公建的设置为一期开发提供了完善的配套。

Vertical Design 竖向设计

According to the original terrain, the plan basically retained the original topography features. The semi-underground garage is arranged by use of low-lying parts, maintaining the original natural state, meanwhile reducing the amount of filling and excavating and garage operating costs. 3# commercial building and sales offices both use the terrain to drop one floor to the side of lake, and sunken square is set up between 3# commercial building to be closer to the lake, forming the center landscape of the entrance area and business. Garden house group is about 4 m higher than business streets. The steps connect Spanish-style gatehouse, combining with environmental design to form rich three-dimensional landscape.

场地按原有地形进行调整规划，基本保留了原始地形地貌特征，利用低洼部位设置半地下车库，在保持原有自然风貌的同时，减小了填挖土方工程量及车库的运营成本。3#商业及售楼处均利用地形向湖面跌落一层，并利用地势，在售楼处与3#商业之间设置下沉式广场，使此处与湖面更亲近，并形成小区入口及商业的中心景观。花园洋房组团地势比商业街高约4 m，通过台阶及西班牙风格的门楼连接，结合环境设计，形成丰富立体的景观。

Typical Elevation Design Sketch 典型高程设计示意图

Facade Design 立面设计

Multi-layer buildings in project adopt Spanish Style and high-rise buildings are modern style. The two styles are connected by similar material and colors. Spanish-style roofs, cornices, moldings, iron ornamental arch, colonnades, pergolas, towers, iconic porch, courtyard, pieces, greening and many other featured detail components, as well as undulating roof outlines, express the characteristics of Spanish architecture and let the whole community fill with rich Spanish flavor.

按建筑单体策划建议，本项目多层建筑均采用西班牙风格，高层建筑为现代风格，二者之间通过类似材质及颜色相互衔接，通过西班牙风格的屋顶、檐口、线脚、铁花拱卷、柱廊、花架、塔楼、标志性的门廊、庭院、小品、绿化等很多风格特征明显的细部构件，以及高低错落的屋顶轮廓线，来表达西班牙建筑的特性，使整个小区洋溢着浓郁的西班牙风情。

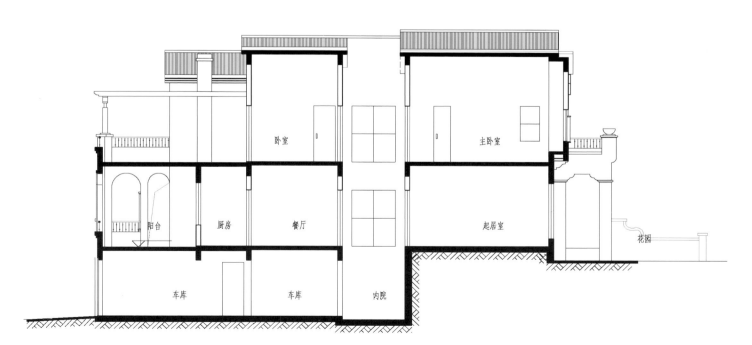

Sectional Drawing 剖面图

House Type Design　户型设计

House type design principles are to create comfortable and practical indoor and outdoor environments for households, considering both orientation and landscape, to maximize the value of the Chujia Lake's landscape. The main bedrooms all face to south or southeast to enjoy plenty of sunshine and beautiful landscape, with north-south light permeability, appropriate depth and good ventilation. Low-rise residence and townhouse both depending on the different area and the proportion set up leisure function rooms underground, public spaces and guest rooms on the first floor, the bedroom area on the second floor and partially terraces on the third floor, fully enjoying the beautiful lake scenery. V1-type low-rise residences more contain indoor elevator, exuding a sense of dignity. Superposed garden houses have four floors. Each unit sets up four households with two duplexes, of which the upper two households have a roof garden on top while the lower two households have ground garden. Each floor has a private semi-underground garage, and ground floor residence is attached with underground multi-functional room. Garden houses have five floors. Each household possesses terrace garden and back platform in each floor, forming rich facade effect. High-rise residences contain indoor garden or home garden, with south-north permeability, great lake view, and wide vision.

户型设计原则为住户创造舒适实用的室内外环境，兼顾朝向及景观，尽量发挥楚家湖的景观价值。各户型主要居室均朝南或偏东南向，享有充足的阳光和优美的景观，南北通透，进深适宜，通风良好。低层住宅及TOWNHOUSE均根据不同面积，按比例设有地下休闲多功能室，一层为公共空间及客人房，二层为卧室区域，低层住宅及部分TOWHOUSE设三层露台，可尽享美丽湖光美景，V1型低层住宅更设有室内电梯，尤显尊贵。叠拼洋房共四层，每单元设四户，每户为复式两层，上两层住户有屋顶花园，下两层住宅拥有地面花园，每层均有私家半地下车库，底层住宅附送地下多功能房。花园洋房共五层，每户拥有露台花园，各层退台，形成丰富的立面效果。高层住宅有户内花园或入户花园，南北通透，湖景绝佳，视野开阔。

Simple, Elegant and Modern High-rise Buildings of Chinese Style
简约、清新、淡雅的现代高层中式建筑

Jiangnan Aristocratic Family　　**湛江江南世家**

Location/ 项目地点：Zhanjiang, Guangdong, China/ 中国广东省湛江市
Architectural Design/ 建筑设计：Shing & Partners / 广州汉森伯盛国际设计集团
Land Area/ 用地面积：51 257 m²
Floor Area/ 建筑面积：225 536 m²
Plot Ratio/ 容积率：3.5

Keywords　关键词

Ecological Residences; Unique Characteristics; Modern Materials
生态人居　特色风貌　现代材料

 "China's Most Beautiful Apartments" Finalist
入围"中国最美楼盘"奖

Designed mainly in modern New Chinese Style, the buildings have combined Chinese traditional architectural elements with modern architectural symbols by using black-white-gray color and point-line-surface composition, abandoning complicated traditional structures, simplifying the combination items, and choosing modern building materials. The exterior wall, bay window, balcony, louver and other decorative items are well combined to create a series of simple and elegant high-rise buildings of modern Chinese style.

项目以现代新中式为建筑风格，引用传统中式元素，组合具有时代感的现代建筑标识，运用黑白灰、点线面的构图手法，简化构成素材，选用现代材料，把建筑的外墙、飘窗、阳台、百叶、装饰构件有机地协调统一起来，构筑成简约、清新、淡雅的现代高层中式建筑。

Overview　项目概况

The project has fully considered the city's characteristics, especially the gateway and the coastal landscape of the city, arranging the ecological axis and designing the building forms carefully to allow the sea views to penetrate into the city. it takes the open coastal spaces into account when building the internal urban spaces, extending the urban spaces to the seaside and strengthening the connection between the sea and the city. It emphasizes the importance of the coastal environment and introduces large-area waterscapes into the community to highlight the coastal lifestyle. With about 18,000 m² green lands, the development is envisioned to be a modern, eco and green community that highlights the theme of "green ecology". Hence, urban green belts and green spaces within the community are well organized to create a harmonious residential environment in Zhanjiang.

项目充分考虑城市风貌特色要求，尤其是城市门户和滨海景观设计，运用渗透、对景、标志点等多种手法，结合海景，对生态轴带、建筑形态等精心布局和设计，强调滨海开敞空间与城市内部空间建设的结合，形成城市内部空间序列在海边的延续，加强海与城的交流，强调滨海空间环境的重要性，小区内更引入大片园林水景，突出滨海生活特色，区内约18 000 m²的景观绿地以表现现代化的生态绿色小区为规划设计的目标，突出"绿色生态"的主题，有机组织城市绿化带与小区绿化空间的布局，营造湛江城市和谐发展的人居环境。

Architectural Style 建筑风格

Designed mainly in modern New Chinese Style, the buildings have combined Chinese traditional architectural elements with modern architectural symbols by using black-white-gray color and point-line-surface composition, abandoning complicated traditional structures, simplifying the combination items, and choosing modern building materials. The exterior wall, bay window, balcony, louver and other decorative items are well combined to create a series of simple and elegant high-rise buildings of modern Chinese style that caters to the new oriental lifestyle. The air conditioner ledge is hidden behind the louver or set in the small decorative balcony, providing the basic functions and ensuring a good facade effect at the same time.

项目以现代新中式为建筑风格，引用传统建筑中式元素符号，组合具有时代感的现代建筑标识，运用黑白灰、点线面的构图手法，摒弃繁复的传统构造，简化构成素材，选用现代材料，把建筑的外墙、飘窗、阳台、百叶、装饰构件有机地协调统一起来，构筑成简约、清新、淡雅的现代高层中式建筑。营造简约东方新人居，空调机位隐藏在装饰百叶后，或结合立面考虑的装饰小阳台上，使功能与造型有机结合在一起，满足住户使用的同时也保证立面设计追求的效果。

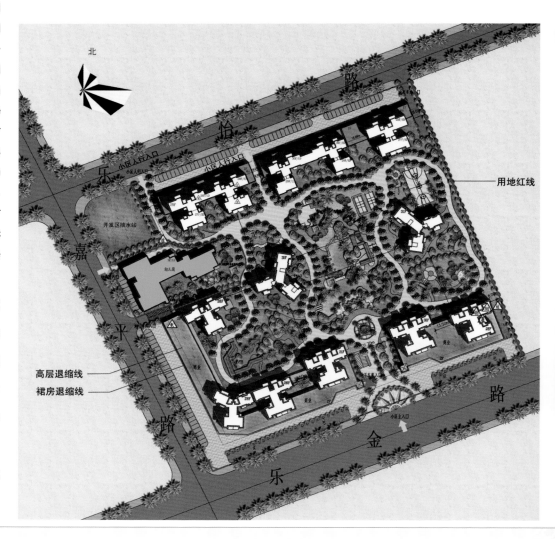

Landscape Design 景观设计

The ground floor and the open green floor are combined to accommodate the entry lobby, which also connect with the central garden to create more open and flowing spaces within the community. It provides the spatial experience of a semi-open private courtyard. The residences are bright and spacious with large-area glass windows in the main rooms to enjoy great views outside. Even the kitchens and bathrooms are well arranged and designed to have enough daylight and natural ventilation.

The planning and design have fully shown the landscape characteristics of a modern coastal city. Chinese-style landscape matches the new Chinese-style facade perfectly to create an ecological and natural residential environment in this modernized coastal city.

首层结合绿化架空层布置住户大堂，与园林充分交融，让小区内部空间更加开放流动，创造出半私家院落的空间感受，住宅厅房宽敞明亮，主要功能房间有大玻璃窗面向景观，视野开阔，每户住宅均明厨明厕，有良好的采光通风。

小区的规划设计展示现代海滨城市的景观特色，中式风格的园林设计配合新中式的立面设计体现了东方新人居的楼盘特色，使沿江滨水区域的城市氛围呈现由现代工业向自然山水的转变，实现真正的生态人居。

建筑细部——坡屋顶

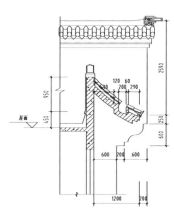
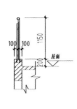
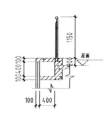
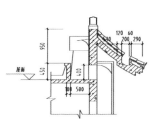

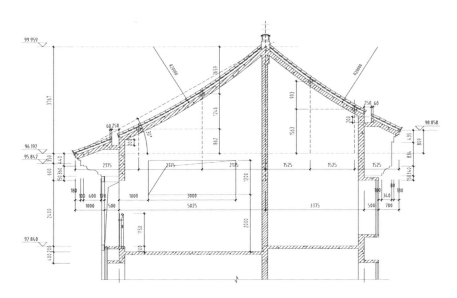

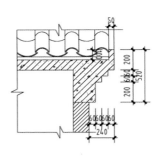
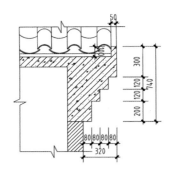

建筑细部——马头墙(一)

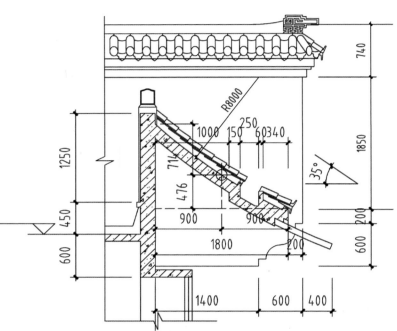

建筑细部——马头墙(二)

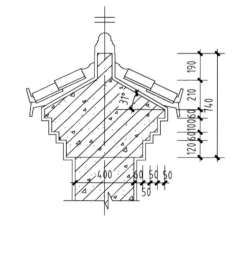
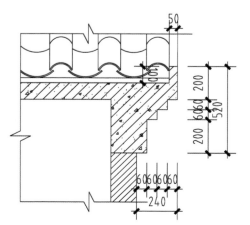

建筑细部——墙身

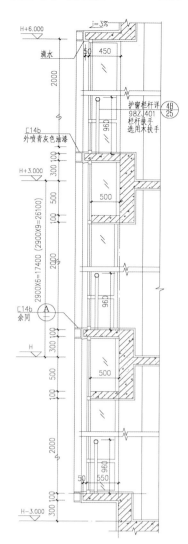
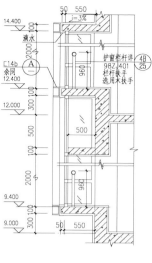
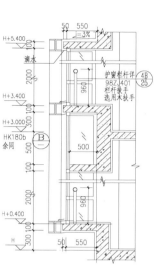

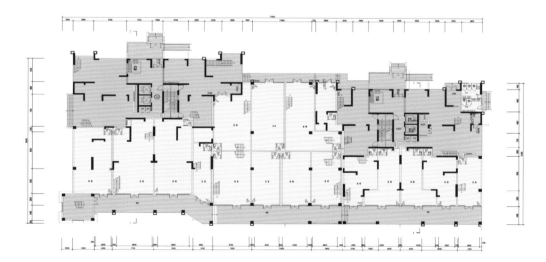

Unit Types Design 户型设计

According to modern lifestyle, every unit is carefully arranged and designed together with the landscapes and the building orientation, providing clear functions, square spaces, pleasant scales and high utilization rate. It also makes full use of the corner unit to get the best orientation and landscape effect. Most of the rooms are opened to the south, southeast or southwest. And the butterfly-shaped floor plan allows the units to extend obliquely and have a maximum view of the central garden and the sea.

住宅户型依据现代人居的生活习性，结合景观与朝向设计，精心布局，分区合理、平面方正、尺度适宜，有较高的使用率，并利用转角户型的特殊设计，达到最佳朝向与景观效果，大部分厅房朝南、东南或西南，住宅蝶形设计使户型斜向伸展，尽可能保证每户有最大限度的视野朝向中心园林和海景。

Innovative, Unique and Large-scale Integrated Community

新颖独特的超大规模综合社区

Times Property Eolia City (Phase IV, V, VI)　　时代山湖海花园（四五六期）

Location/项目地点：Jinwan District, Zhuhai, Guangdong / 广东省珠海市金湾区
Architectural Design/建筑设计：Guangzhou Hanhua Architects + Engineers Co., Ltd. / 广州瀚华建筑设计有限公司
Total Floor Area/总建筑面积：264 500m²

Keywords　关键词

Novel and Unique; Orderly and Elegant; Natural Ventilation
新颖独特　整齐大气　自然通风

"China's Most Beautiful Apartments" Finalist
入围"中国最美楼盘"奖

The buildings are designed in modern simple style to keep unified and present some rhythmic changes. Balconies, bay windows, roof trusses and the floating slabs of the exterior wall are combined to form innovative and unique building facade. In addition with the elegant colors, it will create a graceful and dignified atmosphere.

小区建筑采用现代简约主义的设计风格，设计手法强调统一中的韵律变化，通过阳台、凸窗、屋顶构架及外墙飘板等设计元素的组合，使建筑造型新颖独特，立体感强，凸显建筑的个性化，形成高尚、整齐、大气的建筑氛围。

Overview　项目概况

Eolia City is strategically located in Xihu Town, which is a key government-reserved area on Airport East Road of Jinwan District and also the future western center of Zhuhai City. Adjacent to the planned Jinwan District Government, the development enjoys great mountain views, sea views and golf views. It is a large-scale integrated community providing multi-storey lift apartments, high-rise seaview apartments, large activity center, commercial street and kindergarten. Upon completion, it will be one of the biggest developments in Jinwan District with the best facilities and the highest grade.

时代山湖海项目区域优势突出，位于金湾区机场东路政府重点土地储备区域——西湖城区，是未来珠海西部的中心城区。紧邻规划中的金湾区政府，并且同时拥有山、海、高尔夫等多重景观。项目是由多层电梯洋房、高层海景住宅、大型社区活动中心、商业街、幼儿园组成的超大规模综合社区。建成后将是金湾区规模最大、配套最完善、档次最高的项目之一。

Design Idea 设计理念

Taking advantage of the surrounding landscapes and environment, it tries to provide all units with good views, natural ventilation and enough daylight. The buildings are arranged in row according to the long site to have the best orientations and natural ventilation. And the internal environment is created for a high-end residential community according to the natural topography. Based on the principle of "economical and beautiful", the floor plans are carefully designed to make full use of each corner.

项目设计上力求最大限度利用周边的景观及环境条件，使各栋住宅均有良好的视线、景观及自然通风采光条件营造良好的休息环境。针对用地特点住宅单体呈行列式布局，合理的利用了用地南北方向的进深，最大程度保证了住宅拥有良好的朝向和通风，规划旨在充分利用自然地形，营造内部环境，建设高档住宅小区。设计中仍以经济美观为原则，对住宅的平面进行精心设计，使每寸面积发挥应有作用。

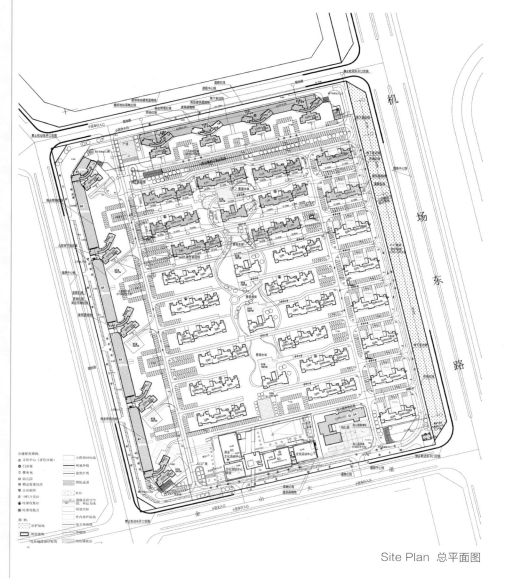

Site Plan 总平面图

Planning Layout 规划布局

Phase IV, V and VI of Eolia City include five 32-storey residential buildings (four units share one lift on each floor), four 33-storey buildings (four units share one lift on each floor), twenty one 6-~9-storey buildings (two units share one lift on each floor), four 11-storey buildings (two units share one lift on each floor) and one 18-storey building (two units share one lift on each floor). Among these buildings, every three 6-~9-story buildings and every two 11-storey buildings will combine to form a building complex. And the buildings surpass 18 stories are independent ones. The building complexes stand in row from south to north or tower in group with one basement floor for equipment rooms and underground parking. The main entrance and a sub entrance are set along Jinshan Avenue according to the overall plan. Residents of all units can reach the lobby on the ground floor directly through the lift or staircase, and the buildings all open to small paths to ensure safety in emergency.

四五六期共布置了5栋一梯四户32层住宅、4栋一梯四户33层住宅、21栋一梯两户6～9层住宅、4栋一梯两户11层高层住宅和1栋一梯两户18层高层住宅，其中6、9层三栋形成一个联体、11层由两栋形成一个联体、18层以上为独栋建筑。住宅联体呈南北条形或点式布局，另设一层地下室，布置设备房及地下车库。总体布局上，靠金山大道一侧为小区主入口和次入口。小区住宅出入口设于宅间小道上，每单元均为一梯两户或四户，电梯及楼梯均直达首层大堂，直通室外，保证消防疏散的合理性。

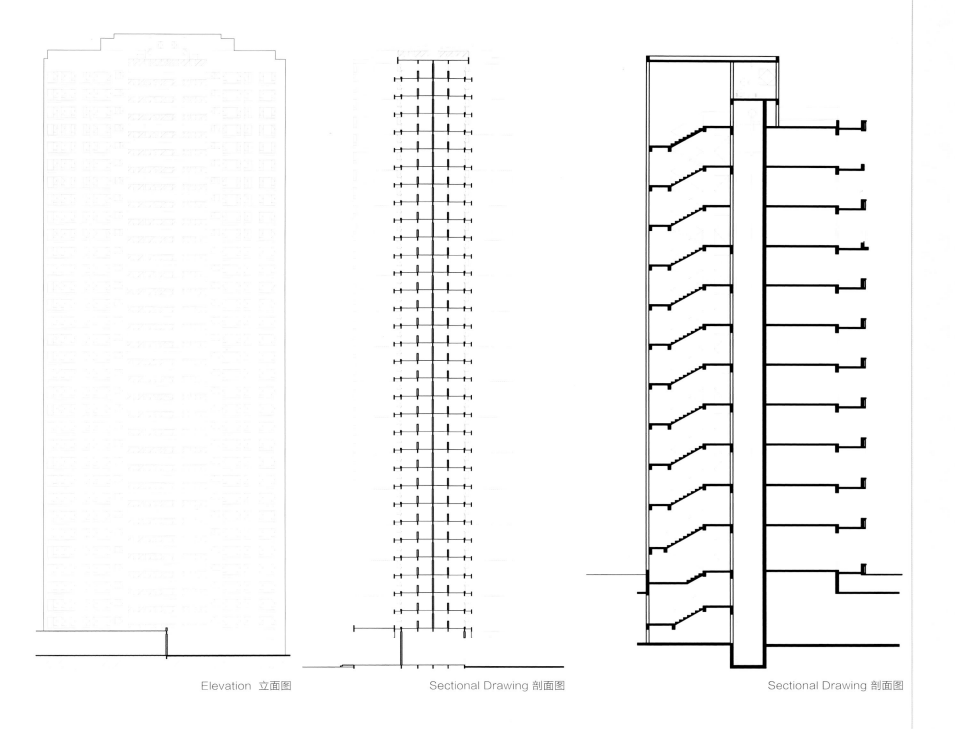

Elevation 立面图　　　Sectional Drawing 剖面图　　　Sectional Drawing 剖面图

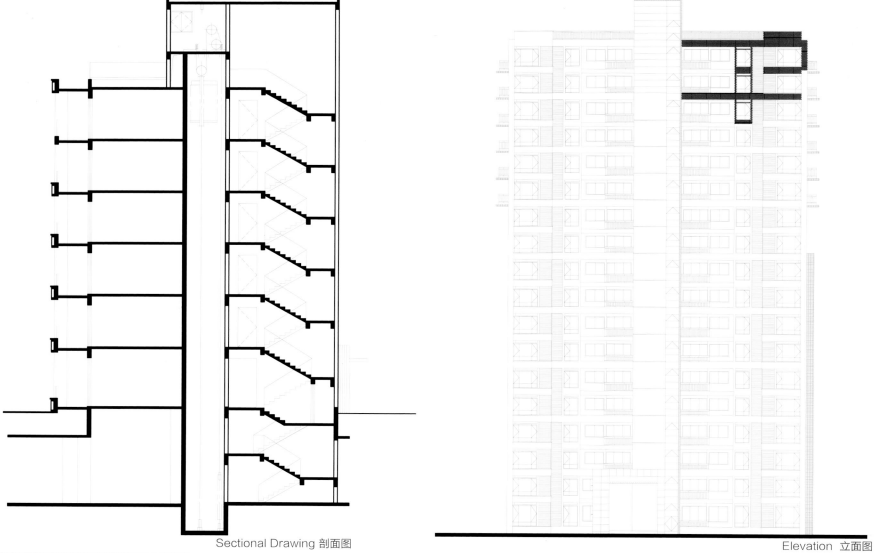

Sectional Drawing 剖面图

Elevation 立面图

Architectural Style 建筑风格

The buildings are designed in modern simple style to keep unified and present some rhythmic changes. Balconies, bay windows, roof trusses and the floating slabs of the exterior wall are combined to form innovative and unique building facade. In addition with the elegant colors, it will create a graceful and dignified atmosphere.

小区建筑采用现代简约主义的设计风格，设计手法强调统一中的韵律变化，通过阳台、凸窗、屋顶构架及外墙飘板等设计元素的组合，使建筑造型新颖独特，立体感强，凸显建筑的个性化，形成高尚、整齐、大气的建筑氛围，典雅的色彩配搭，更加能烘托出整个小区雅致高贵的品质。

Floor Plan 平面设计

The floor plan allows all units to have natural ventilation, natural light, good orientation and views, creating a comfortable interior environment and improving the life quality. It also pays attention to the privacy of the residents by avoiding direct sight.

All these units are well arranged and designed in varied types to meet different requirements. The living room and bedrooms enjoy open and direct views. Many units are designed with home gardens and wide balconies, and all units boast bright kitchen, bright bathroom and spacious living room.

住宅平面设计使各个单元均有良好的自然通风采光及景观，每栋户型具有良好的景观及朝向，营造良好的室内环境和空气品质，最大限度提升其舒适度质量。设计注重各户之间的私密性，尽量避免了相互之间的视线干扰。

各住宅单元布局合理，有多种户型以满足不同需求。客厅及卧室视野开阔，无视线干扰。户型设计中均设置了入户花园及宽阔景观阳台，每户住宅均为明厨明厕，客厅宽敞明亮。

Reconstructing the Memories of Alhambra Palace
重构阿尔罕布拉宫的回忆

Hongbao Villas, Lohas Island, Suzhou 苏州中海·鸿堡

Location/ 项目地点：Suzhou, Jiangsu, China/ 中国江苏省苏州市
Developer/ 开发单位：China Overseas Property / 中海地产
Architectural Design/ 建筑设计：Urban Architecture / UA 国际
Land Area/ 用地面积：39 679 m^2
Floor Area/ 建筑面积：50 600 m^2
Plot ratio/ 容积率：0.58
Greening Coverage Rate/ 绿化率：40%
Types/ 项目类型：Residential Villas / 住宅别墅
Status/ 项目状态：Completed / 已建成

Keywords 关键词
Suzhou Features; Unique Style; Lighting Effects
苏州特色 风格独特 光影效果

"Excellent Engineering" Award
荣获"优秀工程"奖

The project organizes each unit space by lanes and yards, from lanes into yards, then from the yards into doors, constructing spatial texture of Suzhou features. Hongbao creates the spatial mood of going along the lanes through yards and into doors. Molding style selects Islamic forms, taking Alhambra Palace as the motif design.

苏州中海·鸿堡以巷、院来组织各个单元空间，由巷进院，由院入户，构建具有苏州特色的空间肌理。鸿堡营造出一种过巷入门庭、进院落的空间意境。造型风格选用伊斯兰形式，以阿尔罕布拉宫为母题来设计。

Overview 项目概况

The project is located in Dushu Island, south of Dushu Lake in Suzhou. After combing through and interpreting Suzhou City's texture and context, phase II block B organizes each unit space by lanes and yards, from lanes into yards, then from the yards into doors, constructing spatial texture of Suzhou features. Hongbao adopts overall underground garage to separate people and vehicles, which meets the requirements of modern life and meanwhile creates the spatial mood of going along the lanes through yards and into doors. Molding style selects Georgian and Islamic forms, taking Alhambra Palace as the motif design.

项目位于苏州独墅湖南侧独墅岛上，2期B地块通过对苏州城市肌理和脉络的梳理解读，以巷、院来组织各个单元空间，由巷进院、由院入户，构建具有苏州特色的空间肌理。采用整体地下车库进行人车分流，满足现代人生活需求的同时，营造一种过巷入门庭、进院落的空间意境。造型风格选用乔治亚与伊斯兰两种形式。其中鸿堡是以阿尔罕布拉宫为母题来设计的。

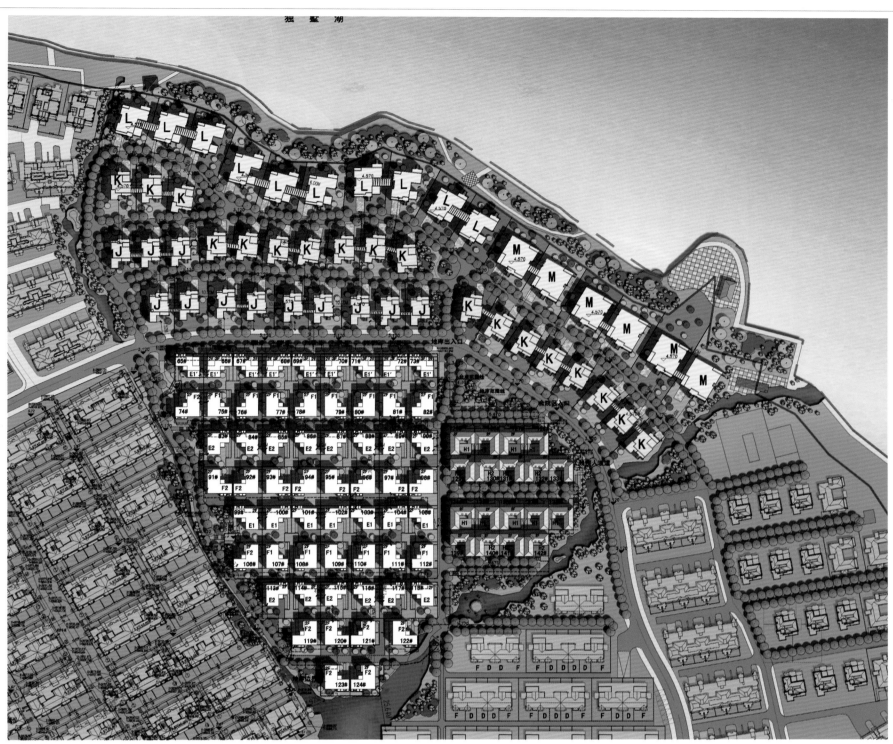

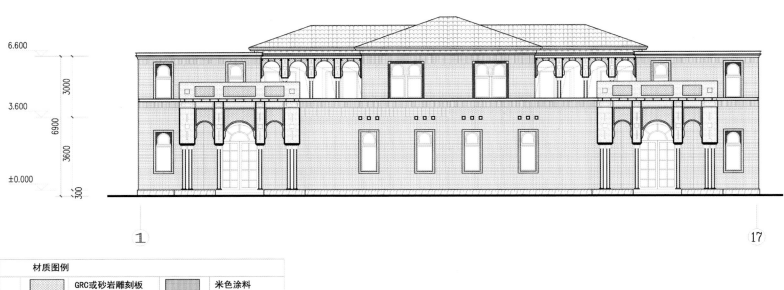

材质图例	
GRC或砂岩雕刻板	米色涂料
米色劈开砖竖贴	暖黄色真实漆
石材	灰色波型瓦
熟褐色涂料	

注：工字钢颜色为深灰色

Side Elevation 轴立面图

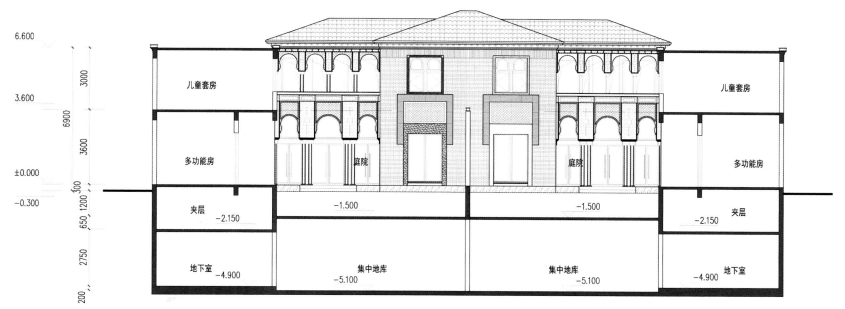

Sectional Drawing 剖面图

Design Thought 设计思路

Hongbao attempts to reconstruct the memories of Alhambra Palace from several aspects, such as courtyards, water, lighting, decoration and materials.

Courtyard

Courtyard is the carrier of Chinese traditional lifestyle. Hongbao refers to the features of Alhambra Palace and forms multi-level spatial construction which is private or open in different size such as the entrance, main courtyard, backyard and sunken garden, defined by porch, enclosure, screen wall and some non-architectural elements of plants and sculptures.

Colonnade constitutes an essential element of Shiquan courtyard to form lighting effects of time transferring between daylight and moonlight. Comparing to Shiquan courtyard, approximately 50 m^2 of main yard is small but complete, and it cleverly borrowed the golden ratio of Shiquan Courtyard. Colonnade arch and column are also divided by golden ratio. The use of colonnade creates a rich courtyard spatial level.

鸿堡试图从院落、水、光影、装饰、材料等几个方面重构阿尔罕布拉宫的回忆。

院落

院落是中国人传统生活方式的载体。鸿堡即是参照阿尔罕布拉宫院落特点，通过廊子、围墙、照壁以及一些非建筑元素如植物、雕塑等界定领域，形成玄关、主院、后院、下沉庭院等大小不一或私密或开放的多层次空间结构。

柱廊是构成狮泉庭院不可或缺的要素，白天和月夜之间能够形成时间流转的光影效果。鸿堡五十见方的主院对比狮泉庭院可谓具体而微，但巧妙在于借用了狮泉庭院的黄金比例，且柱廊拱顶及列柱同样是按黄金分割的比例来划分的。柱廊的采用营造出了庭院丰富的空间层次。

Water

The use of water in this courtyard is of considerable ingenuity. Water overflows out from the water stoup down to the pool in the yard, and then links into a loop along the ditch in the edge of yard to accept sunlight and raindrops falling from above. The pools also play the role of waterscape effect in Myrtle courtyard. Buildings and myrtle bushes reflected on the water together, thus heavy buildings at once look more pleasing, visually reducing the heaviness of architecture. The water that can catch sunlight is medium for buildings, plants and sunlight interactively interweaving to produce charming and colorful effect.

水

庭院中对水的利用颇具匠心。水从水钵中溢出来跌落入下方占了小半个庭院的水池，沿院子边缘的水渠联通成一个回路，接纳从上方落下的阳光与雨水。水池的存在也同样达到了爱神木庭院的水景效果，建筑与爱神木树丛一起倒映水中，厚重的建筑立刻轻盈灵动起来，可减轻建筑在视觉上的沉重感。而水又可以捕捉阳光。水成了建筑、植物、阳光互动的媒介，相互交织产生一种旖旎多姿的效果。

Light and shadow

Alhambra Palace makes clever use of light. To achieve the same effect, Hongbao deliberately sets up an antechamber connecting the hallway. The contrast caused by slight darkness in antechamber and dazzling sunlight in courtyard creates a surprise after startle.

Deep into the courtyard, half in sunlight and half in shadow, the sunlight falls from the oblique eaves down to the pool, and then is reflected to the top of colonnade and arch, creating a mysterious atmosphere of blurred light and shadow. The light color is rapidly changing, showing the magical flow of space and time in quiet.

Decoration

Alhambra Palace's beauty lies in the Moorish decoration which is seen here and there. Simple and abstract pattern is usually used to be infinitely superimposed and replicated to produce complex graphics rich in rhythm and change. Alhambra beautiful checkerboard pattern is engraved on top of the arch, so that the former heavy buildings seem light and lively. Alhambra original complex horizontal decorative band here is reduced to a dark U-shaped sash, revealing a little faint Alhambra intention. Sunlight and lamplight go through the classic Moorish windows and indoor grilles and form the geometric patterns overlapping with similar pattern on the floor.

Materials

Hongbao adopts plenty of contemporary materials to interpret traditional mood. I-beam embedded anticorrosive woods replaces Alhambra white marble pillars, the former intricate frieze is replaced by I-beam and skirting board and walls paved with colorful glazed tiles are also replaced by modern material of split tiles and real stone paint. Thus there are less extravagance and more modern elegance.

光影

阿尔罕布拉宫对光有巧妙的运用。为达到相同的效果，鸿堡刻意设置了一个与玄关相连的前厅，途径微暗的前厅突然迎来庭院耀眼的阳光，光亮和黑暗的反差造成一种错愕之后的惊喜。庭院深深之中，一半阳光，一半阴影，阳光从檐角斜铺下来落在水池中，再被反射到柱廊与拱券之上，营造出光影迷离的神秘氛围，光色瞬息变化，安静闲适之中可感受到时空流转的幻化之妙。

装饰

阿尔罕布拉宫的精美在于摩尔式的装饰。常常利用简单抽象的图案无限复制叠加，产生富有韵律和变化的复杂图形。鸿堡中摩尔式的装饰图案无处不在。萃取阿尔罕布拉宫精美的棋盘格式装饰图案镂刻于拱券之上，使原本厚重的建筑显得轻盈活泼。阿尔罕布拉宫原本复杂的横向装饰带在这里也被简化为一条深色的U型饰带，隐约之中尚能透出些许阿尔罕布拉宫的意向。经典摩尔式镂窗和室内格栅则可以把阳光或灯光过滤成几何形图案叠覆在具有相似花纹的地板上，相映成趣。

材料

鸿堡采用了大量当代材料来诠释传统意境。在这里嵌有防腐木的工字钢替换了阿尔罕布拉宫白色大理石的柱子，原本复杂的横饰带被简约的工字钢所代替，铺砌有绚丽的釉面砖的壁脚板及墙身也替换成了现代材料劈开砖和真石漆，虽少了奢靡之气但不失现代的高贵典雅。

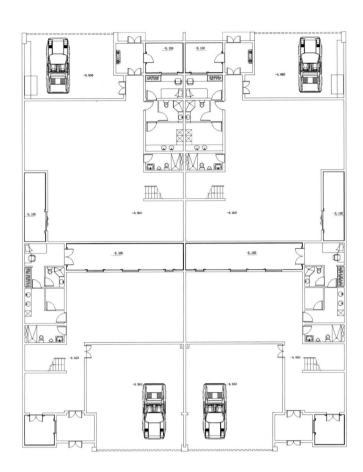

Plan for Basement Floor 地下一层平面图

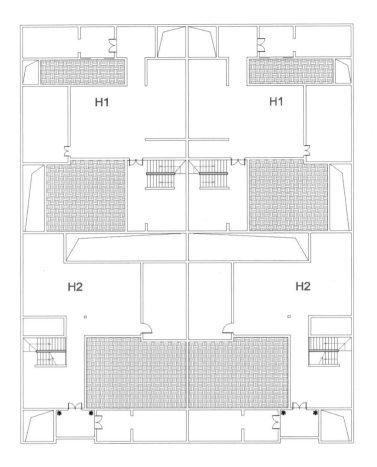

Basement Mezzanine Plan 地下夹层平面图

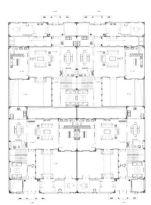

First Floor Plan　一层平面图

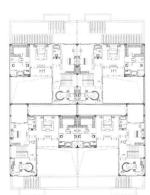

Second Floor Plan　二层平面图

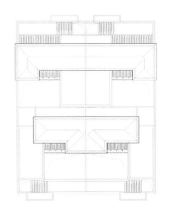

Roof plan　屋顶平面图

Mueju Encyclopedia　美居小百科

Alhambra, located in the hills in the east of Granada city, is Spain's famous imperial palace, which is palace of Granada kingdom established by the medieval Moorish in Spain. Alhambra means red fort in Arabic word. Alhambra Palace is the essence part of all the monuments that Moorish Spain retained in Spain, known as the " the City of Palace" and "wonder of the world". In terms of the architectural history, the Alhambra Palace is a master piece of aesthetics art just like Chinese Epang palace, Tongquetai (Bronze Sparrow Terrace) and many other buildings.

阿尔罕布拉宫，坐落在格拉纳达城东的山丘上，是西班牙的著名故宫，为中世纪摩尔人在西班牙建立的格拉纳达王国的王宫。阿尔罕布拉，阿拉伯语意为"红堡"。阿尔罕布拉宫为摩尔人留存在西班牙所有古迹中的精华，有"宫殿之城"和"世界奇迹"之称。从建筑历史而言，阿尔罕布拉宫与中国的阿房宫、铜雀台等诸多建筑一样，都是美学艺术的集大成者。

Living in Beautiful Landscape
缔造在风景里生活的岁月

Hongju Mediterranean　宏聚·地中海

Location/项目地点：Yuhua District, Changsha, Hunan/湖南省长沙市雨花区
Architectural Design/设计单位：Ming Zhu Design Group/明筑设计机构
Developer/开发商：Hunan Hongju Real Estate Development Co., Ltd./湖南宏聚房地产开发有限公司
Total Land Area (Phase I)/一期总用地面积：102 807.8 m²
Total Floor Area (Phase I)/一期总建筑面积：268 938.2 m²
Residential Land Area (Phase I)/一期住宅用地面积：558 39.7 m²
Green Land Area (Phase I)/一期集中绿地用地面积：3 000 m²
Category/项目类型：High-rise Residences/高层住宅
Status/项目状态：Built/已建成

Keywords　关键词

Exotic Style; Bright Color; Relaxed and Cozy
异域风情　色彩明快　轻松惬意

"Excellent Engineering" Award
荣获"优秀工程"奖

With the theme of "leisure home of South European Style", the facade is designed in classic Mediterranean style with bright colors and lively lines.

项目以"南欧风情的休闲家园"为立意起点，建筑立面选择典型浓郁的地中海风格，使用鲜明跳跃的色彩和明快的线条作为建筑造型的基本语言。

Overview　项目概况

The site is located on Times Sunshine Avenue, Yuhua District of Changsha City, belonging to the most rapidly developed government area which will be the future new CBD of the city. Built as a large-scale and high-end community integrating residences, commerce, hotel, clubhouse and kindergarten, the project is divided into two parts (north and south) which are developed respectively. The residential part offers totally 1,480 units, boasting a plot ratio of 3.0; the hotel part has a plot ratio of 5.5; the kindergarten 0.8.

宏聚·地中海位于长沙市雨花区时代阳光大道，属于长沙目前发展最快的南城核心位置——省府板块，是城市未来的新商业中心区。项目是一个集住宅、商业、酒店、会所和幼儿园于一体的大型高尚社区，分南北两部分地块开发，其中住宅部分容积率为3.0，总户数为1 480户；酒店部分容积率为5.5；幼儿园部分容积率为0.8。

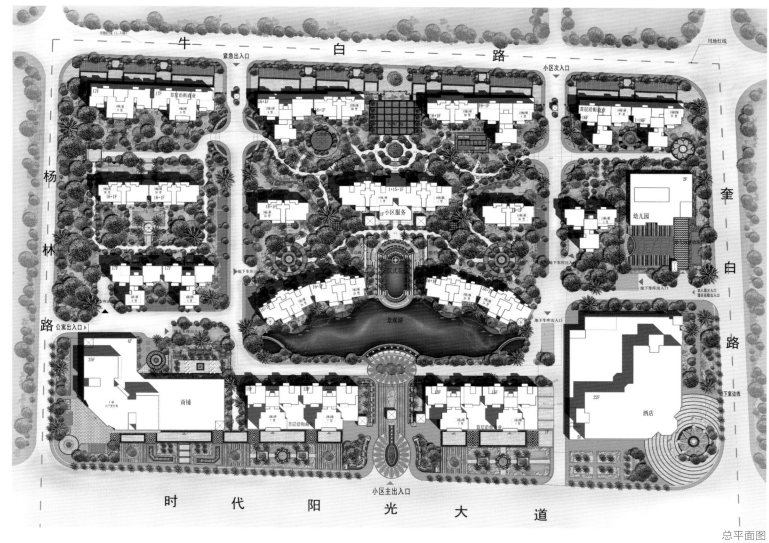

总平面图

Design Idea 设计理念

The design idea comes from the Greek word "CORSO", which means "lakeside" and describes a kind of cozy and romantic lifestyle. Inspired by this idea, the development aims to be the "leisure home of South European Style". There are mainly high-rise residential buildings, commercial facilities, star hotel, central waterscape and swimming pool to draw the outline of the development. The facade is designed in classic Mediterranean style with bright colors and lively lines. In addition with typical Mediterranean elements such as the sandstone walls, iron railings and vermilion imbrex, it creates a relaxed, cozy, sunny and pleasant living environment. Through landscape design, it has presented a Mediterranean style cityscape by using lake, landscape walls, sculptures and Roman colonnades, describing the elegant and wonderful Mediterranean lifestyle and highlighting the theme of "living in the scenic spot".

本案的理念为"CORSO"。"CORSO"来自希腊语，意为"湖滨"，描绘的是一种尽享悠闲的浪漫生活。在"CORSO"理念的指导下，项目以"南欧风情的休闲家园"为立意起点，整个项目主要由高层商品房住宅、配套商业、星级酒店、中心水景和特色泳池为主线的总平面概念跃然纸上，建筑立面选择典型浓郁的地中海风格，使用鲜明跳跃的色彩和明快的线条作为建筑造型的基本语言，并引入砂岩石墙、铁艺栏杆、朱红色筒瓦等地中海元素，营造轻松、惬意、阳光明媚和时尚宜人的生活情境。其景观设计在于缔造地中海式城市风貌，利用湖泊、景墙、特色雕塑、罗马柱廊等讲述地中海风情与简洁时尚的优雅精彩，体现"风景区居住生活"的设计理念。

Overall Layout 总体布局

The project is planned with "two axes and one heart". Taking advantage of the surrounding roads and the topography, the main entrance is set in the south. When entering into the community through the main entrance, two main roads on the left and right run through from south to north, forming "two axles" of the community — the main landscape axis and the lifeline. A 4,500 m² artificial lake is set where two axles meet with the main entrance, becoming the "heart" of the community.

Phase I consists of 17 medium-rise buildings with an average height of 50 m. Built mainly in "south-north" orientation and arranged in groups, these buildings can enjoy natural ventilation, sufficient sunlight and open views, and they are skillfully staggered to form a beautiful skyline.

在总体布局上归纳为"两轴一心"的形学格局。根据周边道路和现有地形的特点，在用地南侧设置小区主入口，从主入口进入小区后，左右两条主干道南北连通整个小区，形成小区的主要景观及生命轴线，即"两轴"；两条主轴与主入口相交，人流最为密集处营造出 4 500 m² 的景观湖面，形成小区独特的主景区，即"一心"。

项目第一期由 17 栋小高层组成，平均高度约 50 m，遵循南北向摆布，采用围合式组团规划，充分满足各楼栋最适宜的阳光照射度，也保证了开阔的景观视野。在获得良好的采光和通风的同时，设计师在设计中注重天际线的高低错落，使整个建筑群的天际线丰富、协调。

Facade Design 建筑立面

With the concept of "leisure home of South European Style", the designers have defined the facade in Mediterranean style, trying to enable residents to experience the poetic holiday lifestyle with pretty sunshine, bright colors and exquisitely designed details and materials. The characteristic cornice, arch, imbrex, cultural stone, iron decorations and wooden frameworks are combined to enhance this kind of atmosphere and create colorful space experience. What's more, with terraces on different levels and building blocks staggered, it has formed a more textured facade which makes the buildings more dynamic and distinguished.

为体现"南欧风情的休闲家园"这一立意，设计师将住宅立面处理为地中海风格，期翼以地中海建筑明媚的阳光感、鲜亮的色彩以及人性化的建筑细节与材质，真正给住户一种休闲度假的写意生活。风格独特的挑瞻、拱券、筒瓦配合文化石及铁艺装饰、木构架，把地中海风格体现得淋漓尽致，使整个立面设计充分展示出层次感和丰富的空间表情，浓郁的地中海风情扑面而来。另外立面通过体量化解、退台、错落等设计手法，使立面形象而更高级的物业类型靠拢，使建筑形象更加生动，更有品质感和协调性。

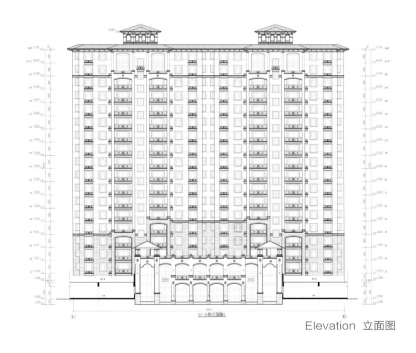

Elevation 立面图

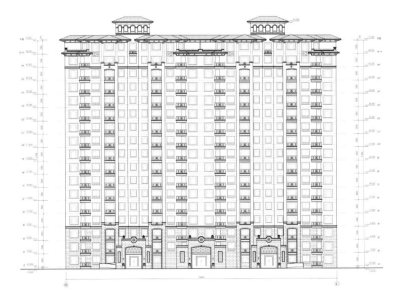

Elevation 立面图

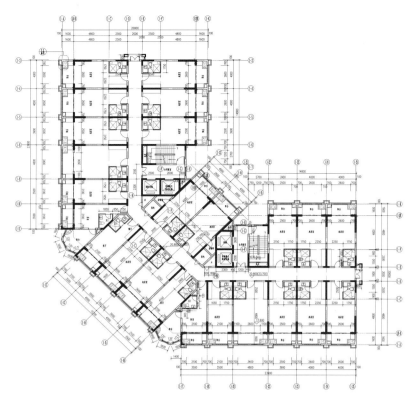

Standard Floor Plan for the Apartment 公寓标准层平面图

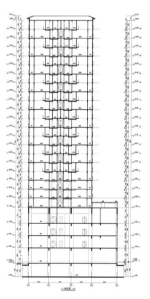

Sectional Drawing 剖面图

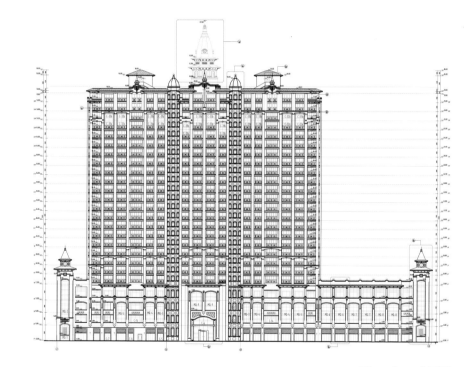

Elevation 立面图

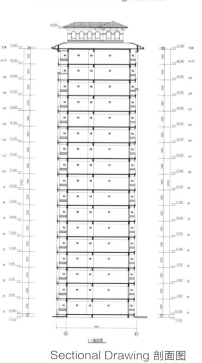

Sectional Drawing 剖面图

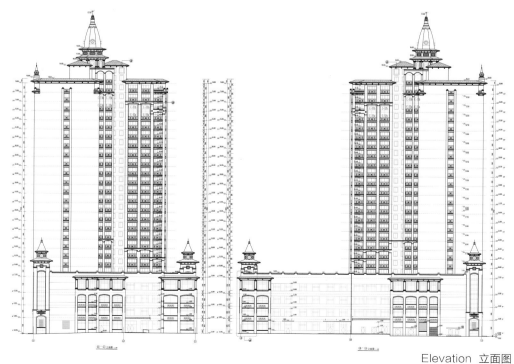

Elevation 立面图

Landscape Design 景观设计

The landscape design aims to present a Mediterranean style cityscape. Taking advantage of the elevation difference and elevating the underground floor, it has created magnificent waterfalls and an endless pool. Buildings enclose a large central lake which is the soul of the landscape system, enjoying luxurious and pleasant living spaces. The sunken pool and the lake are visually connected, increasing the view levels and introducing the landscape into the underground parking. In this way, landscape will change and continue in the indoor and outdoor spaces, adding great value to the project.

The design has focused on the sculptures and the key elements to well present the main theme. The sculptures bring healthy and fresh waterscapes as well as fantastic experience. The landscape walls and other hard landscapes also make reference to traditional Mediterranean style to create a natural, colorful and cheerful atmosphere with wood, natural stone and patterns.

项目景观设计在于缔造地中海式城市风貌，充分利用地面高差，抬高半层地下室，从而打造出宏大的叠水瀑布和下沉式无极泳池。围合式建筑布局围绕出超大水面的湖景，打造最佳空间尺度，营造宜居空间的低调奢华。中央湖景的设计，使水景成为小区园林的灵魂。中央下沉景观泳池与景观湖面在视线上连成一片，增加了水面的进深感，下沉的设计，不仅使整个园林层次感更强，也将景观引入地下车库，使地下车库更通风、更明亮，同时塑造出步移景换的内外部空间效果，更加强了住宅的景观价值，凸显楼王单位的珍贵。

园林设计中，雕塑和主题元素的运用是设计主题的立体重点，力求把主题特色呈现眼前。雕塑品展现梦幻式的感受，并带出健康清新的水景主题。由此取用传统地中海式设计细部风格作立体墙壁和其它硬景元素的设计蓝本，以达立体的空间效果，并配以木材和天然石块、图案设计，强调出天然设计的效果和缤纷色彩的欢乐气氛。

Unit Design 户型设计

Designed with "four units to share one lift", "three units to share one lift" or "two units to share one lift", the floor plans have tried to get the best ventilation, daylight and landscape views, reduce interference, control the depth-width ratio, avoid irregular plane and lower the construction cost for the high-rise buildings.

The unit design has well organized different functional spaces to ensure a comfortable living environment. It has attached great importance to the design of the family activity center, the living room and the dinning room. Both the master bedroom and the living room open to the south to get cross ventilation and sufficient daylight. It's also paid attention to save land, energy and materials, and integrate the residential spaces and with the cultural atmosphere, trying to meet different living requirements.

采用"一梯四户"、"一梯三户"和"一梯两户"的建筑平面组合方式，在满足通风，采光，取景和减少干扰的前提下，控制高宽比，避免平面不规则，以确保高层建筑在经济合理范围内降低建造成本。

户型设计上，合理安排各功能空间，按公私分离、食寝分离、洁污分离的原则，保证居住的舒适性。重视家庭活动中心，起居厅和餐厅的设计。主卧室和起居室均朝南，充分利用良好朝向，有效组织南北的自然通风，做到全明设计。同时注重节地、节能、节材措施，重视住宅群体空间特色设计和住宅尺度设计与整体形象和文化氛围的协调性。满足住户居住生活的多样性和个性化的生活需求。

High-end Residential Community of Exotic Style

富于异域风情的高尚生活小区

CITIC Lake Forest Orchid Valley   中信森林湖兰溪谷

Location/项目地点：Dongguan, Guangdong/广东省东莞市
Developer/业主：CITIC Real Estate Co., Ltd./中信地产
Architectural Design/建筑设计：AIM Group International/加拿大AIM国际设计集团
Land Area/用地面积：158 290 m²
Floor Area/建筑面积：365 000 m²

Keywords 关键词

Exotic Style; User-friendly Design; Green and Ecological
异域风情 人性设计 绿色生态

"Excellent Engineering" Award
荣获"优秀工程"奖

Designed mainly in South Californian Spanish style, Orchard Valley has presented its own architectural languages in accordance to the local residential requirements. The design of the facade has focused on volume, western scale and colors, creating an exotic-style community for the residents.

兰溪谷整体以美国南加州西班牙建筑风格为主题，结合东莞本地居住者在建筑空间与平面布局上的要求，形成独具特色的建筑语言，建筑立面注重体量及西部尺度，色彩处理，为居住者精心打造一个富于异域风情的高尚生活小区。

Overview 项目概况

Located in the south of downtown Dongguan, Guangdong, the development provides Spanish-style buildings. It aims to build a high-quality modern community which features the romance and elegance of Spanish style. Orchard Valley, phase V of CITIC Lake Forest, nestles at the south end of the site, opposite to the eco lake in the north, close to the green world and Dongguan Botanical Garden in the south, and overlooking Mt. Shuilian Park. There are sixteen 11- ~ 33-storey apartment buildings in Orchard Valley, offering totally 2,400 residential units which range from 120 m² to 400 m². These buildings stand around the lake and rise gradually from north to south, allowing most of the units to have a view of the 70,000 m² lake.

项目位于广东省东莞市区的南面，为西班牙建筑风格。全力打造一个充满西班牙浪漫典雅生活气息、高品质的时尚住宅小区。中信森林湖兰溪谷是中信森林湖的第五期产品，位于森林湖项目的正南端，北面正对森林湖原生态湖泊，南侧紧邻绿色世界、东莞植物园，可远眺水濂山公园。兰溪谷总户数约2 400户，是由16栋11~33层的高层洋房组成，产品面积为120 m² ~ 400 m²。兰溪谷绕湖而建，由北向南逐级抬升，绝大部分单位可望森林湖70 000 m²超大湖面。

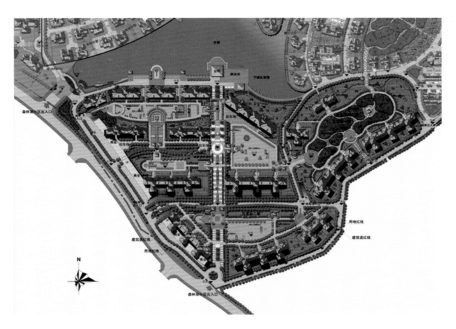

Site Plan 总平面图

Architectural Style 建筑风格

Designed mainly in South Californian Spanish style, Orchard Valley has presented its own architectural languages in accordance to the local residential requirements. The design of the facade has focused on volume, western scale and colors, creating an exotic style community for the residents.

兰溪谷整体以美国南加州西班牙建筑风格为主题，结合东莞本地居住者在建筑空间与平面布局上的要求，形成独具特色的建筑语言，建筑立面注重体量及西部尺度，色彩处理，为居住者精心打造一个富于异域风情的高尚生活小区。

Landscape Design 景观设计

The costly landscape is designed with the idea of "Royal Botanic Garden". It not only creates a "Broadway" of precious trees but also sets different theme for each garden according to the requirements of the owners. Friendly walk paths, children's playground, pavilion and corridor as well as the square for gathering, will encourage people to enjoy the landscapes from far away or closely.

The magnificent landscape axis is about 400 m long (the length of three football fields) and 25m wide (the width of three villa axes). The number of the precious trees and the length of the "Broadway" have surpassed that of CITIC's previous projects. With the long axis running through Orchard Valley, it defines a geomantic pattern that embraces the mountain and lake.

380 m long corridor beside the lake allows residents to fully enjoy the beautiful lake views which are usually presented for lakeside villas only.

园林以"皇家植物园"的理念重金打造，除了是名树百老汇之外，还根据业主需求，每个园林组团都设置不同主题，设置人性化的散步路径，还有儿童游乐设施、休息的亭廊、聚会的广场，让人充分参与到园林中去，能"远观"，也可以"近享"。

恢宏大气的景观中轴，长约 400 m，宽约 25 m；长度相当于 3 个足球场的超阔距离，宽度相当于别墅中轴的 3 倍，比别墅更大气；名树数量、规划长度均超越中信以往项目。长轴贯穿兰溪谷形成一个山湖相望的风水格局。

380 m 湖滨长廊，让人充分享用湖面所带来的惬意，一线湖居别墅才能私享的湖岸线，在兰溪谷就可以尽情享有。

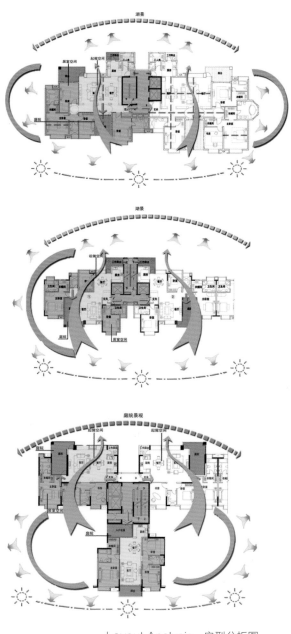

Layout Analysis 户型分析图

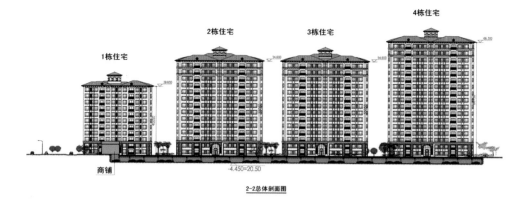

2-2总体剖面图

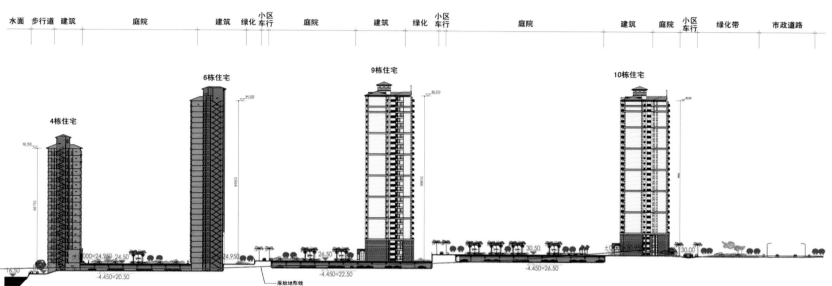

Overall Section 总体剖面示意图

China's Most Beautiful Residential Landscapes
中国最美人居景观

Aesthetics	美学性
Livability	宜居性
Commercial Value	商业价值性
Humanity	人文性
Sustainability	可持续性

Top Mansion in the Urban Area
内环内品牌轨交豪宅

Poly Phili Mansion, Shanghai　　上海保利翡丽云邸

Location/ 项目地点：Shanghai, China/ 上海市
Developer/ 开发商：Poly Property Group/ 保利置业
Architectural Design/ 建筑设计：Urban Architecture /UA 国际
Planning & Landscape Design 规划 / 景观设计：De Design Co., Ltd. / 上海地尔景观设计有限公司
Land Area/ 占地面积：19 284.3 m²
Floor Area/ 建筑面积：2 887 m²
Plot Ratio/ 容积率：2.0
Green Coverage Ratio/ 绿化率：41%

Keywords　关键词

Courtyard Pattern; Ancient Orchestra - Bayin; New Asian Style
院落模式　古代八音　新亚洲主义

"China's Most Beautiful Residential Landscapes" Award
荣获"中国最美人居景观"奖

Landscape designers integrate the connotation of Bayin, metal, stone, string, bamboo, gourd, clay, leather and wood, with the courtyard design ingeniously to play a beautiful symphony for the city.

项目将古代八音——金、石、土、革、丝、木、匏、竹的内涵巧妙融入进院落，使之与未来的居住场景有机融合，诚然为这座城市谱写了一首交响曲。

Overview　项目概况

Located in the urban area of Shanghai City and close to Lujiazui and CBD of the Bund, only 324 m away from Dalian Road Station of Metro Line 4 and Line 12, Phili Mansion, developed by Poly Real Estate, is a comprehensive project that collects residence, commerce and offices. It is designed in New Asia style; while for landscape design, corridors, private clubs with bottom overhead and eight landscape themes are carefully designed. Phili Mansion is an intelligent comprehensive community that has passed the LEED certification Level 3, including solar energy, permeable pavements and natural lighting underground garage.

项目位居上海市中心、内环内，邻近陆家嘴、北外滩CBD，距离4、12号线大连路站324 m。项目是集住宅、商业和办公于一体的综合性项目，由保利置业集团打造。翡丽云邸建筑风格采用新亚洲主义；在景观设计上，选用风雨连廊、底层架空私家会所、八大景观主题设计等风格；翡丽云邸是绿色三星认证项目，项目是集太阳能热水入户、透水路面、自然采光车库等为一体的智能化综合体社区。

Site Plan 总平面图

Architectural Design 建筑设计

Adopting the New Asian style, the architecture of the project is designed with concise masses, graceful proportion, bright colors and exquisite details which have been the features of the community. Through careful control and application on the materials and colors, and selection on the modern materials and technology, architects have succeeded in an elegant, warm and romantic atmosphere in the development and revealing the characteristics of modern times.

项目的建筑风格以新亚洲现代建筑风格为主，建筑以体块简洁、比例优美、色彩明快、注重细部处理的手法为其鲜明的特征，造型现代大方，通过对细部的雕琢和材质及颜色恰当地控制和运用，营造出优雅统一以不失温馨浪漫的社区氛围，适当运用新技术新材料，更好地反映时代的特征。

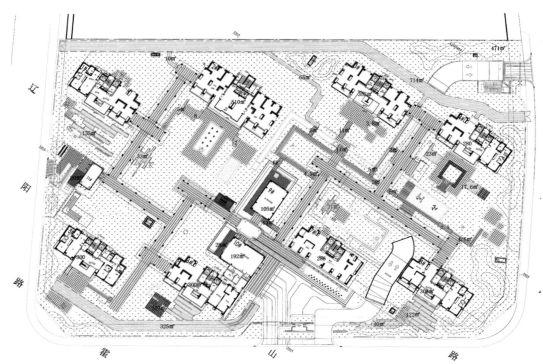

Landscape Design Concept 景观设计理念

Courtyard Pattern
With a house and a yard in a courtyard is the expression of richness and dignity of users.

As an important part of Chinese architecture, courtyard is enclosed space which is the typical living model in China with profound culture heritage. The landscape of courtyard is designed with New Asian style including the elements of pavilion, shed, enclosure, corridor, path and furnishings.

Bayin Elements
Bayin stands for eight categories of musical instrument in ancient orchestra. In the book of Zhouli Chunguan Dashi, it says that Bayin refers to metal, stone, string, bamboo, gourd, clay, leather and wood. The musical instrument is not only for playing music, but also performed as entertainment tools and decorations, they even can be used for communication in the ancient time. Landscape designers integrate the connotation of Bayin with the courtyard design ingeniously to play a beautiful symphony for the city.

院落模式
院有宅必有园，始为富贵所在。

院落是中国建筑的重要组成部分，是中国的典型居住模式，它植根于深厚的传统文化，其基本形式是围合空间，以院子的形式而存在。景观以新亚洲院落风格为主，包括间、亭、榭、围墙、回廊、花木、路径、陈设等元素。

八音元素
八音是中国古代对乐器的统称。《周礼·春官·大师》云："皆播之以八音，金、石、土、革、丝、木、匏、竹。"中国的乐器除了主要作为演奏音乐之用外，其实还有种种其他功能。它不但是各个时期的娱乐用器和装饰摆设，更是重要的礼仪及传讯用器。将古代八音的内涵巧妙融入进院落，使之与未来的居住场景有机融合，诚然为这座城市谱写了一首交响曲。

Revealing Gardens in Tang Dynasty Poem Through Interesting Landscape
趣味景观彰显园林唐朝诗韵

Poly Oriental Mansion　北京保利东郡

Location / 项目地点: Chaoyang District, Beijing, China / 中国北京市朝阳区
Developer / 开发单位: Poly (Beijing) Real Estate Development Co., Ltd. / 保利（北京）房地产开发有限公司
Landscape Design / 景观设计: L&A Design Group / 奥雅设计集团
Land Area / 用地面积: 46 188 m²
Landscape Area / 景观面积: 34 414 m²
Plot Ratio / 容积率: 2.8
Green Coverage Ratio / 绿化率: 32%

Keywords　关键词

Enclosed Gardens; Artistic Courtyards; Zen Habits
围合造园　诗意花园　禅意趣味

"China's Most Beautiful Residential Landscapes" Award
荣获"中国最美人居景观"奖

Adopting new Chinese garden as the design concept, designers are inspired by the popular poem Ti Po Shan Si Hou Chan Yuan, written by Chang Jian of Tang Dynasty. Following the tour path and conception of the poem to form different theme gardens, the project is eager to create an artistic garden with exquisite landscape.

从新中式园林的设计理念出发，项目追寻唐朝诗人常建脍炙人口的《题破山寺后禅院》诗句中游览的路径及意境的描述塑造出不同院落的意境主题，营造一幅具有景观趣味和意境的诗意花园。

Overview　项目概况

Located in the expanded area of Chaoyang District CBD, the project enjoys outstanding natural and transportation resources. Carrying out the corporate value of "Harmony Builds Good Virtue" of Poly Real Estate throughout the project, architects meant to express the cultural heritage of "harmony" and value criterion of "kindness" and to create an oriental private mansion with Sino-west mix and strong characteristics.

项目地处北京朝阳区CBD辐射区；交通和绿化条件优越。整体设计贯彻了保利品牌"和者筑善"的理念，体现保利地产"以和为道"的文化底蕴和"以善为达"的价值准则，建成一个中西合璧、深具灵魂品格的东方私人庭院。

Planning Layout 规划布局

Placing main entrances at the north and south gateways and setting the buildings facing the inner heart to form "enclosed gardens" which is exactly the planning idea of the project; buildings are placed in form of point to maximize the landscape area. The architectural style of the project is neoclassicism cooperating with design language of exotic building skin and Chinese culture. Buildings are designed varying from 10-storey to 18-storey to reveal rich architectural outlines and expressions. The facades are steady and generous which can be a landmark and without being too assertive, the simple and concise design makes the project full of urban atmosphere and modern sense.

　　住宅北侧和南侧各设小区的主要出入口，通过建筑围合，形成"围合造园"的规划理念；小区内部以点式住宅布置，为项目提供了最大化的绿化景观。整个社区建筑风格为新古典主义，中西合璧，充分运用国际外表，中式灵魂的设计语言。建筑从10层到18层错落布置，显现出丰富的建筑轮廓和表情，建筑立面稳重大方，既具备标示性又不过分张扬，简洁的设计使其更具有城市氛围与时代感。

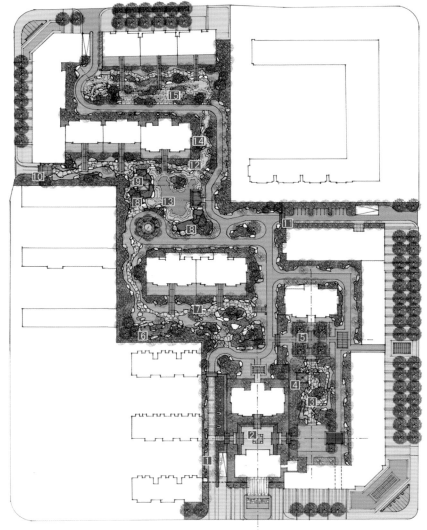

1，林荫道——"清晨入古寺"
2，会所
3，绚秋林——"初日照高林"
4，木亭
5，静明庭——"禅房花木深"
6，健身场地
7，山石苑——"山光悦鸟性"
8．日式修剪灌木群
9，金亭
10，东入口大门
11，西入口大门
12，平桥
13，龙潭——"潭影空人心"
14，叠云溪
15，龙吟庭——"惟闻钟磬声"

Site Plan 总平面图

Unit Design 户型设计

The project is designed with large units. 10 buildings in total, and residential buildings of No. 2, No. 4, No. 6, No. 7 and 8 are designed with 11 units per storey, while office buildings of No. 1 and 3 and commercial building No. 5 are designed with 5 units per floor. And only 4 units type are offered, Unit A is 315 m², Unit B is 285 m², Unit C is 190 m² and Unit D is 165 m², with the floor height of 3.10 m to highlight the dignity of users.

大宅社区，户型纯粹。项目共设计 10 座楼栋，2#4#6#7#8# 住宅楼共有 11 个单元，1#3# 办公及 5# 商业楼共有 5 个单元。仅有四种户型，A 户型 315 m²；B 户型 285 m²；C 户型 190 m²；D 户型 165 m²。层高 3.10 m 聚汇高端人群显赫层级。

1. 入口特色水景　　7. 荷花池　　　　13. 门廊
2. 巨石 logo　　　 8. 特色景墙　　　14. 景观桥
3. 会所　　　　　　9. 种植台地　　　15. 铜灯笼
4. 中心特色水景　　10. 树池　　　　 16. 入口
5. 住宅　　　　　　11. 种植池
6. 连廊　　　　　　12. 景观置石

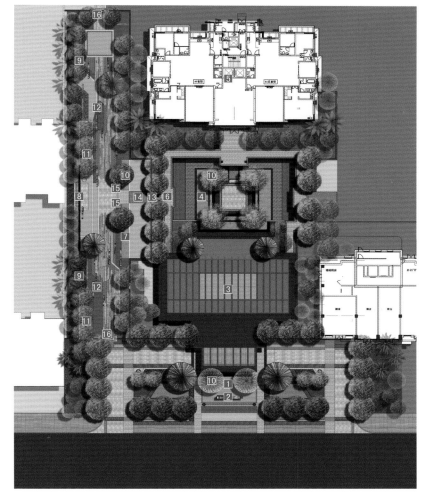

Landscape Design 景观设计

Trying to adopt brand new design idea to explore traditional conception in modern ways, landscape designers learnt from the local history and multi-culture of Poly Real Estate to create a project that is unique and original, and they are also eager to get breakthrough in the landscape design with the highlights on mind purification and sublimation. Not only to meet the requirement of enjoyment, landscape should also be able to reach a higher level of spiritual yearnings that can make both environment and users' soul "empty, smart and quiet".

项目以全新的设计思想来探索传统意境的现代演绎，从本案地域历史和保利自身丰富的多元文化中挖掘题材，打造设计的独创性和唯一性，并试图在景观概念上有所突破，着眼于心灵净化、思想升华。在环境营造上不单单满足人的观赏性，更多的是表达人更高层次的精神诉求，使环境和心灵都达到"空、灵、静"的唯美境界。

Artistic Themes 意境主题

Adopting new Chinese garden as the design concept, designers are inspired by the popular poem Ti Po Shan Si Hou Chan Yuan, written by Chang Jian of Tang Dynasty. Following the tour path and conception of the poem, the project is designed with an axis from south to north connecting six different theme gardens which are "temple", "forest", "flowers", "birds", "pond" and "bell" respectively to create an artistic garden with exquisite landscape. Two pavilions standing for Zen Habits are added to the garden, which are Golden Pavilion and Wooden Pavilion respectively, with the same landscape structure to offer the same functions with different appearance while Golden Pavilion is in yellow color like the gold and Wooden Pavilion uses wood. It reflects the preference and choice regarding vanity and plainness in users' inner hearts.

设计单位从新中式园林的设计理念出发，追寻唐朝诗人常建脍炙人口的《题破山寺后禅院》诗句中游览的路径及意境的描述，使其成为一个主轴从项目南侧串联到北侧，"清晨入古寺""初日照高林""禅房花木生""山光悦鸟性""潭影空人心""惟闻钟磬音"成为不同院落的意境主题，营造一幅具有景观趣味和意境的诗意花园。在整个园区中设计单位加入了一个具有反思性的禅意趣味景观——金亭、木亭。相同形式的景观构筑物都能满足遮风避雨，品茶休憩的功能，不同的是材质金与木的区别。浮华与质朴在每个人的心中都有一种倾向与选择。

First Impression Building of the Capital
首都第一印象建筑

Wangjing SOHO 望京 SOHO

Location/ 项目地点：Chaoyang District, Beijing, China/ 北京市朝阳区
Developer/ 开发单位：SOHO China Ltd. /SOHO 中国有限公司
Landscape Design/ 景观设计：ECOLAND Plan and Design Co., Ltd. /ECOLAND 易兰规划设计院
Cooperative Design/ 合作设计：Zaha Hadid Architects/ 扎哈·哈迪德建筑事务所
Land Area/ 占地面积：115 393 m²
Floor Area/ 建筑面积：521 265 m²
Green Coverage Ratio/ 绿化率：30%
Plot Ratio/ 容积率：8.14

Keywords 关键词

Dynamic Beauty; Huge Garden; New Green Building
动态美感 超大园林 新绿色建筑

"China's Most Beautiful Residential Landscapes" Award
荣获"中国最美人居景观"奖

The unique surface design has made it possible for the buildings to show the dynamic beauties and elegance from every angle. Aluminium plates and glass curtain walls are adopted for the facades, which have integrated with the blue sky to reveal the artistic conception of mountains in the cloud and mist.

项目建筑独特的曲面造型使建筑物在任何角度都呈现出动态、优雅的美感。塔楼外部被闪烁的铝板和玻璃覆盖，与蓝天融为一体，象征了山中云雾缭绕的意境。

Overview 项目概况

Situated in the center of Wangjing, which will be the second CBD of Beijing, Wangjing SOHO enjoys board view and can also overlook Wangjing. It connects with Futong West Street in the east, Fu'an East Road in the south, Wangjing Street in the west and Fu'an West Road in the north.

项目位于未来北京的第二个CBD——望京核心区，视野开阔，俯瞰望京。东至阜通西大街、南至阜安东路、西至望京街、北至阜安西路。

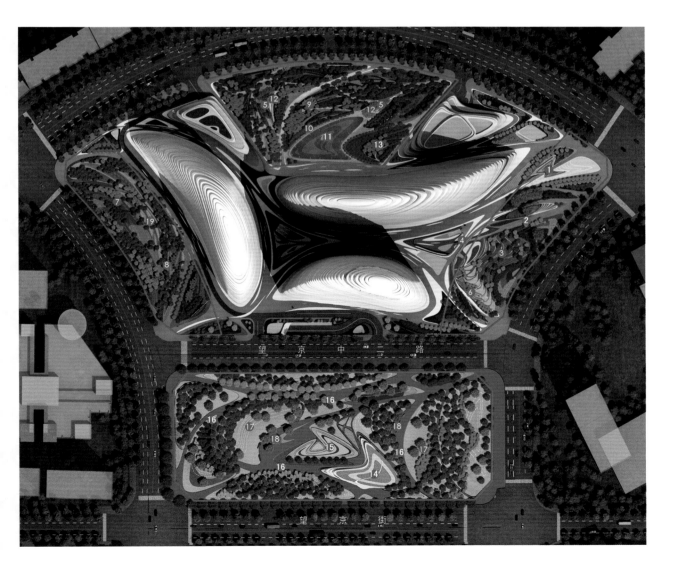

图例

① 音乐喷泉广场
② 步行坡道
③ 疏林慢坡
④ 步行廊桥
⑤ 现状大树
⑥ 跌水花园
⑦ 野趣花岛
⑧ 起伏康体步道
⑨ 林荫漫步花园
⑩ 梨花水岸
⑪ 中心喷泉
⑫ 活力广场
⑬ 林下氧吧
⑭ 公园管理中心
⑮ 中心水景花园
⑯ 漫跑道
⑰ 阳光活力草坪
⑱ 林荫休闲广场
⑲ 花境漫步休憩道

Location Advantage 区位优势

It enjoys sophisticated transportation while it is nearby Wangjing Station of Metro Line 15 and Line 14 that is under construction, close to Wangjingxi Station of Line 13, Sanyuanqiao Station of Line 10 and Airport Express, and it's 25 minutes' drive away to Beijing Capital International Airport, International Trade, Asian Sports Village and Olympic Village.

周边轨道交通便利，地铁14号线（在建）、15号线交汇于望京站，可快速到达地铁13号线的望京西站、10号线和机场快线的三元桥站，25分钟即可抵达首都国际机场、国贸中心、亚运村和奥运村。

Positioning Strategy 定位策略

The project was originally positioned as an urban sub-center that's close to Beijing Fifth Ring Road with the main function as a residential area that can accommodate 600,000 people, while with the development of the city, Wangjing has become an area collecting a lot of IT corporations. The settlement and development of the north area of Dawangjing have made the project become an urban business district of Beijing that is closest to the airport.
Plot B29 is set on the planned central axis and parallel to the Airport Express, lying on Dawangjing and opposite to Wangjing Central Park in the south. The central location and close relationship with surrounding buildings have worked together to make Plot B29 the center and landmark of Wangjing.

项目最初规划是一个紧靠五环，以居住为主的能容纳60万人口的城市副中心。随着城市发展，望京逐渐成为IT公司聚集的区域。北侧大望京的确立和兴建，使项目顿然成为距离机场最近的北京城市商务区。

B29地块处在项目规划的中轴线上，与机场高速平行。背靠大望京，南侧面对望京中央公园，这一特殊的中心位置和与周边楼宇的关系，决定了B29地块必然成为望京的核心和地标。

Planning Layout 规划布局

The project is comprised of three high-rise buildings for offices and commerce, and three low-rise commercial buildings, with the highest building of 200 m. Wangjing SOHO has become the "First Impression Building of the Capital" after its completion in 2014, it is the first outstanding high-rise landmark building on the way from Beijing Capital International Airport to the urban area.

项目由3栋集办公和商业一体的高层建筑和三栋低层独栋商业楼组成，最高一栋高度达200 m。自2014年建成后望京SOHO将成为了"首都第一印象建筑"，是从首都机场进入市区的第一个引人注目的高层地标建筑。

Design Concept 设计理念

Designed by local top-level design company ECOLAND, and Zaha Hadid Architects, design style and strengths of these two companies have been combined and expressed perfectly on both architectural and landscape design. The unique surface design has made it possible for the buildings to show the dynamic beauties and elegance from every angle. Aluminium plates and glass curtain walls are adopted for the facades, which have integrated with the blue sky to reveal the artistic conception of mountains in the cloud and mist.

由国内顶级设计机构易兰规划设计院与扎哈·哈迪德（Zaha Hadid）建筑事务所倾力合作，从建筑设计到景观设计，双方设计风格和实力得到了完美的结合和充分展现。项目建筑独特的曲面造型使建筑物在任何角度都呈现出动态、优雅的美感。塔楼外部被闪烁的铝板和玻璃覆盖，与蓝天融为一体，象征了山中云雾缭绕的意境。

Landscape Design 景观设计

Following the design concept, ECOLAND combines the function and shape together in the landscape planning and design, making the huge landscape garden of 50,000 m² echo with the buildings. With a high green coverage ratio of 30%, designers create a unique office environment that's like an urban landscape garden. In order to reflect the changes of seasons, four landscape themes of leisure theatre, outdoor sport space, artistic sculptures and water features are carefully designed for Wangjing SOHO. The creative musical fountain and landscape make the project as a garden full with sunshine and complementing with the buildings. All of these have worked together to reach LEED standards including architecture, landscape and construction, to create an energy and water saving, comfortable and smart green architecture.

延用这一建筑设计理念，易兰在项目景观规划设计中将功能与形式充分结合，使项目5万m²的超大景观园林，与极具视觉和空间动感效果的建筑形成对话，相得益彰。绿化率高达30%，形成了独树一帜的都市园林式办公环境。为了体现四季更迭变化，易兰设计团队为望京SOHO特别打造了休闲剧场、场地运动、艺术雕塑、水景四大主题景观。其独具匠心的音乐喷泉和园林景观，远远望去如同洒满阳光的花园，与楼群相辅相成。这一切使得整个项目在建筑、景观和施工组织等方面都达到美国绿色建筑LEED认证标准，打造出一个节能、节水、舒适、智能的北京新绿色建筑。

Meiju Encyclopedia 美居小百科

LEED (Leadership in Energy and Environmental Design), is a green building certification program. Its purpose is to reduce the negative impact on environment and residences in the design process. It is eager to build a complete and accurate green building concept, to avoid the overflow of green building. LEED is developed by the U.S. Green Building Council (USGBC) in 2003; it is a legal mandatory standard in some states of USA and some countries.

LEED（Leadership in Energy and Environmental Design），是一个评价绿色建筑的工具。其宗旨是在设计中有效地减少环境和住户的负面影响。其目的在于规范一个完整、准确的绿色建筑概念，防止建筑的滥绿色化。LEED 由美国绿色建筑协会建立并于 2003 年开始推行；在美国部分州和一些国家已被列为法定强制标准。

Landmark "Living Room" of Nanjing

金陵新城地标"会客厅"

Poly Joytown 保利·堂悦

Location/ 项目地点：Nanjing, Jiangsu, China/ 江苏省南京市
Developer/ 开发单位：Ploy Real Estate Group/ 保利地产集团
Landscape Design/ 景观设计：International Greenview Landscape Design Limited /GVL 国际怡境景观设计有限公司
Floor Area/ 建筑面积：76 722 m²
Land Area/ 占地面积：60 000 m²
Green Coverage Ratio/ 绿化率：30%
Plot Ratio/ 容积率：4.0

Keywords 关键词

Chinese Flavor; Landscape Obstacle; Connotation of Nanjing
中式韵味 绿化障景 秦淮意蕴

"China's Most Beautiful Residential Landscapes" Award
荣获"中国最美人居景观"奖

Large scale and grand fountain, modern lamps, delicate theme sculptures and cloths, as well as landscape corridor with lush plants of display square are working together to reflect the beauty of Joytown against colorful lights. The contrast of quietness in the sunshine and dynamism at night represents brand new sensorial impact to visitors, which is highly interesting and attractive.

展示广场大气整齐的旱喷涌泉、线性现代的折线灯带、精致韵味的主题雕塑及布品、斑驳成影的绿道通廊在繁华灯光的烘托下映衬着堂悦的惊艳。阳光下的静雅与夜幕下的灵动为来客呈现着全新的感官冲击，令人流连忘返。

Overview 项目概况

Located in the G48 plot of Qinhuai District, Nanjing, which is the central zone of Southern New Town, it is positioned as a new center of commerce, office and residence with a land area of 76,722 m². It is designed with modern methods while decorating with classic elements to build a new city skyline, while creating an international community, it merges into connotation of ancient capital to carry historic heritage and to create a landmark in Nanjing.

项目位于南京秦淮 G48 地块，属于南部新城中心地段，占地 76 722 m²，开发定位为集商业、办公、居住于一体的新区域中心。全区以中魂西技为设计理念，运用现代的手法提取古典元素，在展示城市新风貌，打造南京国际社区的同时，注入古都蕴含的气韵，以此传承城市历史记忆，树立区域地标。

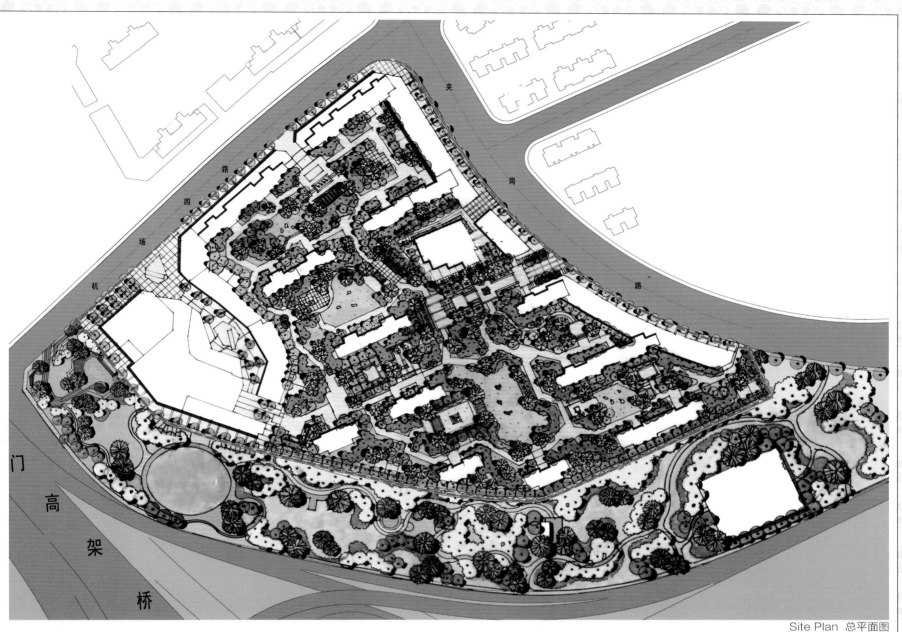

Site Plan 总平面图

Plot Value 板块价值

The project is situated in the core zone of Southern New Town, Qinhuai District, which is the last undeveloped plot in the downtown. While Qinhuai District is the cradle of Nanjing culture, also Qinhuai is an alternative name of ancient Nanjing. It is prosperous in ancient, and more convenient in nowadays, with sophisticated supporting facilities and transportation network. Furthermore, it enjoys bright planning prospects and will be the most important window of Nanjing's image. The project is the first complex project of Poly Real Estate after entering into Nanjing City, it will also be the Poly Jiangsu Headquarters in the future.

项目所处南部新城的核心地段，是主城内仅存的可供规模开发的区域，隶属秦淮。而秦淮则是南京文化的发源地，是南京的古代别称之一。旧时繁华，现时便捷，目前生活配套齐全，出行四通八达。该地段作为展示南京城市形象的重要门户，规划前景明朗。这也是保利进驻南京、深耕发展后的首个综合体项目，后期也将成为江苏保利总部所在地。

Positioning Strategy 定位策略

The project is situated in the core zone of Southern New Town, which is the last undeveloped plot in downtown, with sophisticated supporting facilities and transportation network, it will be regarded as the "urban living room". It is the first complex project of Poly Real Estate after entering into Nanjing City, integrating modern elements with Art Deco style together for the display area, cooperating with changeable spatial layout and exquisite details to create high-end delicate space.

项目地处南部新城核心地段,是主城内仅存的可供规模开发的区域,未来将被规划定义为"城市会客厅",该地段生活配套齐全,出行四通八达。这是一直深耕商品住宅市场的保利地产进入南京后打造的首个综合体项目,现代中式结合 ART DECO 风格的展示区,空间开合变化丰富,注重细节的处理,打造出高品质的精致空间。

Planning Layout 规划布局

Poly Joytown is positioned as an urban complex, with 200,000 m² for high-rises and 100,000 m² for commerce which includes serviced apartment, top-level office building and shopping mall; it is also equipped with public supporting facilities such as kindergarten and culture center.

保利·堂悦定位城市综合体,20 万 m² 高层,近 10 万 m² 商业;商业包含酒店式公寓、甲级写字楼、购物 MALL;并配建幼儿园、文化活动所等公建配套。

Design Concept 设计理念

Modern elements and Art Deco style are adopted for the project, concise lines, changeable spaces, exquisite hard landscaping volume, delicate details and well-proportioned Chinese sculptures are working together to reveal the connotation of thousands years of Qinhuai, and to tell the Zen of Chinese courtyards quietly.

在设计风格上采用现代简约与中式 ART DECO 相结合，简约的线条、开合变化的体验空间、反复推敲的硬景体量、精心雕琢的细节组合、错落其间极具中式韵味的主题雕塑等共同营造出千年的秦淮意蕴，也娓娓细说着中国院子的幽幽禅意。

Landscape Design 景观设计

When entering into the project, the Logo wall coming into visitor's eyes cooperates well with the water feature, the grand and concise seasonal flower base reveals the unique characteristics of the project. While walking on the paths, the landscape varies from step by step, a landscape wall carved with Song of Qinhuai written by Qianlong, an emperor of Qing Dynasty, says, "it's hard to find out the stories of last dynasties, all the splendor will fade away while the only thing does not change is the beauty of Yangtze River". It reminds visitors of the history of Nanjing, like a landscape tunnel running through the time, space, history and culture and promoting visitors' curiosity to walk deeper.

A corridor with horizontal grass belt is connected with display square, space and well-proportioned landscape representing the feelings of high expectation, delicate landscape and refined details are telling about the uniqueness of the "urban living room". Walking to the end of the corridor, visitors will stand in a "front room of landscape" enclosed by three exquisite landscape walls, strong Chinese flavor can be easily felt through carved iron wall, while the landscape walls that set symmetrically form an invisible "gate" to embrace the beauty into its arms, which will maximize the visitors' expectation regarding landscape.

初识堂悦，呈现在来客面前的是结合水景精心设计的异形LOGO景墙，简洁大方的时花草坪景观难掩场地的独特气质。沿着参观流线步移景异，一面题刻着乾隆《秦淮歌》的异型景墙映入眼帘，"六朝往事难寻迹，王谢燕飞谁氏宅。风流将令倦游归，唯见长江依旧碧。"金陵往事依稀浮现，仿佛穿越了时间、空间、历史、人文的景观隧道，使来客满怀期待的心情探寻前行。

连接展示广场的是平行镶嵌着导向草带的折线通廊，空间的收敛与疏密有致的绿化障景酝酿着来客无限向往的情绪，精致的绿化组合与细节处理，无一不体现着"会客厅"的独有气质。沿着通廊到达尽头，来客便置身于三面精致景墙围合而成的"景观前室"，精心雕琢的铁艺景墙散发着浓厚的中式韵味，对称而设的组合景墙仿佛形成了无形的"门"，将园外美景收纳其中，让来客的景观期待推向最高潮。

New and Luxury Expression of Longfor Landscape
龙湖景观的华丽新阐释

Longfor Chunjiang Central　杭州龙湖春江郦城

Location/ 项目地点：Hangzhou, Zhejiang/ 浙江省杭州市
Developer/ 开发单位：Longfor (Hangzhou) Real Estate Co., Ltd. / 杭州龙湖地产
Landscaped Design/ 景观设计：Weimar Group/ 上海魏玛景观规划设计有限公司
Land Area/ 占地面积：128 383 m²
Floor Area/ 建筑面积：600 000 m²

Keywords　关键词
Magnificent Pavements; Landscape Folded Walls; Linear Space
华丽铺装　折面景墙　轴线空间

"China's Most Beautiful Residential Landscapes" Finalist
入围"中国最美人居景观"奖

Although with the same design language throughout Chunjiang Central, users will enjoy different psychological feelings in different spaces. The square on the Jiangnan Avenue shows the welcoming atmosphere, and the parking lot will rise up the users' expectation and curiosity about the development. While the linear spaces created in the sales center will offer the sense of dignity to users directly.

春江郦城的设计语言高度统一，但传递出的心理感受在不同的空间却大不相同。江南大道上的广场给人以浓浓的邀请之感；停车场给人以满满的期待之感；而售楼处的轴线空间则给人以实在的尊贵之感。

Overview　项目概况

Located in the Binjiang District, Hangzhou City, project Longfor Chunjiang Central is a model of urban life of Longfor third generation that integrating TOP commercial brand of "Paradise Walk" with high end residential brand of "Central" together.

龙湖春江郦城，位于杭州市滨江区，是将龙湖商业 TOP 品牌"天街"系和高端城市豪宅"郦城"系融合为一体打造的龙湖地产第三代城市生活范本。

Location Advantage　区位优势

The development is set along Jiangnan Avenue, the urban arterial road of Binjiang District, with complete and perfect transportation and other supporting facilities in this sophisticated area including commerce, politics, hospital, education, sport and ecology. It is connected with urban area through Third and Forth Bridges, only 1 km away from Binjiang District Government and Binhelu Station of Metro Line 1, and also close to Jianghanlu Station of planned Line 6.

项目位于滨江城市主干道江南大道旁，周边生活、交通配套完善，区域发展成熟，商、政、医、学、体、生态六心合一，可享周边醇熟配套。通过三桥、四桥可快速到达主城区，距区政府仅 1 km，距地铁 1 号线滨和路站 1 km，和规划中的地铁 6 号线江汉路站零距离。

Unit Planning　户型规划

With a planned gross floor area of 600,000 m^2, high-rise residential buildings are equipped with compact three bedrooms of 90 m^2, pleasant four bedrooms of 138 m^2 as well as luxury four bedrooms of 168 m^2.

项目总建筑面积约 60 万 m^2，高层住宅户型以精致三房约 90 m^2，舒适四房约 138 m^2，奢华四房约 168 m^2 为主。

Landscape Design 景观设计

Entrance

Jiangnan Avenue, the urban arterial road of Binjiang District, is parallel to the Qiantang River closely, while the entrance of the development is attractive and unique from others. Project is eager to create an urban interface that is full of modern sense, the magnificent pavements, tough landscape folded walls, and the stronghold of the spirit that has integrated with landscape walls have been the rare elements of traditional Longfor products. The whole project is designed become welcoming and generous.

Parking Lot

Designers are paying a lot attention to the unification of the design language for the project, including the parking lot. Despite the traditional definition of preserving vehicles only, designers extend the landscape space into the parking lot by forming echoing between geometrical tree-like pond and the square at the corner, revealing the sense of dignity through slant facades and gorgeous cobblestones road. Besides, vitality can also be easily expressed by the ground pavements.

入口界面

江南大道，无疑是杭州滨江最主要的一条道路，平行于钱塘江而相距甚近；体验区的入口因势而为，开阔、引人注目、给人眼前一亮；设计单位打造出一个现代感的城市界面，华丽的铺装、硬朗的折面景墙和与景墙融为一体的精神堡垒成为龙湖传统产品中少有的元素；整个空间的感觉充满了包容与欢迎感。

停车场

即使是停车场，设计单位在这个项目中也注重设计语言的整体统一。设计单位摒弃了停车场作为单纯停车功能的定义，有效地延展了景观空间。几何式的树池造型与转角广场形成强烈的呼应，倾斜的外立面与优美的石纹路给人高贵的感觉；地面铺装亦展现出十足的张力。

Sales Center 售楼处

Although with the same design language throughout Chunjiang Central, users will enjoy different psychological feelings in different spaces. The square on the Jiangnan Avenue shows the welcoming atmosphere, and the parking lot will rise up the users' expectation and curiosity about the development. While the linear spaces created in the sales center will offer the sense of dignity to users directly. The symmetrical structure and exquisite layout cooperate together to express the language of the building facades in a better way, and the network of ground pavements and water features have followed the style and characteristics of the square at that corner.

春江郦城的设计语言高度统一，但传递出的心理感受在不同的空间却大不相同。江南大道上的广场给人以浓浓的邀请之感；停车场给人以满满的期待之感；而售楼处的轴线空间则给人以实在的尊贵之感；对称的结构和严谨的布局更好的烘托售楼处建筑的整体立面感受；而网状的地面铺装纹理、水景形式延续了春江郦城转角广场的气质。

Adopting COSMO Concept to Reorganize the Urban Life Elements
COSMO 理念重组都市生活要素

COSMO, Hangzhou   杭州东润原筑壹号

Location/ 项目地点: Jianggan District, Hangzhou, Zhejiang, China/ 中国浙江省杭州市江干区
Developer/ 开发单位: Dongrun (Hangzhou) Real Estate Co., Ltd. / 杭州东润世纪置业有限公司
Landscape Design/ 景观设计: Palm Landscape Architecture Co., Ltd. / 棕榈园林股份有限公司
Land Area/ 占地面积: 82 041 m²
Floor Area/ 建筑面积: 200 000 m²
Landscape Area/ 景观面积: 62 127 m²
Green Coverage Ratio/ 绿化率: 30%
Plot Ratio/ 容积率: 2.2

Keywords   关键词
Obstructive Method; Leisure Courtyard; Waterfall
障景手法 休闲庭院 跌水景观

 "China's Most Beautiful Residential Landscapes" Finalist
入围"中国最美人居景观"奖

According to the architectural styles, five building groups are divided for the residences, various elements including landscape walls, water features, sculptures and trees are working together to form changeable entrances and present different landscape.

住宅区域根据建筑多变的空间形态，将住宅区域划分出五个不同小组团，每个组团的入口通过景墙、水景、雕塑、树池等不同元素的组合形成多变的入口感觉，让每一个组团呈现不一样的景色。

Overview   项目概况

Situated in the east of the urban area, which is the center of the newly constructed Rail Way East Station business district, the development is close to Desheng Express and Qibao Station of Metro Line 1, connecting with Tongxie Road on its right and occupying strategic location of Shanghai-Hangzhou Express.

原筑壹号坐落于杭州主城区城东新城，处新火车东站区域核心，紧邻德胜快速路，右接同协路，与地铁一号线七堡站咫尺之遥，占据沪杭高速等核心枢纽位置。

Site Plan 总平面图

Location Advantage　区位优势

As 20 new towns and 100 urban complexes are in the planning, as well as a large-scale urban complex in the Rail Way East Station is under construction, Jianggan District has become the most promising district of Hangzhou. The development sitting on the east of Desheng Road in which is the developing area of Hangzhou, will be a new beginning of urban life. Developing as the first urban complex of Jianggan District, COSMO is the landmark of the east plot of Hangzhou, and will start the urban life which is high-efficient and convenient.

　　随着杭州20大新城，100个城市综合体的规划以及火车东站大型城市综合体的建设进行，江干区荣升为城市发展的明星板块。德胜路东这一顺应城市发展规律的主城区核心属地，成为新都市生活的原点。项目以区域首个都市综合体的姿态，成就城市东大门地标性建筑群，启幕平衡高效的都市生活新篇章。

Positioning Strategy　定位策略

Studying and investigating the living habit of targeted users, designers adopt the leading concept of COSMO for the project to reorganize the elements of urban life, including high-end apartment, hotel, office buildings, boutique residences, etc., with a gross land area of approx. 200,000 m².

设计师潜心考量现代人的生活习惯，以国际前沿的 COSMO 理念重组都市生活要素：高级公寓、国际酒店、独院办公、风尚写字楼、精品公馆等生活形态，缤纷呈现近 20 万 m² 现代都市生活集。

Planning Layout　规划布局

Bringing leading concept of COSMO into the project, designers create this complex with the functions of living, commerce, leisure, entertainment and hotel, and the modern brands including apartment, Yue LOFT, Enjoy Space and Intercontinental Hotels Group, to carefully express the exquisite urban life in the 200,000 m² plot.

项目全面引入国际主流的 COSMO 理念，是一个将都市居住、商务、休闲、娱乐、酒店等各类功能复合、高度集约的建筑综合体，融合风尚公寓、悦 LOFT、原筑悦庄和洲际假日品牌酒店等现代生活形态，缤纷呈现约 20 万 m² 现代都市生活。

Design Concept　设计理念

The project is comprised of hotels, apartments, high-rise residences and villas, while different architectural types are designed with different landscape themes.
As the guests of hotel are mainly for business and travel and not familiar with the hotel, the signage and wayfinding system should be perfectly designed. Besides, a welcoming entrance of the hotel is also essential to attract more potential guests.
Users of villas are pursuing for the spaces for private courtyards, which is the main difference for that of apartments. Aesthetic tastes are changing from one person to another, hence several kind of landscape elements are selected to form different landscape atmosphere to meet users' requirements. While designers should search for the unification among these changes to avoid any disorder may be caused.

项目由酒店、公寓、高层住宅、别墅住宅组成，设计中根据不同的建筑类型营造不同的景观。这也是设计团队景观设计的根源。

酒店的使用者为商务旅游休闲的人群为主，对于酒店而言只是匆匆过客，每个使用者对酒店都是陌生的，所以标示引导系统必须做到位，同时酒店入口应做到让人心动，让潜在顾客有进入的冲动。

别墅住宅的使用者更注重私人室外活动庭院的需求，这是别墅与公寓的主要区别。而不同的人会有不一样的审美情趣，所以可以通过不同的景观元素营造不同的景观氛围，满足不同人的需求。然而变化是有限度的，过分的变化会带来混乱，所以设计中应注意变化中寻求统一。

Landscape Design 景观设计

Landscape for Hotel

With the highlights on the entrance square and inner court, landscape of entrance is set the fountain as the center to form a welcoming entrance. The entrance of underground garage is close to the inner court which will affect the beauty of the court. A 3.5m height waterfall wall is designed to separate the underground garage and inner court, meanwhile, the sound of the waterfall will also help to decrease the noise of the vehicles.

酒店景观

酒店区域着重打造酒店入口广场以及酒店休闲内庭。酒店入口广场以喷泉水景为中心，通过动感的喷泉、跌水，营造热闹的入口氛围。酒店休闲内庭紧邻地下车库出入口，严重影响到内庭的使用。因此景观设计通过设置3.5 m高的跌水景观形成对景，通过障景的设计手法隔绝地下车库出入口的不利因素，同时通过跌水的声响掩盖车辆出入的噪音。

Landscape for Residence
According to the architectural styles, five building groups are divided for the residences, various elements including landscape walls, water features, sculptures and trees are working together to form changeable entrances and present different landscape. Modern and concise design methods are adopted to reveal the power and rhyme of the courtyards so that the courtyards and buildings are integrated well with the sense of layer and order. For the courtyards of villas, designers regard water, grass, stones and trees as the landscape concepts respectively to build a beautiful picture of courtyards that enjoy unique landscape. Designers use the same gate doors and fences for the villas to make a unified landscape effect.

住宅景观
住宅区域根据建筑多变的空间形态，将住宅区域划分出五个不同小组团，每个组团的入口通过景墙、水景、雕塑、树池等不同元素的组合形成多变的入口感觉，让每一个组团呈现不一样的景色。以现代简约的设计手法，使庭院景观表现遒劲而富于节奏，使庭院和建筑立体形式有条不紊地融为一体，更具序列和层次感。别墅庭院设计以水、草、石、木等不同的景观元素为主题衍生出不同的景观概念，营造一庭院一风景的美好愿景。通过相同的入户门头、分户围墙，形成统一的景观效果，在变化中寻求统一。

Landscape for Apartment
The apartment is comprised of the entrance square and inner courtyard, while the entrance use circles as the main element to form a strong centripetal visual effect, and an artistic sculpture setting in the center becomes the visual focus of the square. For the inner court, designers make full use of the topographical height difference cooperating with plants in different heights to create a modern and natural court. Learning from ancient gardens, designers are aimed at building landscape atmosphere that's like people are walking in a painting so that to feel the changes of seasons in the court.

公寓景观
公寓区域由公寓入口广场以及公寓休闲内庭组成。公寓入口广场以圆为构图元素，形成强烈的向心视觉效果，在圆心中设置艺术雕塑形成整个广场的视觉焦点。公寓内庭通过丰富的地形高差变化结合高低错落的植物群落，形成现代自然式的休闲庭院。藉古典造园之法，旨在营造"房流于林影，人行于画中"的场景氛围，以此感知庭院四季更替的变化。

The Top Poetic Dwellings of Zhengzhou

郑州第一诗意栖居名盘

Qinghua·Remember the Southern China　　郑州清华·忆江南

Location/项目地点：Zhengzhou, Henan, China/河南省郑州市
Developer/开发单位：Zhengzhou Qinghua Garden Real Estate Development Ltd./郑州清华园房地产开发有限公司
Landscape Design/景观设计：Greentown International Design & Engineering Group (HK) Ltd./重庆绿城景观设计工程有限公司
Land Area/用地面积：1 433 808 m^2
Floor Area/建筑面积：824 662 m^2
Landscape Area/绿化面积：760 000 m^2
Plot Ratio/容积率：0.575
Green Coverage Ratio/绿化率：53.7%

Keywords　关键词

Dynamic Corridor; Green Gallery; Landscape for Every Unit
活力梯道　自然绿廊　户户有景

"China's Most Beautiful Residential Landscapes" Finalist
入围"中国最美人居景观"奖

The lake running through the residential area is lying on the mountain and following the terrain of the slope, being sinuous, or a pond or running under the bridges, and feeding into the river outside the residential area to form a green gallery with changeable landscape to integrate the functions of entertainment, recreation, fitness and communication.

居住区以一条主要水系贯穿，这条主要水系依山就势在坡地中流淌、时而曲折蜿蜒、时而变换为清澈见底的水潭、时而又穿流过桥，最后汇入组团外的河道，形成一条自然绿廊，同时也形成一个富于变化的景观系列、一个集娱乐、休闲、健身和交谈的活力梯道。

Overview　项目概况

Situated on the south bank of Yellow River and close to Peach Valley, Qinghua·Remember the Southern China has a planned gross land area of 1,430,000m^2, which is a garden community that integrates villas, garden houses, schools, a senior entertainment center and so on in various styles.

清华·忆江南位于黄河南岸、桃花峪旁。规划总占地面积约143万 m^2，是集别墅、花园洋房、学校、老年活动中心等功能齐全、风格各异的江南园林式社区。

表1 用地分配表

用地类别	面积（M²）	百分比（%）
别墅用地	467797	32.6
花园洋房用地	279935	19.5
高层住宅用地	183821	12.8
公共建筑用地	175224	12.2
湿地用地	327030	22.9
总用地面积	1433808	100

表2 用地平衡表

用地类别	面积（M²）	百分比（%）
别墅用地	424581	29.6
花园洋房用地	209994	14.6
高层住宅用地	150921	10.5
公共建筑用地	169691	11.8
湿地用地	285987	19.9
道路用地	192627	13.6
总用地面积	1433808	100

表3 经济技术指标

住宅类型	建筑面积（M²）
独立别墅	89100
双拼别墅	21600
联排别墅	61600
花园洋房	206912
小高层	100100
大高层	270800
住宅总面积	750112

续表3 经济技术指标

公共建筑	建筑面积（M²）
幼儿园1	2350
售花店（前期售楼处）	220
老年公寓	5400
度假公寓	9550
小木屋	2600
高尔夫会馆	6000
培训中心	3260

续表3 经济技术指标

公共建筑	建筑面积（M²）
幼儿园2	2350
步行街	11360
培训中心	7090
净菜市场	4270
假日酒店	17500
会所	2600
公建总面积	74550

续表3 经济技术指标

总建筑面积	824662M²
总用地面积	1433808M²
容积率	0.575
绿化率	53.7%

注：如果大高层不建，容积率为0.46

图例说明 拟建中学

① 主入口
② 假日酒店（康乐中心）
③ 会所（内含物管、邮局、银行等）
④ 培训中心
⑤ 商业步行街
⑥ 停车场
⑦ 垂钓中心（游船俱乐部）
⑧ 高尔夫会馆
⑨ 高尔夫练习场
⑩ 插花庙公园
⑪ 幼儿园
⑫ 幼儿园（临时售楼处）
⑬ 庄园式别墅
⑭ 翠竹居
⑮ 怀秀居
⑯ 倚潮苑
⑰ 秀瞻居
⑱ 凌云居
⑲ 丰泽苑
⑳ 文洋苑
㉑ 翰风居
㉒ 逸仙居
㉓ 徽南居
㉔ 平远居
㉕ 听涛苑
㉖ 翠峰居
㉗ 长岛
㉘ 夕阳红康复中心
㉙ 潭溪
㉚ 长河
㉛ 花园洋房
㉜ 假日沙滩
㉝ 休闲沙滩
㉞ 天潭
㉟ 游泳池
㊱ 度假公寓
㊲ 度假木屋
㊳ 湿地探索小径
㊴ 花港观鱼
㊵ 净菜市场
㊶ 滤水池
㊷ 河畔观鸟屋
㊸ 煤气调压站
㊹ 步行栈桥
㊺ 游船停靠码头
㊻ 垃圾回收点
㊼ 变电站
㊽ 商业步行街
㊾ 电瓶车点
㊿ 停车场
51 休闲茶室
52 煤气小广场
53 泥滩观鸟区
54 树林及草地

Site Plan 总平面图

Location Advantage　区位优势

Being close to the large-scale ecological wetland of Yellow River, a forest of 2,000,000 m², and a lake of 333,000 m², the development enjoys the scenery of bridges, a large lake, ambulatories and pavilions, while the large-scale wetland makes the whole project more idyllic and attractive. Rivers and the lake are crisscrossed in the plot to provide spaces for recreation and entertainment including yachts, boating, fishing, waterfront paths and commercial pedestrian street, which make this plot in the north of China full of the charm of gardens of Southern China and create a modern countryside that is close to nature.

清华·忆江南紧临万亩的黄河生态湿地、200万m²的自然山林和33.3万m²的伴山湖畔，处处小桥流水、回廊秀亭，大型湿地公园使整座园林富有诗情画意、引人入胜。园内河道、湖泊纵横交错，在水边形成了游艇、划船、垂钓中心、滨水步道和商业步行街等，使本案虽处于北国疆域却富含江南园林的韵味，成为现代人亲近、感受、体验自然的现代田园。

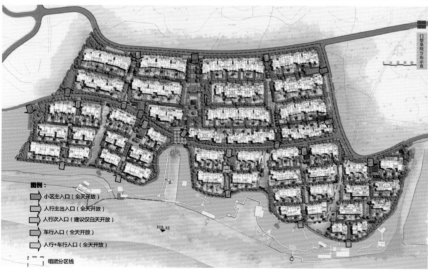

Design Concept 设计构思

With the design concept of the poem Yi Jiang Nan Ci San Shou, written by poet Bai Juyi of Tang Dynasty, landscape designers express the pursuit and recall of gorgeous landscape of Southern China through modern landscape design method so that users are able to experience the beauty in the development. Working together with the themes of "Riverbank, leisure, vacation and comfort", also following the ideas of "being natural, ecological, healthy and casual", designers decorate the landscape area that is over 80,000 m² as a garden that is close to water and mountain, being colorful and modern. By adding some landscape elements which are imbued with local characteristics and the unique charm of gardens in the Southern China, designers use the modern methods to redefine a "retreat away from the world".

项目景观规划设计主题以白居易的《忆江南词三首》（"江南好，风景旧曾谙；日出江花红胜火，春来江水绿如蓝。能不忆江南！江南忆，最忆是杭州；山寺月中寻桂子，郡亭枕上看潮头。何日更重游！"）为设计蓝本，把人们对江南秀美山水的梦想、向往和追忆以现代景观设计的手法阐释、营造出来，让人去畅游、感悟。以"梦里江南云水情，人间清华是重游"，同时结合楼盘的"水岸、休闲、度假、乐居"理念，秉承"自然的、生态的、健康的、休闲的"思路，将小区 8 万多平方米的园林景观设计成为丰富多彩的现代、亲水、休闲的坡地园林。在此基础上挖掘富有江南园林风情与民俗风味的景观元素，用现代白话的语汇重新组织新语境下"世外桃源"。

Design Philosophy 设计原则

Adopting water as the dynamic element, landscape designers follow the design philosophy and purpose of leaving every corner with green, sunshine and fresh air and making each unit can enjoy the landscape when sitting in their rooms. Modern and concise styles are selected for the landscape design, which is echoing with the traditional garden design philosophy of Southern China that is as natural as possible. Besides, the concise layout is just meeting the requirement of modern ways of life, being fully considered by using abstract art to embellish the landscape elements. The natural decorations, although have been carved, make the landscape style perfectly match the architectural style.

　　以"水"为景观活跃要素，使得"户户有景"，让每个空间充满绿色、阳光和新鲜的空气是本次设计的基本原则和目的，运用简洁现代的园林景观构图来组织景观秩序，这与设计单位崇尚江南园林"虽由人做，宛自天开"的传统造园理念更为吻合。简练的构图是现代城市生活的迫切需要，看似简单却独具匠心，用抽象的艺术来修饰各景观要素，用天然材质，虽由人工雕琢，使小区的景观风格与建筑协调一致。

Landscape Design 景观设计

The lake running through the residential area is lying on the mountain and following the terrain of the slope, being sinuous, or a pond or running under the bridges, and feeding into the river outside the residential area to form a green gallery with changeable landscape to integrate the functions of entertainment, recreation, fitness and communication.

The development is connected by water which has fully represented the beauty of Southern China, while the lake is changing all the way, being wide or narrow, rapid or slow, forming a large lake or shaping as a sinuous creek. Every residential or landscape area it runs through has been designed with different styles and unique characteristics, working together to form a harmonious and unified landscape with special water features as well as being changeable. The landscape design has strategically improved the taste and quality of the development, and provided high end living environment for users, representing the beauty that surpasses that of the Southern China.

居住区以一条主要水系贯穿，这条主要水系依山就势在坡地中流淌、时而曲折蜿蜒、时而变换为清澈见底的水潭，时而又穿流过桥，最后汇入组团外的河道，形成一条自然绿廊，同时也形成一个富于变化的景观系列、一个集娱乐、休闲、健身和交谈的活力梯道。

设计以水为线，充分体现江南水乡的含蕴，水系在景观中时小时大、时宽时窄、时放时收，有时是宽阔大气的湖面，有时是曲折蜿蜒的小溪。水系流过的每个组团、每个景区形成不同风格、各具特色的水景园林景观，整体构成既统一、和谐、又有丰富变化的水景特色园林景观，极大地提高小区的格调与品质，同时为业主提供高品味的休闲居住环境，不是江南却胜似江南。

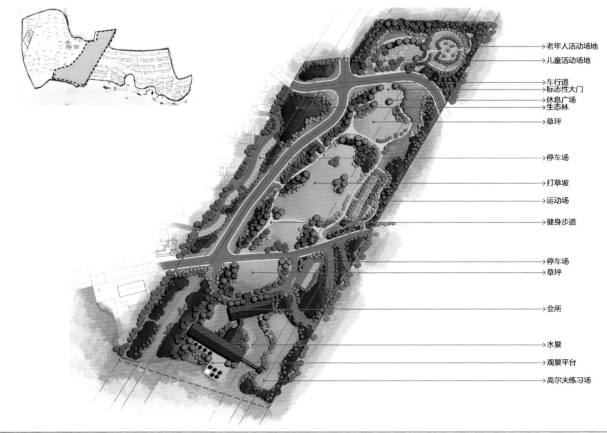

→老年人活动场地
→儿童活动场地
→车行道
→标志性大门
→休息广场
→生态林
→草坪
→停车场
→打草坡
→运动场
→健身步道
→停车场
→草坪
→会所
→水景
→观景平台
→高尔夫练习场

Green Transition on the Grey Site

灰色地块上的绿色蜕变

Civic Cultural Square at Park Residence, Changchun 长春柏翠园市民文化广场

Location/项目地点：Chaoyang District, Changchun, Jilin, China/ 吉林省长春市朝阳区
Client/项目委托：Jiahu (Changchun) Real Estate Co., Ltd. / 长春嘉湖房地产开发有限公司
Landscape Design/景观设计：Palm Landscape Architecture Co., Ltd. / 棕榈园林股份有限公司 / Palm Design Co., Ltd. / 棕榈景观规划设计院
Land Area/占地面积：50 000 m²

Keywords 关键词

Ecological Revivification; Site Memory; Plant Community
生态还原　场地记忆　植物群落

"China's Most Beautiful Residential Landscapes" Finalist
入围"中国最美人居景观"奖

In order to make the site become "green" instead of "grey", designers adopt the most simple and effective way of creating multi-layer plant communities. Meanwhile, the plant communities have helped to divide the functional spaces, hide the messy spaces and provide shades to users.

为了实现场地从"灰色"到"绿色"的蜕变，项目直接营造层次丰富的植物群落并自我演替。同时，场地各功能空间需要浓密的植物群落对场地进行二次划分，消隐零乱的痕迹，提供良好的遮阴。

Overview 项目概况

The project is a public space sitting in the Nanhu Plot, the center of Chaoyang District, Changchun City, which is the former site of 288 Factory. It is a grey land with a concrete coverage ratio over 50%, along with rare landscape resources, a lot of tall and graceful pine trees and poplars planted with the tree height up to over 10 m.

项目位于长春市朝阳区核心——南湖板块，原为原长春228厂旧址，是对外开放的公共空间。整个场地混凝土覆盖面超50%，是标准意义上的灰色地块，同时，场地内有大量的松树和杨树，部分高达十余米，树干挺拔，姿态优美，是不可多得的景观资源。

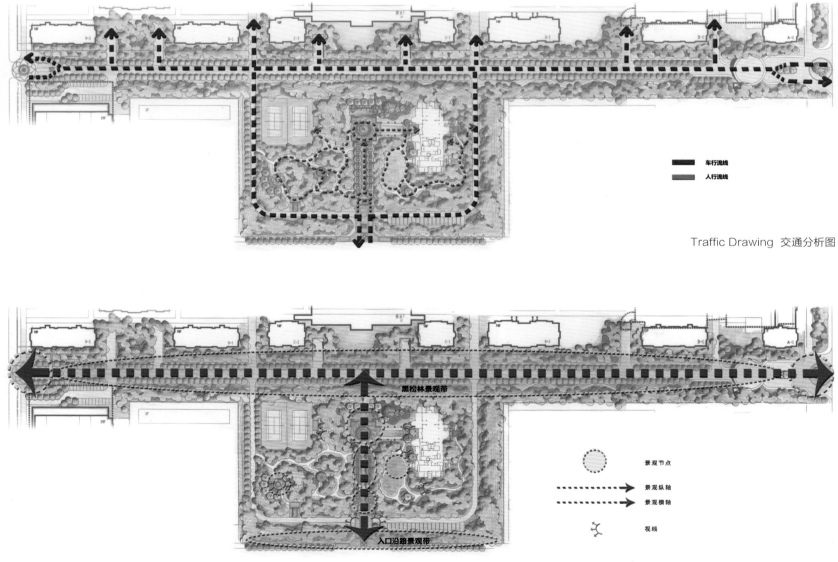

Traffic Drawing 交通分析图

车行流线
人行流线

Landscape View Analysis 景观视线分析图

景观节点
景观纵轴
景观横轴
视线

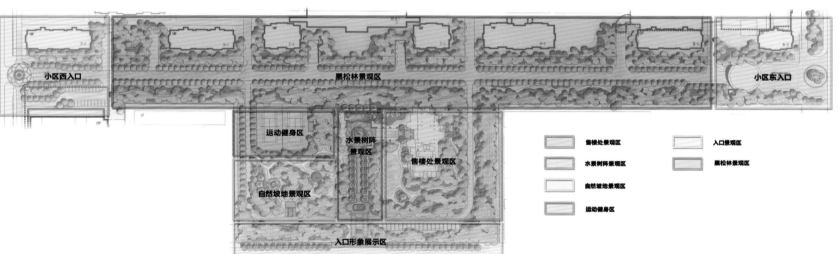

Landscape View Analysis 景观视线分析图

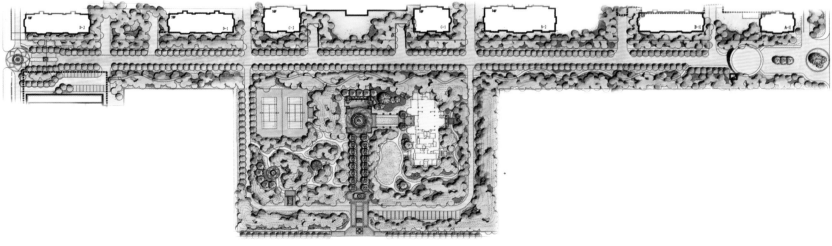

Site Plan 总平面图

Location Advantage 区位优势

Nanhu Plot is the fifth largest business circle of Changchun City. As the development of Changchun City, especially when the government office building moved to the southern plot, commercial atmosphere of Nanhu Plot is enhanced dramatically, and because it enjoys large-scale potential area, Nanhu Plot has represented the positive posture for becoming a new business circle and urban agglomeration. According to the geographic position, Nanhu Plot is planned to be divided into "one center, four streets and seven plots".

项目所在地南湖板块是长春市的第五大商圈。随着近年的不断发展，特别是长春市政府南迁以后，南湖地块的商业氛围愈加浓厚，可开发利用空间较大，已具有发展成为新的商圈和集聚区的良好态势，纳入重点培植。根据区域地理位置，南湖商圈可按"一心、四街、七大板块"功能格局进行统一规划。

Positioning Strategy 定位策略

Planned to transform from grey land to green land, it is essential to change the physical materials, as well as taking human's living needs into consideration so that to make a transition for the site functions. Starting from the living styles and demands of users, designers keep the trees of the former site, trying to make people much easier to remember the plot, which keeps the features of traditional Chinese gardens to receive the rebirth of the site.

项目地块要完成从"灰色"到"绿色"的蜕变，不仅需要物质的改变，更要规划人的综合需求，使场地功能的彻底转型。项目从人们的生活方式和需求出发，保留场地内看似"随意"的大树，锁定场地的记忆，并延展中国传统造园的印记，成功使场地完成蜕变和重生。

Planning Layout 规划布局

Based on the conditions of the site, designers reorganize and divide the site, and pay more attention to the potential value of the site and the overall landscape layout, cooperating with the transportation and buildings around to create an attractive and welcoming entrance, and present the site to users through a central axis with landscape on both sides. A water feature set in the central along with symmetrical waterfalls is designed at the entrance, and cooperating with the water feature at the beginning of the landscape axis to lead people walking to the club which is the main building in the center of the site.

基于场地的基础条件，首先对场地和空间重新进行了梳理和划分。重点考虑地块未来最佳的使用功能，以及整个景观界面的布局，结合周边交通及建筑的功能，划分出具有展示功能的入口空间和界面，并通过中轴景观序列排布，进一步展示场地。以对称的跌水和中心水景，界定入口，结合端头水景作为景观中轴，以此在空间上形成导向性，将人流引至场地中的主体建筑——会所。

Design Concept 设计理念

According to the nature of the site, the design should not only pay attention to the redivision of functional spaces, but also to people's activity demand and the landscape details, trying to the transform the gray land into a green land.
Setting the phenomenology meaning of landscape design as the start point, landscape designers are eager to provide spaces for communication to meet the requirements that beyond substances and functions. The project is exactly a rebirth on the phenomenology meaning from an abandoned plot to green site.

　　基于场地的性质，本案设计既要兼顾功能空间的重新划分，又要关注人的活动需求以及景观细节的体现，完成灰色地块向景观绿地的蜕变重生。

　　设计师从景观设计现象学意义上出发，力求体现该场地中人与人的经验交织，以实现超越物质和功能的需要。本案的设计正是将废弃地块向绿色自然地块一个景观意义上的蜕变重生。

Landscape Design 景观设计

The position and needs of the site have changed dramatically from factory to club and green space. In order to make the site become "green" instead of "grey", designers adopt the most simple and effective way of creating multi-layer plant communities. Meanwhile, the plant communities have helped to divide the functional spaces, hide the messy spaces and provide shading to users. Hence, designers have kept some trees on the original site (i.e. Pinus Thunbergii Parl), and picked some local trees (i.e. Mongolian oak and Prunus armeniaca) for the project, respecting the natural growth rules of plants and placing them in the most suitable places.
From the very beginning, landscape designers insist on the strict requirements on the construction by transplanting the trees without cutting any roots or branches to keep their graceful gesture and make them grow up easily. Flowering shrubs in the middle layer are combining with ground-cover plants and flowers, evergreen and deciduous plants, to make the site be a picture of scenery all the year around. The plant communities will help to speed up the transition of the site, promoting the rebirth of green and ecological plot.

　　从厂区到会所和绿地，场地的定位和需求发生了巨大的变化。实现场地从"灰色"到"绿色"的蜕变，最直接最有效的方法就是营造层次丰富的植物群落并自我演替。同时，场地各功能空间需要浓密的植物群落对场地进行二次划分，消隐零乱的痕迹，提供良好的遮阴。故对原有乔木进行选择性的保留（如黑松等），适当选用当地本土树种（如蒙古栎、山杏等）进行合理搭配，尊重植物生长的生态法则，适地适树，促其演替。

　　设计伊始，设计单位对施工工艺进行限定，采用大树全冠移植，保证其完整优美姿态的同时，又可尽快还原生态。中层花灌木结合地被、草花，落叶和常绿搭配，做到处处有花可观，时时有景可看。丰富的植物群落在短时间内完成地块性质的转变，促使绿色生态的重生。

Humanistic and Natural Landscape in French Style
人文、生态双重景观的法式演绎

Eastern Provence 东方普罗旺斯

Location/项目地点：Nanhu New District, Jiaxing, Zhejiang, China/浙江省嘉兴市南湖新区
Developer/开发单位：Zhejiang Dongfanglanhai Properties Co., Ltd. /浙江东方蓝海置地有限公司
Landscape Design/景观设计：Botao Landscape (Australia) Co., Ltd. /澳大利亚·柏涛景观
Landscape Area/占地面积：235 000 m²
Floor Area/建筑面积：407 000 m²
Plot Ratio/容积率：1.67
Green Coverage Area/绿化率：38%
Completion/建成时间：2014 年

Keywords 关键词
Unique Site; Natural Boundary; Large Depth Space
场地独享 自然界限 大纵深空间

"China's Most Beautiful Residential Landscapes" Finalist
入围"中国最美人居景观"奖

The water system in the villa area is the natural boundary to separate each villa. The designers strive to use humanistic art structures to create exquisite landscape details for the villa area.

别墅区的水体形成了分割其他别墅的自然界限，设计师力图利用有人文情节的艺术小品为别墅区的景观创造精巧的细节。

Overview 项目概况

This development is located at the Nanhu New District of Jiaxing City, 5 km to the south of urban area and only 15 minutes ride to downtown.

项目地处嘉兴市南湖新区，位于市中心南向 5 km 处，从项目用地到达中心城区仅 15 分钟车程。

Site Value 板块价值

The site is connected with Shanghai-Hangzhou Express and high-speed railway on the south. With the city developing to the south, the regional traffic advantage is significant as the project area is the main direction of the city development.

项目向南可与沪杭高速公路及高速铁路交通枢纽连接，随着城市的北控南进。项目所属区域为城市发展的主要方向，区位交通优势非常明显。

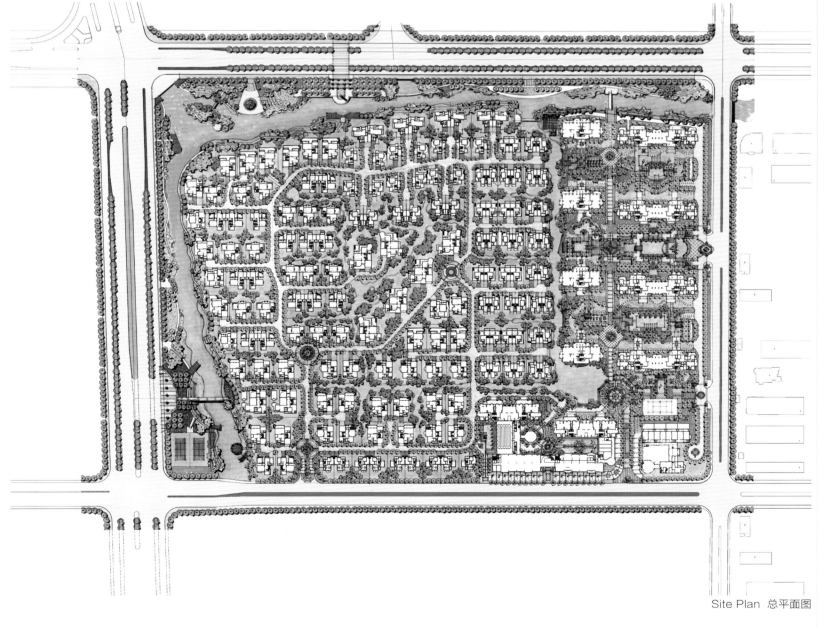

Site Plan 总平面图

Positioning Strategy 定位策略

The development is designed with Mediterranean style. With the earlier stage of product positioning, the planning and design of the project aim at creating a classic mansion with French Provence artistic conception. Through several years, Provence has become a magnificent humanistic beauty with a variety of architectural styles and cultural relics, and it is endowed with a colorful past. As time goes by, the ancient and modern fashion are perfectly blended together in Provence and create a fascinating heaven on earth.

项目是地中海风格的延续,策划和设计经过前期的产品定位,意在打造法式普罗旺斯意境的经典豪宅。历史上时光的磨砺给普罗旺斯留下了一个多建筑风格多文化遗迹的瑰丽人文美景,同时也赋予普罗旺斯一段多姿多彩的过去。岁月流逝,普罗旺斯将古今风尚完美地融合在一起,从而沉积出一个让人流连忘返的人间乐土。

Planning Layout 规划布局

Based on the space concept of the architecture planning and design, the landscape is divided into two landscape sites according to the two types of architecture, high-rises and villas.

基于建筑规划与建筑设计的空间概念,景观按建筑产品高层与别墅两大类,分成两大块景观场地。

Design Concept 设计理念

The landscape design of the development emphasizes the combination of humanistic cultural landscape and natural ecological landscape, which are inseparable, as two wheels to the bike and wings to the bird. Humanistic cultural landscape uses landscape techniques such as carving, copy and forge to reflect the profound history and culture, the relaxation and elegance of Provence needs the support of heavy and exquisite landscape details. Meanwhile, natural ecological landscape emphasizes the wild and harmony of nature, the artificial intervention is just a reasonable control of pastoral style so as to achieve the state of simple but not vulgar.

项目的景观设计强调人文文化景观与自然生态景观，如车之双轮、鸟之双翼不可割分，人文文化景观需要通过雕塑、摹刻、锻打等景观手法，体现深厚的历史与文化，那种普罗旺斯式的轻松与优雅正需要沉重与精致的景观细节支撑和衬托。自然生态景观强调自然界的那份野趣与和谐，人为的介入只是将田园风范合理的控制，从而达到质而不野的境界。

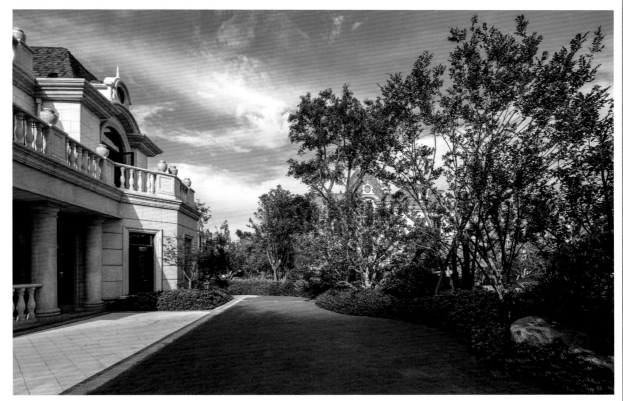

Landscape Design 景观设计

High-rises

The site of the high-rise area has large spatial span and depth of space, and the population density is higher than that of the villa area, so a wilder public activity space is needed. Therefore, the landscape designers propose to create three plazas with different themes displaying a western fountain, an obelisk and a large pavilion as its core respectively. The north side of the site is long and narrow and extends from west to east. The site of the commercial podium on the northwest corner and high-rise apartment is irregular, so the designers use natural waters, lush planting, winding riverbank, uplifted hills and other landscape elements to design the landscape with natural scenery theme as landscape green transition zone to connect the high-rise area. People go into the zonal progressive landscape veranda from here. The landscape is divided into four parts, which takes small scale landscape elements as the core, such as pavilion, corridor, sculpture, small fountain and flower pots.

Villas

In the relatively narrow villa area, the specialty of the design determines the private temperament of the site. The landscape designers take the courtyard as theme and small space landscape as design elements, such as flower pots, private revetment, elegant fence, lighting fixture, sculpture and feature planting to create the courtyard landscape spaces with different themes. The designers emphasize the exclusive use of space in this area and decrease the usage of public space. The water system in the villa area is the natural boundary to separate each villa. The designers strive to use humanistic art structures to create exquisite landscape details for the villa area.

高层区

高层地块的场地，空间跨度较大，有更大的纵深空间，高层区的人口密度大于别墅区的人口密度，需要更宽阔的公共活动场地。因此，景观设计师创作了三个不同主题的广场，分别以西洋喷水池、尖方碑和大型柱亭为核心。北侧的场地较狭长，由西向东延续发展。西北角的商业裙楼与高层公寓的场地边缘较参差，设计师设计了体现自然风光主题的景观，作为连接高层的景观绿色过渡区。这里有景观元素的自然水体、密植的树林、波折的水岸、隆起的丘陵等。从这里人们就开始进入景观游廊的地带。这里的景观以凉亭、连廊、雕塑、小喷泉、花钵等中小体量的景观元素为核心，分为四个单元，带状递进的景观走廊。

别墅区

在相对狭窄的别墅区，产品的特性决定了场地的私有气质。景观设计师在这里以庭院为主题，以小空间的景观元素，如花钵、私家驳岸、优雅的围墙、灯具、情景雕塑、特色植栽为核心，创造不同主题风格的庭院景观空间。在这里，设计师更强调的是场地独享，降低公共空间的使用度，别墅区的水体形成了分割其他别墅的自然界限，设计师力图利用有人文情节的艺术小品为别墅区的景观创造精巧的细节。

World-class Resort Town
世界山居度假小镇

Peacock City·Badaling　　孔雀城·八达岭

Location/项目地点：Beijing, China/北京市
Developer/开发商：CFLD/华夏幸福基业
Planning & Landscape Design/规划/景观设计：Terra Group (Beijing)/北京丹瑞建筑工程设计顾问有限公司
Floor Area/建筑面积：300 000 m²
Landscape Area/景观面积：250 000 m²

Keywords　关键词

Multi-layered Garden; Ecological Life; Fifth-layered Landscape System
立体花园　原生态生活　五重景观体系

 "China's Most Beautiful Residential Landscapes" Finalist
入围"中国最美人居景观"奖

It has the design purpose of revitalizing and representing the essence of world-class resorts for targeted users. In every unit, one can enjoy the beautiful landscape while sitting in the room, and large scale gardens are featured in the site. The pursuit for ecological lives is reflected in the mountain landscape and exquisite architectural and landscape designs are created on the hillside. Fifth-layered landscape system is carefully designed for the site, and the landscape designers follow the terrain of mountain to build featured gardens.

"复兴世界山居精髓，开启山居庭院时光"为目标，园区规划布局内户户瞰景，并有超大立体花园，山景中承载着原生态的生活向往，塑造出半山上的精美建筑设计、繁花树影。项目的设计精心打造五重景观体系，依山势营造出特色山地园林。

Overview　项目概况

Situated between Badaling Great Wall and Guanting Lake, the development is an artificial and low density community lying on a mountain. It is 40 degrees north latitude, which is the internationally recognized golden land for grapes and is one of three greatest grape golden lands along with Bordeaux of French and California of United States. It has a grape-growing area of more than 10,000 Mu and it is also the production base of Great Wall red wine. There are 85 scenic spots around the developmentwithin 30 min drive, including Badaling Great Wall, Shuiguan Great Wall, Badaling Wildlife Park, Longqingxia, Guanting Lake and Wild Duck Lake National Wetland Park, which make the development to be a world-class resort town.

　　项目位于八达岭长城和官厅湖之间，依山造镇，手工打磨经典的山居低密庭院。小镇位于北纬40°。这是国际公认的葡萄种植"黄金地带"，与法国的波尔多、美国的加州并称世界葡萄种植三大黄金地带，区域种植葡萄面积达十几万亩，也是著名的长城葡萄酒生产基地。项目周边半小时车程内有八达岭长城、水关长城、八达岭野生动物园、龙庆峡、官厅湖、野鸭湖等85处风景名胜旅游区，"世界山居度假小镇"名副其实。

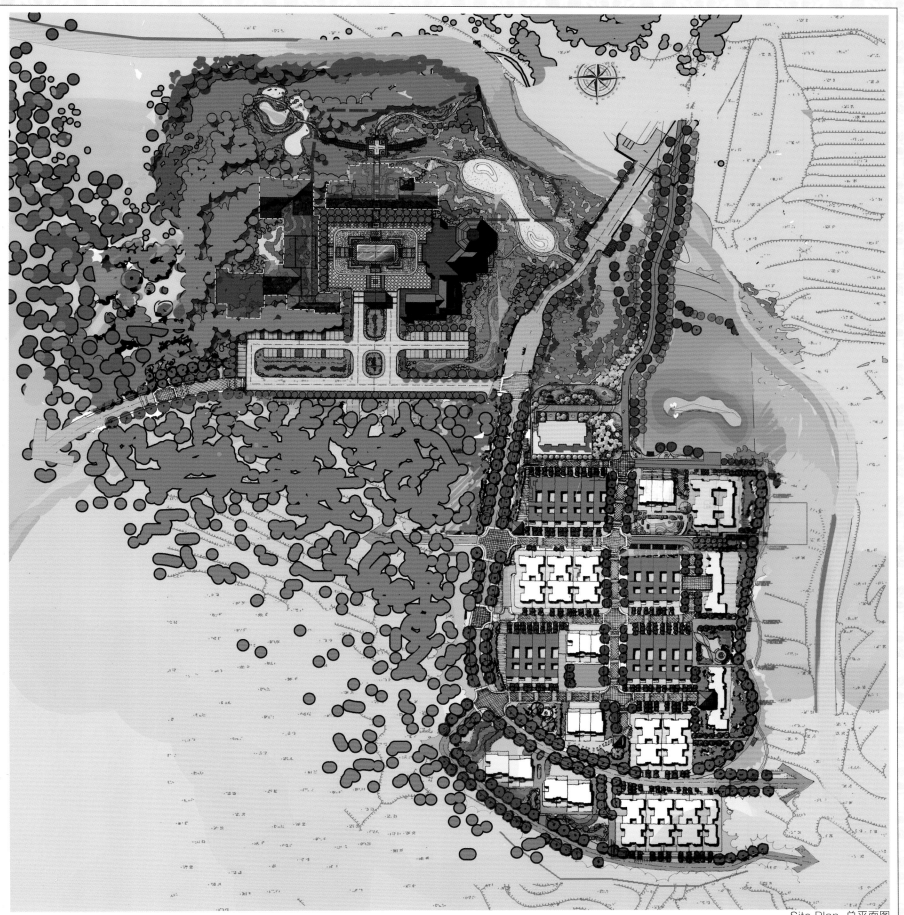

Site Plan 总平面图

Planning Layout 规划布局

Several architectural types are designed for the site including duplex houses, townhouses, low-rise villas and individual villas. It has the design purpose of revitalizing and representing the essence of world-class resorts for targeted users. In every unit, one can enjoy the beautiful landscape while sitting in the room, and large scale gardens are featured in the site. The pursuit for ecological lives is reflected in the mountain landscape and exquisite architectural and landscape designs are created on the hillside.

项目产品包括双拼、联排、叠拼、类独栋等多种建筑，设计以"复兴世界山居精髓，开启山居庭院时光"为目标，园区规划布局内户户瞰景，并有超大立体花园，山景中承载着原生态的生活向往，塑造出半山上的精美建筑设计、繁花树影。

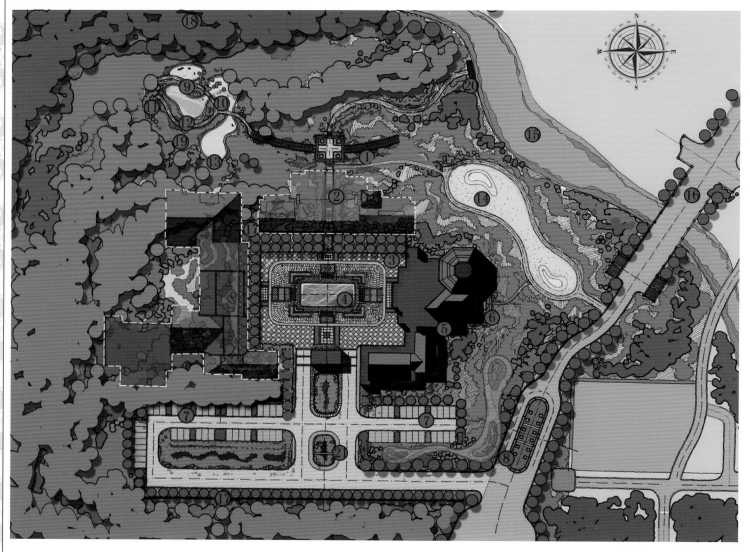

① 观景平台
② 特色阶梯
③ 中心广场
④ 中心冰场
⑤ 酒庄
⑥ 酒庄观景平台
⑦ 停车场
⑧ 入口景观
⑨ 儿童游乐区
⑩ 绿色瑜伽
⑪ 休闲自行车
⑫ 风筝放飞区
⑬ 排球
⑭ 高尔夫
⑮ 冲沟
⑯ 桥
⑰ 挡墙
⑱ 防风林
⑲ 电瓶车路
⑳ 亲水平台
㉑ 景观LOGO墙

Landscape Design 景观设计

Fifth-layered landscape system, including gorgeous country roads, community landscape, garden groups, courtyards and inner gardens, is carefully designed for the site, and the landscape designers follow the terrain of mountain to build featured gardens. When wandering in the site, users will be able to enjoy the incomparable scenery of mountains, flowers and courtyards, and the natural and ecological environment also reflects the thoughts and feelings of lives. The elegant and beautiful landscape is blended with 85 scenery spots around like Longqinfxia and Wild Duck Lake.

项目的设计精心打造五重景观体系，依山势营造出特色山地园林。五重园林景观包括美丽乡村路、社区绿化景观、组团园林、庭院园林、入户花园。漫步园区内，清月叠水，林荫夹道的山间美景，花海烂漫、星缀其间的山间院落，纯粹亲近自然的生态环境，无不深蕴环览群山、意会人生的情怀。园区内丰盈优雅的风情景致，与周边龙庆峡、野鸭湖等85处美景融为一体。

Natural Landscape Painting Nearby Yushan Lake
敔山湖畔的现代山水自然画卷

Thai Hot Cathay Courtyard 泰禾江阴院子

Location/项目地点：Jiangyin, Jiangsu, China/江苏省江阴市
Developer/开发商：Thai Hot (Jiangyin) Real Estate Co., Ltd./江阴泰禾房地产开发有限公司
Landscape Design/景观设计：L&A Design Group/奥雅设计集团
Land Area/占地面积：83 000 m²

Keywords 关键词

Life With Landscape; Hanging Courtyard Villa; Landscape Varying From Each Step
山水生活 空中院墅 步移景异

"China's Most Beautiful Residential Landscapes" Finalist
入围"中国最美人居景观"奖

After the analysis on district planning, architectural style and site condition, designers adopt the design philosophy of modern garden and combine with space design of irregular garden and gorgeous natural landscape to create a living model like a "retreat away from the world" that blended with nature, and to paint a natural painting leading users to step away from the hustle and bustle and walk into the natural, graceful and pleasant lives.

景观设计通过对项目的地域规划、建筑风格、场地条件的分析，运用现代园林的设计哲学，结合自然园林的空间手法、优美的自然景观，以与自然融为一体的居住形式"桃花源"为模型，打造出敔山湖畔的现代山水自然画卷，引领人们远离烦嚣的城市，走向自然、优雅、舒适的山水生活。

Overview 项目概况

As the Chinese ancient saying goes, it's the best to live near the water and mountain, better to live in seclusion and the last to live in the countryside. The development sits in the Yushanwan District of Jiangyin City, Jiangsu Province, enclosed by Yu Mountain, Ding Mountain and Yangtou Moutain with abundant ecological forest and water resource. The vision of the project is to make full use of the natural resource and location advantage to create an attractive and glamorous resort community.

中国古人云"居山水间者为上，村居次之，郊居又次之"。该项目位于江苏省江阴市敔山湾区域，由敔山、定山、羊头山三山合围而成。区域内原生态山体资源丰富，山水自然环境优越。项目的愿景是充分利用好周边自然资源和地理优势，打造一个令人向往的休闲度假社区。

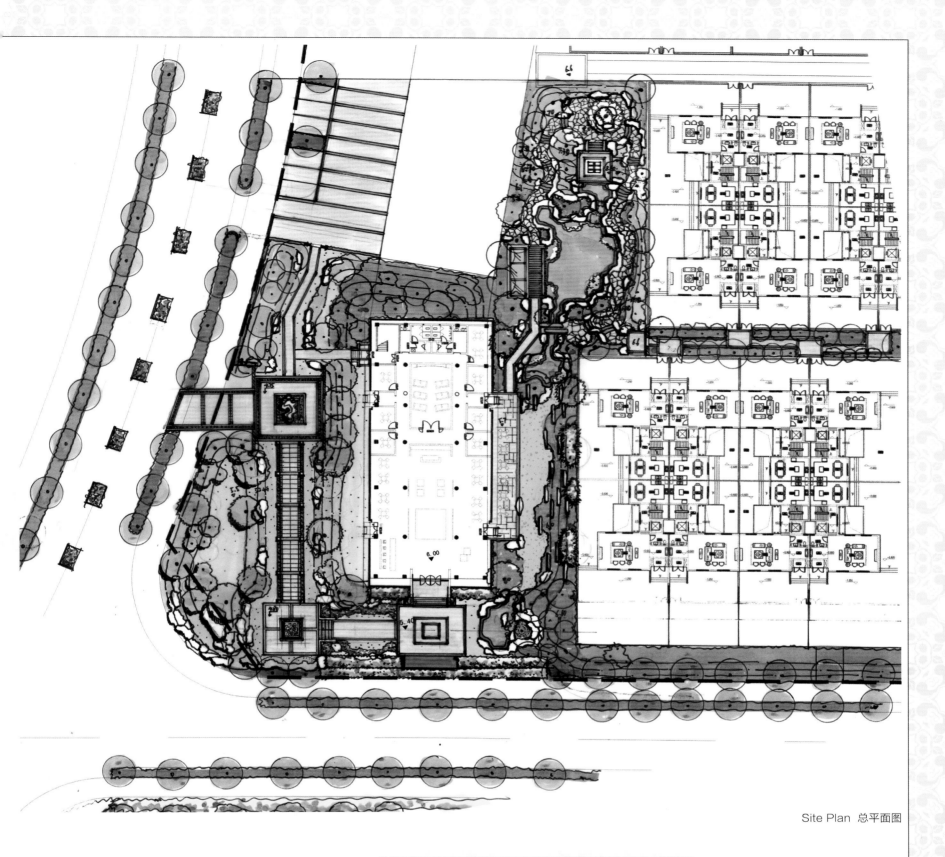

Site Plan 总平面图

Location Advantage 区位优势

The project is located in the Yushanwan New Town, the central area of Jiangyin's planned urban development. The government is trying to build Yushanwan New Town as an urban garden that integrates three features of a modern downtown, a gorgeous ecological area and a leisure scenery spot. Thai Hot Cathay Courtyard is just sitting in the most convenient plot, connecting with downtown which is only 6 km away, as well as three developed towns of Yunting, Zhouzhuang and Shanguan, through Changshan Avenue and Furong Avene. It is also close to several roads such as No. 41, No. 201 and 206.

项目位于江阴敔山湾新城，是江阴未来发展核心要地，江阴市着力将本区打造为"现代化的新城区，山水特色的生态区，休闲度假的风景区"三大功能于一体的"花园中城市，城市中花园"。泰禾·江阴院子位于本区山水交通最优地块，距江阴市中心仅6km。由长山大道和芙蓉大道两大城市主干道与主城区及云亭、周庄、山观三大经济强镇相联系。另有41、201、206等多路公交便捷通达。

Positioning Strategy 定位策略

As the second generation of "courtyard" series of Thai Hot, following the essence of top-level villas of the sister project "courtyards on the canal bank", this development adopts courtyard villas as the main residential building structure to introduce brand new living experience for Jiangyin. Besides, phase I will provide "high-rises lying on the mountain" with higher storey height, and the building height ranging from 130 m to 140 m. Units of high-rises are designed as presidential suites with several balconies to bring sunshine and landscape into the rooms which make them to be "hanging villas" and to open the door of "rich lives on upper level" of Jiangyin.

作为第二代"院子系"产品，项目袭姊妹篇"运河岸上的院子"顶级豪宅血统，以庭院式为主的建筑结构，带来江阴人居新体验。另外，泰禾·江阴院子一期将推出的"伴山高层"采用罕有的 130～140 m^2 跃层结构与时尚的观景平层户型，多阳台总统套房设计，迎纳更多阳光与美景，使之成为名副其实的"空中院墅"，开启江阴的"上层涵养时代"。

Planning Layout 规划布局

Residences are being placed carefully, while the low density luxury villas have ensured the privacy and exclusiveness of users. With green coverage ratio up to 45%, a 23,370 m² garden and waterfront landscape with three lakes, it's free for users to walk in this beautiful picture accompanied by sunshine, oxygen and beauty all around. Furthermore, in order to meet the requirements of daily lives, it also leaves an area of 7,000 m² for community commerce to complement the commerce of Yushanwan New Town.

  项目在区域内部规划上进行了精心打造，低密度的住宅部分充分保证了高端墅区的私密性和生活的专属性；45% 的超高绿化、23 370 m² 的内部园林及 3 大水系组成的内部亲水园林景观，让阳光、氧气和美景处处与业主相伴，业主随时可漫步在优美画卷里。另外，为了满足整个社区的日常生活所需，项目还自带 7 000 m² 的社区商业，与区域内商业形成有效互补，满足社区内日常商业需求。

Design Concept 设计理念

After the analysis on district planning, architectural style and site condition, designers adopt the design philosophy of modern garden and combine with space design of irregular garden and gorgeous natural landscape to create a living model like a "retreat away from the world" that integrates with nature and to paint a natural painting leading users to step away from the hustle and bustle and walk into the natural, graceful and pleasant lives.

景观设计通过对项目的地域规划、建筑风格、场地条件的分析，运用现代园林的设计哲学，结合自然园林的空间手法、优美的自然景观，以与自然融为一体的居住形式，"桃花源"为模型，打造出敢山湖畔的现代山水自然画卷，引领人们远离烦嚣的城市，走向自然、优雅、舒适的山水生活。

Landscape Design 景观设计

The built showflat area uses landscape methods of traditional Chinese garden to create landscapes that follow the terrain and vary from every step. The showflat area is lying on the mountain and facing to the lake, which are dragon vein and water vein respectively. Artificial hills are designed windingly as a "mountain dragon" in the center to form an enclosure partially, meanwhile, a "water dragon" is created in the east of the area to form an interesting picture that the matureness of "mountain dragon" and spirituality of "water dragon" embrace with each other.

目前已建成的项目的样板区，采用中国传统园林的造景手法营造因形就势、步移景异的景观视觉感受。此区域位置背山面水，山是龙脉的延伸，湖是水脉，设计的时候在地块中部利用假山在园内做部分围合，形态蜿蜒取意"山龙"，同时在地块东部挖方理水，引入一条"水龙"。山龙之稳健与水龙之灵动互相环绕，饶有趣味。

Beauty of Elegance and Nature Created Through Neoclassicism
新古典主义缔造优雅与自然的精致美

CIMC Zijin Wenchang　中集紫金文昌

Location/ 项目地点: Hanjiang District, Yangzhou, Jiangsu, China/ 中国江苏省扬州市邗江区
Developer/ 开发单位: CIMC Dayu (Yangzhou) Property Co., Ltd. / 扬州市中集达宇置业有限公司
Planning & Landscape Design/ 规划 / 景观设计: Yalantu Landscape Engineering Design Co., Ltd. / 深圳市雅蓝图景观工程设计有限公司
Land Area/ 占地面积: 51 263 m²
Floor Area/ 建筑面积: 160 000 m²
Landscape Area/ 景观面积: 42 104 m²
Green Coverage Ratio/ 绿化率: 39.2%
Plot Ratio/ 容积率: 2.2

Keywords　关键词
Environment-friendly; Neoclassicism; Intelligent Life
绿色环保　新古典主义　智能化生活

"China's Most Beautiful Residential Landscapes" Finalist
入围"中国最美人居景观"奖

After careful study on the project, the landscape design of the community is in neoclassicism style, eager to create a brand new cultural landscape for commerce, office and residences that is elegant and exquisite.

根据对项目的分析，项目的景观设计主要是以新古典主义风格为主题，极力打造一个全新的"古典优雅，精致从容"的商业办公及居住豪华、高档的人文景观环境。

Overview　项目概况

Zijin Wenchang, invested by CIMC (China International Marine Containers), is developed by CIMC Dayu (Yangzhou) Property Co., Ltd. Occupied the west plot of government office building that is in the center of Wenchang, it has seven 17+1 storey high-rise residences, three townhouses, a 13+1 storey office building (CIMC Zijin Plaza) and the underground garage, which is a AAA intelligent community collecting international high-end high-rise residences, environmental friendly townhouses, an intelligent office building and a five-star club together.

　　中集紫金文昌，由中集集团倾力打造，扬州市中集达宇置业有限公司开发建设。项目雄踞文昌中轴核心市政府西侧，由七栋17+1层住宅、三栋联排别墅、一栋13+1层中集·紫金广场（写字楼）、地下停车库组成，是集国际化高端高层住宅、绿色节能环保联排别墅、优质智能写字楼、五星级会所为一体的AAA智能社区。

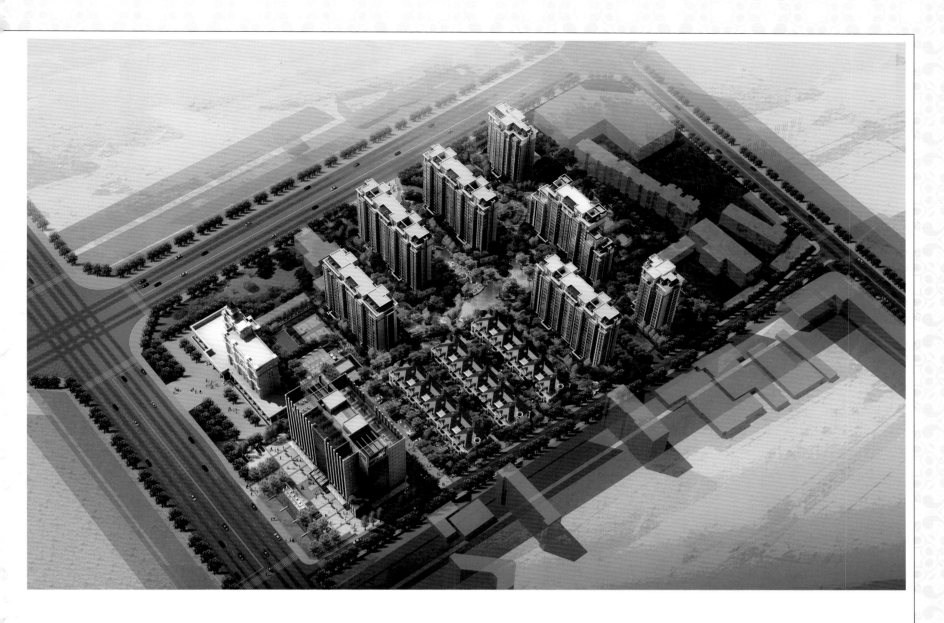

Location Advantage　区位优势

The project will be a multi-level space which is ecological, environment-friendly and healthy after its completion. Strategic location and perfect supporting facilities are working together to reveal the scarcity value of the urban high-end residential plot.

　　项目建成后将形成立体的、多层次的生态、绿色、健康空间。得天独厚的地理优势，充足便利的生活配套，彰显着城市高尚居住核心领地的稀缺价值。

Positioning Strategy 定位策略

It insists on environmental protection concept to create a low-carbon life environment. A lot of eco-friendly materials are selected for the architecture such as improved aluminium alloys and double layer low-e glass.

项目坚持秉承先进环保开发理念，积极创造低碳生活环境。在建筑材料上，采用断桥铝合金、双层LOW-E中空玻璃等众多环保材料。

Planning Layout 规划布局

Including high-end residences, a club and a five-star office building, etc., the project adopts advanced technology and process decorating with intelligent and environment-friendly elements, and takes full consideration of the harmony with surrounding buildings and interaction with Yangzhou City by featuring exquisite design methods such as double ground level, two lobbies, hanging gardens and roof gardens.

项目包括高尚居住社区、休闲会所、5A级金融商务办公楼等。使用了新技术、新工艺，引入了"智能化、绿色环保"元素，设计充分考虑了与周边建筑物的协调及城市的互动，采用双首层、双大堂、空中花园、屋顶花园等设计手法。

Design Concept 设计理念

Following the concept of CIMC of "self-improvement, challenge the limit", the project adopts dignity and refinement as the main thought of landscape design, neoclassicism style as the design concept, and design philosophy of "culture, refinement, elegance, nature and privacy" for the creation of every single corner.

项目设计秉持中集集团的"自强不息,挑战极限"之道,以尊贵与精致为景观设计的主要思想,以新古典主义景观风格为设计理念,遵循"文化、精致、优雅、自然、私密"的设计原则,臻于每一处点滴的创作与营造。

Landscape Design for Office 办公景观设计

1. Through the study on the architectural style of commercial and office space, modern architectural image is revealed through pavements, water features, lampstandards, sculptures, LOGO wall, leisure chairs, landscape spaces and landmarks of Yangzhou

2. Transparence of the commercial plaza is required for the project to represent its fashion and grand characteristics to users, plants are set in dot or line while adding some rare plants to the local plants to create exotic atmosphere. The project is divided into five zones on the landscape design that are central garden zone, sport and recreation zone, club zone, entrance landscape zone, and commercial and residential zone, to express neoclassicism landscape.

1. 商业办公空间通过对建筑风格的充分把握,在铺装形式材料、水景、灯柱、雕塑小品、商业LOGO墙、休闲的伞架椅、城市的景观空间划分及城市的标识物等各方面延伸大气沉稳的现代建筑形象。

2. 城市商业广场要求通透性,以展示其时尚高贵大气的一面,植物以点、线状为主,不宜过多,同时会以本土树种为主,合理布置珍稀植物树种来营造异域风情;居住空间结合项目的规划布局,景观设计主要分为五大区域构成:中心花园区、运动休闲区、会所景观区、主入口景观区、商业区和别墅区等来表达新古典主义风情园林景观。

Landscape Design for Residence 住宅景观设计

1. Luxury and grand main entrance square is designed, simple flower garden is set around the walls of the community, waterfall and landscape stones are set to attract people's eyes
2. Large scale lawn and high trees are working together to cut off the hustle and bustle outside so that residents can enjoy quietness at home
3. Children play ground is set for the children to play and excise, and to have a happy and free childhood
4. Adding plants and flowers to the pavements, and cooperating well with architectural space, the landscape will lead residents to the garden and enjoy the fun of life
5. Light well is designed integrated with landscape, paying attention to ventilation and natural light and meeting the sustainable requirement
6. Respect the needs of fire protection in order to represent a pleasant and safe living environment for users, to create a high-end luxury community
7. The spaces between walls around the residential zone and buildings are planted with lush plants to cut off the noisy and ensure a quiet living environment for residents

1. 豪华大气的主入口广场，简明的花圃围合社区标志墙，跌水与景观置石将更多的空间留给了人们的视线而不是脚步；
2. 大片绒毯般柔软的庆典草坪，散发泥土和青草混合的芬芳，如同屏风般的高大乔木，隔绝喧哗，尽可体味别样的静谧；
3. 专为孩子们准备的孩趣天地，给孩子们充分撒野、玩耍、锻炼的机会，享受恣意的童年时光；

4. 通过铺装模块中加入自然、随机的种植空间、以及巧妙的与建筑空间相结合等手法，使人们不约而同的因美景而聚在一起，欢享生活的情趣盎然；
5. 采光井引入时尚的设计理念，将采光井的设计与景观设计融为一体，设计注重良好的通风和良好的自然采光，符合当前国际提倡的低碳环保理念；
6. 设计以满足消防车道要求为主，以求提高住宅区内的居住舒适环境和安全环境，营造一个大气高档豪华住宅区；

7. 住宅区围墙内到建筑的周边一带为噪音缓冲区，以浓密种植为主，减少外界噪音对住区的干扰，提高区内安静的居住环境。

Modern Simplified European Style Landscape Create the City Classic

现代简欧风情景观打造城市传世经典

Zhengzhou Haima Park, Phase I 郑州海马公园（一期）

Location / 项目地点：Zhengzhou, Henan, China / 中国河南省郑州市
Developer / 项目委托：Haima Real Estate / 海马房地产
Landscape Design / 景观设计：CSC Landscape / 深圳市赛瑞景观工程设计有限公司
Land Area / 用地面积：63 200 m²

Keywords 关键词

Park Theme; Double Landscape Axis; Vertical Stereo Greening
公园主题 双景观轴线 竖向立体绿化

 "China's Most Beautiful Residential Landscapes" Finalist
入围"中国最美人居景观"奖

As a park-like garden with inspiration and beauty, its luxury Art Deco style integrates into the natural landscape; and the elegant, modern simplified European style landscape creates the Eco-environment of the community.

灵感加美感的公园式园林，奢华的 ART DECO 风格融入自然景观；优雅的现代简欧风情景观风格营造社区生态环境。

Overview 项目概况

Haima Park is located in the commercial and residential logistics park of Zhengdong New District, Zhengzhou. It's a residential complex, including commerce, residence, office, education and scientific research and training. The plot is to the east of Dongfeng East Road, to the west of Xinyi Road, to the south of Yulin South Road and to the north of Shangdu Road. The construction land area within the red line is 333,000 m². The terrain is smooth and the planning floor area is 1,160,000 m².

海马公园项目位于郑州市郑东新区的商住物流园区。用地性质是以居住为主的商住混合用地，包含商业、住宅、写字楼、教育、科研培训等业态，地块西靠东风东路。东临心怡路、北依榆林南路、南接商都路，红线内建设用地为 333 000 m²，且地势平坦，规划建筑面积为 1 160 000 m²。

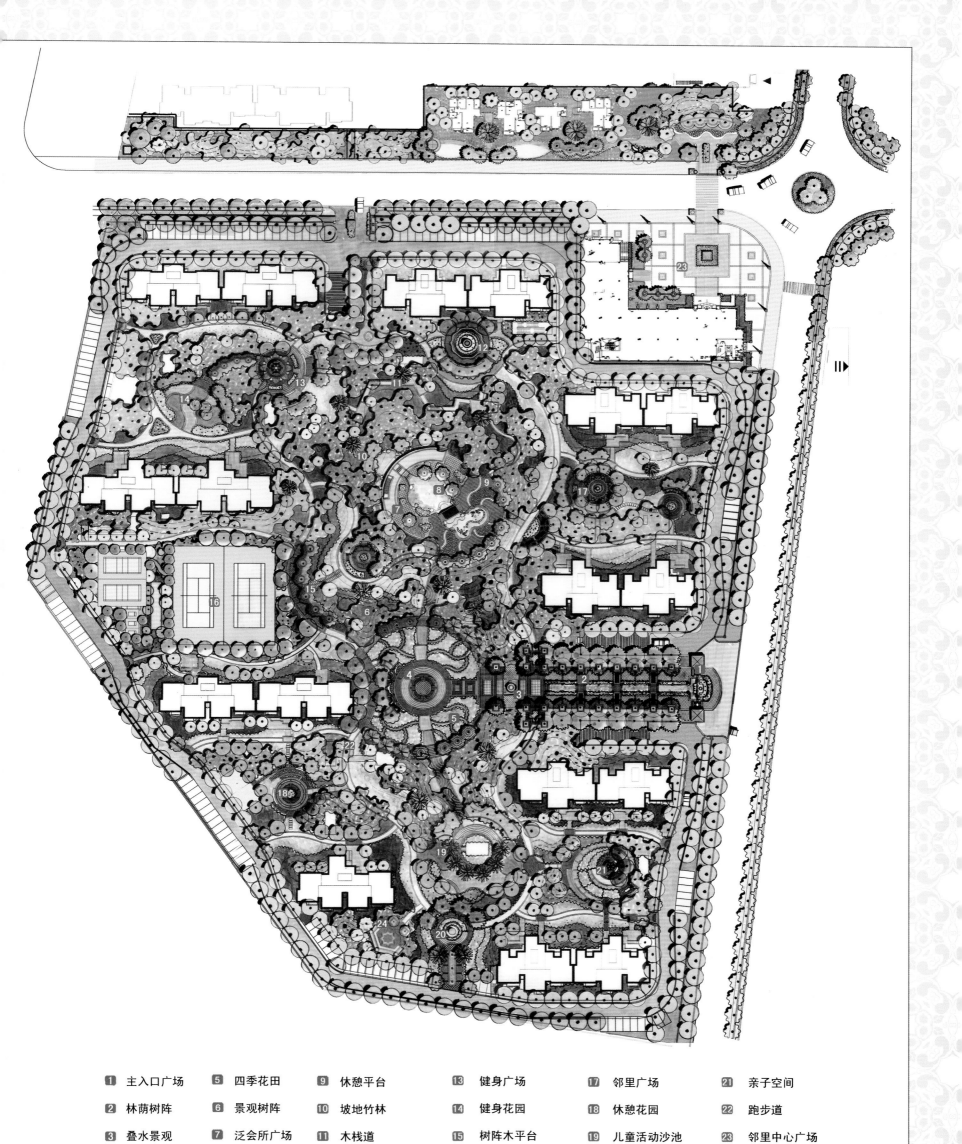

1	主入口广场	5	四季花田	9	休憩平台	13	健身广场	17	邻里广场	21	亲子空间
2	林荫树阵	6	景观树阵	10	坡地竹林	14	健身花园	18	休憩花园	22	跑步道
3	叠水景观	7	泛会所广场	11	木栈道	15	树阵木平台	19	儿童活动沙池	23	邻里中心广场
4	主题广场	8	九曲湖	12	风情雕塑广场	16	运动场	20	雕塑广场	24	朗琴园

Site Plan 总平面图

Planning 规划布局

With an advanced planning, humanized and scientific design, it's a urban complex project that gathers various functions such as residence, education, class-A office, star hotel, leisure and shopping as a whole.

项目规划超前,设计人性化、科学化,是一个集居住、教育、甲级办公、星级酒店、休闲购物等多功能为一体的城市综合体项目。

Design Concept 设计理念

The project uses the modern simplified European style, which is concise and graceful, closer to nature and in accordance with the Chinese aesthetic appreciation. It makes the residents' living expectation and required living environment as planning purpose, and holds the design principle of creating a diversified living space which is full of vigor and vitality.

项目采用现代简欧风格,简欧大气、更贴近于自然,也符合中国人的审美。项目主要以满足业主居住期望和需求的居住生活环境为规划目的,以打造一个多元性的充满生机与居住活力的生活空间为设计原理。

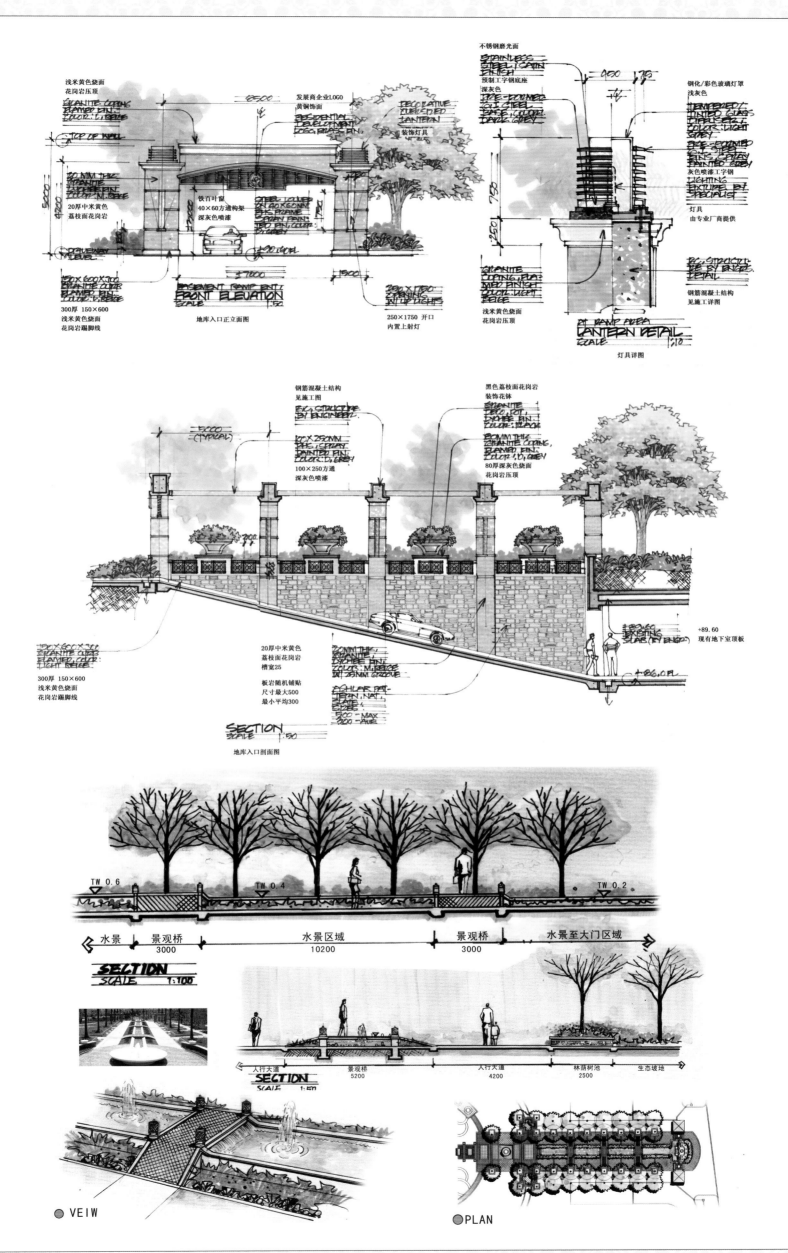

FOREST & WOODEN DECK	FEATURE TRELLIS RELAXATION DECK	LANDSCAPE LAKE & WATERFALLS & BRIDGE	RELAXATION PLAZA
果林 木栈道	特色廊架 休憩平台	景观湖 叠水 木栈桥	休憩广场

SECTION 1:150

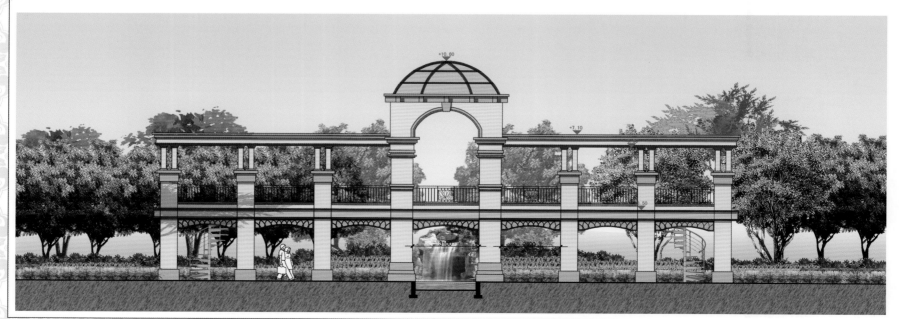

Landscape Design 景观设计

Park Theme

The project design makes park as theme and builds the distinctive community image, and it uses central park, neighborhood park, parent-child park and exercise park to create a strong sense of location and natural belonging. It builds the living space with the "modern simplified European style landscape" design concept, pays attention to the beauty and visual enjoyment of the landscape in the community and puts more emphasis on the interaction between people and the landscape. It uses the modern simplified European style to create urban classic; and the finely arranged grasses, trees make life bloom in this beautiful scenery. The residents can breathe the fresh air and keep proper distance with the prosperous city so that to avoid the hustle and bustle of the city and enjoy certain tranquility.

公园主题

项目设计主要是以公园为主题，建立起鲜明的社区形象，通过中央公园、邻里公园、亲子公园、健身公园，创造强有力的场所感和自然归属感。以"现代简欧风情景观"的设计理念构筑生活空间，注重小区景观的美感及视觉享受，更注重人与景观的互动关系。以现代简约欧式艺术的姿态打造城市传世经典，精雕细琢的一草一木、一树一林，让生活在这优美的风景里盛放。呼吸着新鲜的空气，与繁华保持恰当距离，无都市之喧嚣，无尘世之烦扰。

Double Landscape Axis

The vertical and horizontal two main landscape axes make the central axis landscape area, like a rhythmic ribbon passing through the deep of the buildings; each landscape node on the axis is like a pear inlaid on the ribbon and is connected by the winding sidewalk system; through the organization of landscape dot, line and area, a series of central axis landscape node is formed and an activity place for communication and interaction is built; water is brought in to design waterscape, flower pond, sculpture square of the central landscape area, strengthening the integrity, continuity and diversity of the landscape environment of the public activity space.

双景观轴线

纵横两条主要景观轴线，交汇出中轴景观区，犹如一条律动飘带在楼宇间的纵深部穿过，轴线上的各个景观节点犹如镶嵌在这条飘带上的颗颗珍珠，通过蜿蜒曲折的步行道路系统将其串联起来，通过景观点、线、面的组织，形成一系列中轴景观区的节点，构筑了人群集散交往的活动场所，将水系引入其中，并设置中央景观水景、花田、雕塑广场等，强化公共活动空间的景观环境的整体性、连续性以及元素多样性。

Theme Waterscape

Water is the most affinitive element with people in nature. The designers combine the two large landscape axes and arrange crisscrossed waterscape. From the entrance waterscape to the cascade waterscape and then to the quiet lake surface, which are layer upon layer, static, dynamic and full of change, and all make you enjoy the different feelings such as excitement and tranquility and fascination brought by different rhythms while you are strolling.

主题水景

水是自然界最与人亲和的元素，设计者结合两大景观轴线，布置纵横交错的水景景观，从入口水景、叠水水景到平静的湖面……层层叠叠，有静有动，富于变化，让业主在散步过程中享受到不同节奏带来的不同心理感受：或跳跃、或宁静、或神往。

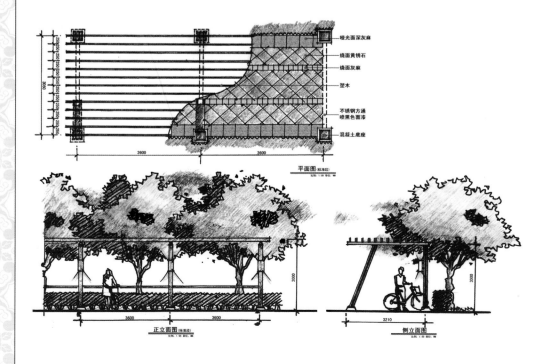

Vertical Stereo Greening

The designers use naturalized design to integrate the landscape into the nature and make it affinitive with the nature. And they adopt split-level planting with plants in different height to enhance the layering of the landscape horizontally; use small trees, shrubs and vegetation as the main landscaping and greening approaches to create stereo greening space; define the characteristics of the roads through the species of the plants, plant (separately or in group) trees, shrubs and ground cover plants in various features such as evergreen, deciduous, symmetric, regular, covering, fragrant, decorative and colorful, and soften the landscape hard facilities to enhance the interest of the landscape.

竖向立体绿化

设计师以自然化的设计将景观融合于自然、亲和于自然，设置不同高低层次的植物错落种植，增加横向关系上的景观层次感；以小乔木搭配灌木与植被为主要造景绿化手法，营造立体绿化空间；以植栽种类界定道路特色，种植（单植或群植）常绿、落叶、对称性、规则性、覆盖性、具象性、装饰性、多彩等不同的乔灌木及地被植物及软化景观硬质设施，增加景观的趣味性。

Exquisite Exotic Experience and Natural Sunshine Garden
别致的异域风情体验、自然清新的阳光花园

Jinke Central Park City　金科大足中央公园

Location/ 项目地点：Dazu District, Chongqing, China/ 重庆市大足区
Developer/ 开发单位：Jinke Group (Chongqing) / 重庆市金科集团股份有限公司
Landscape Design/ 景观设计：Oceanica International Landscape Planning Design Co., Ltd. / 深圳市奥森环境景观有限公司
Land Area/ 占地面积：108 765 m^2
Floor Area/ 建筑面积：450 000 m^2
Plot Ratio/ 容积率：3.2
Green Coverage Ratio/ 绿化率：35%

Keywords　关键词
Countryside Life; Dotted Water Feature; Wander System
田园生活　点式水景　漫步系统

"China's Most Beautiful Residential Landscapes" Finalist
入围"中国最美人居景观"奖

Echoing with architectural style and theme, it makes full use of existing natural resources around to create Tuscany landscape style that is blended into architecture to represent nature, harmony, art and interest to users.

设计手法上与建筑风格及主题相呼应，利用周边的生态自然创建与建筑融为一体的托斯卡纳风格，同样雅致，体现"自然、和谐、艺术、风情"风格。

Overview　项目概况

Situated in the central of southern part of Longgang, Dazu District which is a newly-developing plot supported by government, Central Park City occupies a large area up to approx. 110,000 m^2 and offers high-rise residences and villas.

项目位于大足区龙岗组团城南中心区，该区域是大足区政府大力发展的新兴板块。项目占地近 11 万 m^2，项目类型为高层洋房。

Location Advantage　区位优势

As a newly-developing plot supported by government and positioned as sub-center of Longgang, the plot should meet the requirements of commerce, culture, entertainment, tourism service and residence. The plot is with residential area in the north, Dazu Tianjiabing Middle School in the west, a residential area for sale in the east and a central park planned to be built in the south. It is a flat plot with a maximum terrain height difference of 15 m.

项目所在区域是大足区政府大力发展的新兴板块，定位为龙岗组团城市的副中心城区，须满足大足城市商贸、文化、休闲娱乐、旅游服务和居住功能。项目北面为以已出让住宅用地，西侧为大足田家炳中学，东侧为待出让住宅用地，南侧为规划建设的中央公园。地块内部地势较为平坦，最大高差约为 15 m。

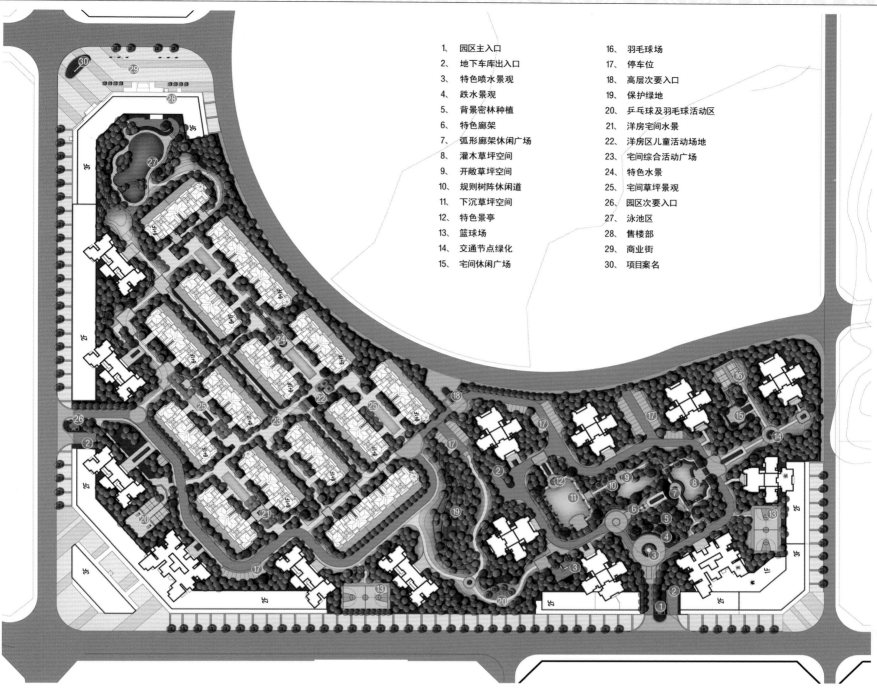

Site Plan 总平面图

1、园区主入口	16、羽毛球场
2、地下车库出入口	17、停车位
3、特色喷水景观	18、高层次要入口
4、跌水景观	19、保护绿地
5、背景密林种植	20、乒乓球及羽毛球活动区
6、特色廊架	21、洋房宅间水景
7、弧形廊架休闲广场	22、洋房区儿童活动场地
8、灌木草坪空间	23、宅间综合活动广场
9、开敞草坪空间	24、特色水景
10、规则树阵休闲道	25、宅间草坪景观
11、下沉草坪空间	26、园区次要入口
12、特色景亭	27、泳池区
13、篮球场	28、售楼部
14、交通节点绿化	29、商业街
15、宅间休闲广场	30、项目案名

Positioning Strategy 定位策略

Positioned as a multi-functional area for commerce, culture, entertainment, tourism service and residence, the project is carefully located in the central of "Shengji Lake urbanism demonstration area" and surrounded by Shengji Lake Central Park, north of a grade A hospital, close to Chengdu-Chongqing Express in the south, lying on Dazu No. 2 Middle School and tourist area in the east and west respectively.

　　项目定位大足区城市商贸，文化休闲娱乐，旅游服务和居住功能区。项目择址"圣迹湖都市生活示范区"中心位置，环抱圣迹湖中央城市公园，北临三甲医院、南近成渝复线高速；东依大足二中，西靠"休闲旅游区"。

Planning Layout 规划布局

It is comprised of commercial shops, villas and high-rises, highlighting on making each part with unique characteristics and reaching harmony of the overall project, to create a harmonious and pleasant residential community.

　　项目由商业街区、洋房区、高层区组成，项目重点考虑打造各个部分、形成各部分特色的同时又兼顾独特性和协调性的统一，创造一个和谐温馨的住宅园区。

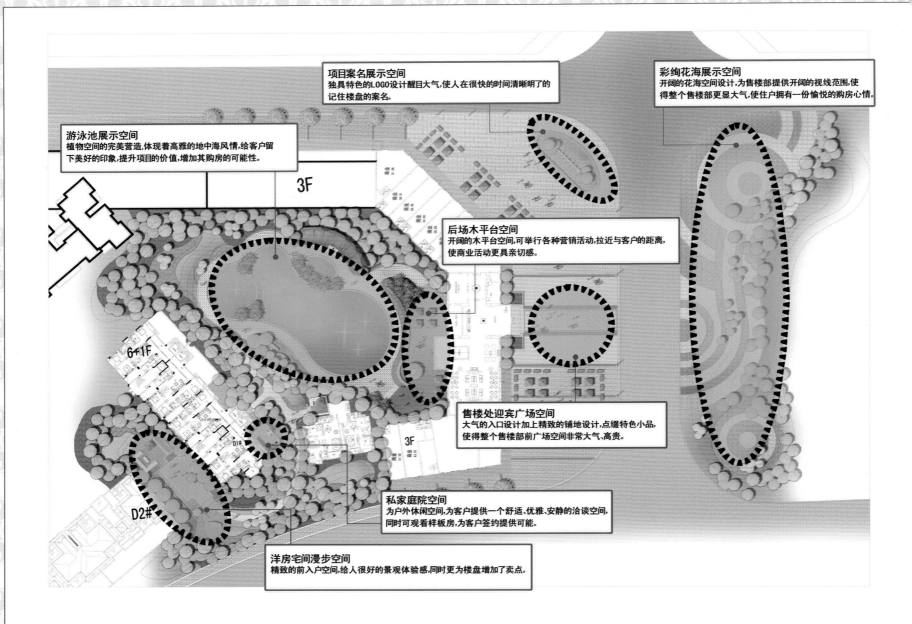

Design Concept 设计理念

The design theme for landscape is to provide "exquisite exotic experience and natural sunshine garden" Tuscany style for users. Echoing with architectural style and theme, it makes full use of existing natural resources around to create Tuscany landscape style that is blended into architecture to represent nature, harmony, art and interest to users. While Tuscany style is rural and simple and elegant, integrating with nature and reminding of hillside, country villas and vineyard in the sunshine. Landscape designers take full advantage of topography and are eager to create a high end community that reveals humanistic spirit and community culture, also integrates with linear layout and featured elements to build wonderful lives in the countryside, so the way that users go back home will be a garden path. Meanwhile, they also take practicability of landscape as primary that reveals its humanity.

项目景观设计主题为"别致的异域风情体验、自然清新的阳光花园",设计风格为"托斯卡纳风格景观的自然营造"。设计手法上与建筑风格及主题相呼应,利用周边的生态自然创建与建筑融为一体的托斯卡纳风格,同样雅致,体现"自然、和谐、艺术、风情"风格。托斯卡纳风格是乡村的、简朴的,更是优雅的,它与大自然形成有机的整体,是色彩的融合,置身其中,便会情不自禁想起沐浴在阳光里的山坡、农庄、葡萄园。设计师充分利用地形与地貌,将自然资源进行合理整合,旨在构建一个体现人文精神和社区文化的现代高档社区,通过直线型的构图、特色元素的融入,打造如田园般的纯美生活,使住户的回家之旅仿佛置身公园。同时注重实用性,体现景观的人性化。

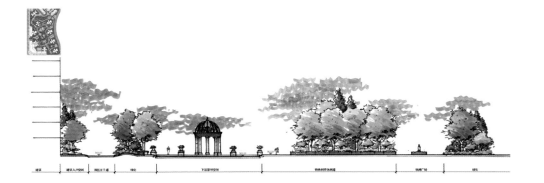

Architectural Design 建筑设计

Gardens of high-rise residences are multi-functional spaces that includes sport, recreation and wander. Courtyard spaces between each residence are carefully designed with pleasant wander system to connect every node space to make users feel be involved to take activities in their back yards after dinner. For villas, landscape designers adopt linear layout, dotted water features and plants and lawns to cooperate with landscape design between villas, to create high end and elegant living environment.

高层组团花园区空间形态多元化，运动、休闲、漫步等一应俱全。设计以绿化和地形进行宅间庭院空间的营造，通过合理、舒适的漫步系统连接各个节点空间，让住户尽情地参与其中，进行各种康乐活动，体验业余饭后的美好时光。洋房区旨在打造高档优雅的住宅环境，通过直线的构图方式、点式水景的点缀、疏林草地配合宅间合理的植物空间设计，打造住宅品质。

Landscape Design 景观设计

Guidelines of landscape design are listed below: promoting sense of layer of space by plants to brighten residential groups and attract users to walk outside; selecting plants that are both beautiful and practical to integrate greening and economic benefit together. For the landscape design of commercial street, plants are set in standard with straight and tall trees to create grand and simple commercial atmosphere, while the lower layer employs bright ground cover and lawn to form colorful tapes. Planting pattern of "tree + shrub + ground cover or lawn" is employed for landscape in villa area, with different trees for different areas to form various landscape themes. Landscape design of high-rises uses evergreen plants as foundation decorating with seasonal plants to provide wide lawn, open wander paths and private communication spaces for users.

　　绿化设计总则——利用植物划分空间增加层次感，使居住建筑群更显活泼，吸引住户进行户外活动；选择观赏性和实用性兼备的植物进行布置，做到绿化、美化和经济效益相结合。商业街区植物景观设计以规整式配置手法为主，乔木运用树干通直、分支点高的观树形大树，打造大气、简洁、强调线条感的现代商业城氛围，下层采用颜色亮丽的地被和草坪形成线条优美的色带。洋房区绿化设计采取"乔木＋灌木＋地被"或草坪的种植方式，不同区域选择不同的主打树种，形成不同的植物景观主题曲和差异化的植物景观。高层区绿化设计以自然式为主，以常绿植物为基调，点缀季相植物，通过植物的层次、疏密创造开阔的阳光草坪空间、开放式运动漫步空间和私密性的静坐交流空间。

Modern and Fashinable Working Environment
极具时尚现代气息的花园式办公环境

Landscape of China Merchants Baoyao Area (Net Valley) 招商创业宝耀片区景观（招商网谷）

Location/ 项目地点：Nanshan District, Shenzhen, Guangdong, China/ 广东省深圳市南山区
Client/ 委托单位：China Merchants Property Development Co., Ltd. / 深圳市招商创业有限公司
Landscape Design/ 景观设计：Dongda Landscape Design Co., Ltd. / 深圳市东大景观设计有限公司
Area/ 项目规模：45 000 m²

Keywords　关键词

Humanistic Concern; Modern and Fashinable; Flowing Water Feature
人文关怀　现代时尚　流动水系

"China's Most Beautiful Residential Landscapes" Finalist
入围"中国最美人居景观"奖

Modern and concise design style is adopted for the project, inserting humanistic concern into the environment and expressing through various sensory judgments including gustation, touch, hearing and vision, and small spaces such as elegant stop, tactual wall, time showcase, spray square, dreamland and rhythm garden, to create a modern and fashion working environment.

项目采用现代简洁的设计手法，在自然的环境中注入更多的人文关怀，通过味觉、触觉、听觉、视觉的不同感官体验，运用品位小站、触觉体验墙、时代橱窗、水雾广场、幻境花园和韵律之声等小空间，打造出一个多元化具有时尚现代气息的花园式办公环境。

Overview　项目概况

The project located in the center of Shekou Net Valley is the landmark building of Net Valley. After upgrading and reconstruction, Baoyao area is the central zone collecting three core public technologies of internet, foundation and application of e-commerce and Internet of Things, creating high-tech and dynamic zone for Shekou Industrial District and increasing its attraction to build a new model regarding working and living style for Shekou.

宝耀片区项目位于蛇口网谷核心区域，是网谷片区的标志性建筑。片区通过升级改造，打造互联网、电子商务基础及应用、物联网技术及应用示范三大核心主体园区及公共技术平台，成为"蛇口网谷"的中心区域，为蛇口工业区打造出高端活力湾区，增强了片区整体吸引力，竖立了蛇口地区工作及生活方式的新样板。

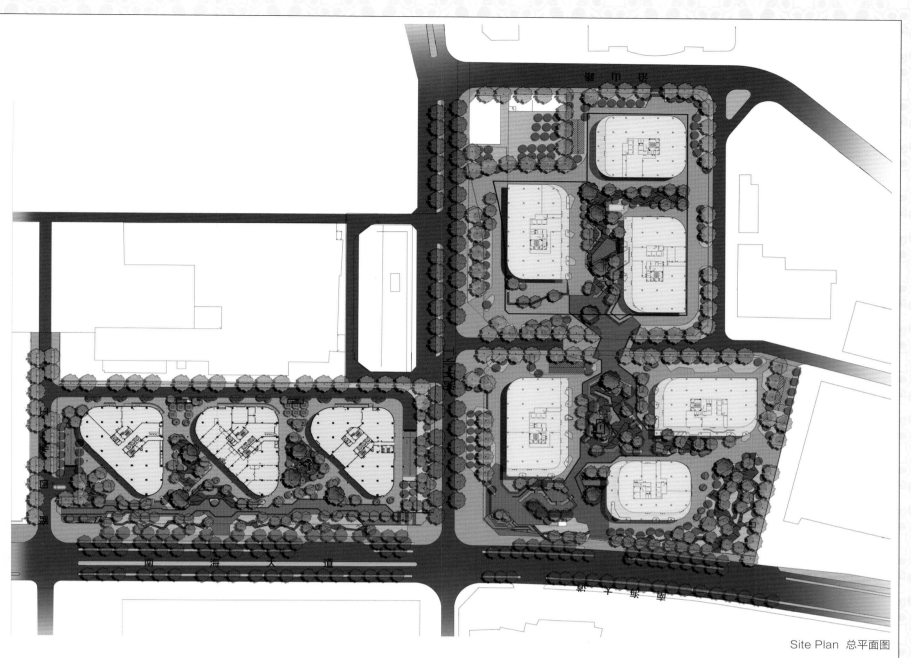

Site Plan 总平面图

Architectural Design 建筑设计

On the basis of functional demands, architectural design presents pleasant sense of space and cosy scale for users. Designed according to local conditions, specially featured hanging platforms and gardens make it possible to reach harmony between environment and architecture. With a modern, concise and unique design style, the project reveals the rich connotation of electronic industry.

项目的建筑设计在符合功能性要求的基础上，体现宜人的空间感及舒适的尺度。项目规划因地制宜，空中平台及空中花园的特色设计使环境和建筑和谐对话。整体设计风格现代、简洁、新颖独特，展现出电子产业文化的丰富内涵。

Landscape Planning 景观规划

Modern and concise design style is adopted for the project, inserting humanistic concern into the environment and expressing through various sensory judgments including gustation, touch, hearing and vision, and small spaces such as elegant stop, tactual wall, time showcase, spray square, dreamland and rhythm garden to create a modern and Fashinable working environment.

项目采用现代简洁的设计手法，在自然的环境中注入更多的人文关怀，通过味觉、触觉、听觉、视觉的不同感官体验，运用品位小站、触觉体验墙、时代橱窗、水雾广场、幻境花园和韵律之声等小空间，打造出一个多元化具有时尚现代气息的花园式办公环境。

Landscape Structure 景观架构

One Framework

The framework is landscape of urban interface formed by Nanhai Avenue and Industry 5th Road. The project reserves the existing landscape of municipal roads and takes all the roads as a whole to build an external display framework.

Three Corridors

Frontage Display Corridor: tropical plants set in linear cooperating with winding paths are designed as the main design elements of this corridor, it is designed with large-scale theme sculptures echoing with other areas of Net Valley which make the display corridor as the first window to the outside and also to cut off the noise of Nanhai Avenue.

Central Time Corridor: adopting trace of history as a landscape element, integrated ground with imprint is highlighted by LED lamps to tell the history of Baoyao area. Several places are designed on the ground for communication.

Landscape Corridor: cooperating with planned roads and landscape style, the corridor connects the project which has been divided by the municipal roads as a whole.

一个框架

框架由南海大道及工业五路交汇所形成的城市界面景观。设计当中保留了蛇口现状市政道路的原有绿化风貌，对总体街道方案的统一化处理，形成对外展示的整体风貌。

三条廊道

临街风貌展示廊：带状的热带植物配与蜿蜒曲折的小路形成廊道的主要设计元素，南海大道一侧，结合外围绿化，大型主题雕塑的设计与网谷其他片区相互呼应，使得此展示廊道成为第一道园区对外风貌展示的窗口，并借此隔离南海大道对园区的影响。

中心时光通廊：运用历史时代的印迹作为景观元素，设计整体的印迹地面并加以夜景灯带的强化提示，诠释宝耀片区的历史。在这片历史的印迹中，设计若干个场所，提供驻足交往空间的可能。

空间缝合景观廊：结合规划道路及整体景观设计风格缝合由于市政规划道路割裂项目的完整性。

Six Places

Elegant Stop: sunken enclosed small space is designed with quiet water features, while chairs and bars are set outdoors to provide public dining space.

Tactual Experience: a space with sense of texture is created by protruding texture of road and vertical landscape wall, various kinds of materials with different texture are cooperating with plants to attract people to touch the materials and to create a diverse leisure space.

Time Showcase: the landmark sculpture wall is telling the development of Baoyao area and showing some important evens (i.e. Huayi Aluminium Factory) on the facade, the special treatment for the texture has created a unique facade landscape.

Spay Square: offering a cooler outdoor environment for people in the hot summer, also gathering them to the square to look for relaxation, the square is designed with linear water front space and artificial fountain to be the landscape that is attractive.

Dreamland: just like a shallow pool with dark titles, it excites people's curiosity, when night falls down, lamps under the water are in different colors and shadows are reflecting on the water; while goldfishes are swimming in the water to form a double-layer space, just like walking into a dreamland and experiencing the fun and interests of a difference space.

Rhythm Garden: a place that adopts sound as its theme, a flowing water feature is designed on the land with height difference, with the background of water feature, a stage with multi-layer is created for small-scale concert and Jazz performance, furthermore, audible device is placed in each step, even there is no music performance it still provides leisurely sound to bring bright sound and dynamism to the place.

六个场所

品位小站：采用下沉形式的围合小空间、点缀的静态水景，在围合的小空间中，设置户外小座以及穿插的吧台，提供公共餐饮空间。

触感体验：地面的肌理纹路较为突出及立面推拉景墙形成一个肌理纹路比较强的场所空间。不同质感的硬质材料与植栽配合，引人触摸感受的欲望，创造质感丰富的休憩空间。

时代橱窗：标识性的雕塑墙面给宝耀工业区提供展示园区发展历程以及重要时代印迹（华益铝厂）在立面空间上的诉说平台，通过肌理的质感处理，形成独特的立面景观。

水雾广场：广场在室外酷暑环境中提供一种清凉的气息，也将人群聚集到这里来放松自我。带状的亲水空间、人造水雾喷泉塑造了主要入口的聚集人气，吸引人流的提示景观。

幻境：如同地面的深色浅水池，吸引游人一探究竟，当夜晚天幕降下时，水面中的灯柱散发各色光芒，向水面望去，自己的影像将倒影在水面上；水下又有动感的金鱼游动，双层空间，使人们走进幻境，体验另一类空间的乐趣及欣赏角度。

韵律之声：这是一个以声音为主题的场所，通过高差设计一处流动的水系，在流动的水声背景下，并设计富有高差的表演舞台，提供了小型的音乐会和爵士乐表演的平台，在没有音乐表演时，每级水阶下设有音响装置，依然会发出悠然的声音，为此场所带来愉悦的声音和活力。

The First Ecological and Low-carbon Office District Demonstration Area of Guangzhou-Foshan Economic Circle
广佛经济圈首个绿色低碳产业商务总部集群示范区

Ecological Office District·Guangzhou-Foshan Basement 中企绿色总部·广佛基地

Location/项目地点：Lishui Town, Nanhai District, Foshan, Guangdong, China/广东省佛山市南海区里水镇
Client/委托单位：Genyuan Investment Co. Ltd. /广东广佛现代产业服务园开发建设有限公司
Landscape Design/设计单位：Guangzhou BAC Landscape Design Co., Ltd. /广州市柏澳景观设计有限公司
Land Area/占地面积：300 000 m²
Floor Area/建筑面积：500 000 m²
Landscape Area (Apartment)/景观面积（公寓部分）：25 000 m²
Project Type/项目类型：Commercial and Residential/商住
Design Content/设计内容：Landscape & Construction Drawing Design/景观方案及施工图设计

Keywords 关键词

Entering the Hall Before Walking Into the Room; Transition; Personalized Connotation
登堂入室 起承转合 个性化情愫

"China's Most Beautiful Residential Landscapes" Finalist
入围"中国最美人居景观"奖

Inspired by Chinese spatial concept of "entering the hall before walking into the room", landscape design of the project is simple and concise, following the order of "introduction, elucidation, transition and summing up" to express distinct concept and reflection.

景观设计以中式"登堂入室"空间概念作为景观创意蓝本，以起、承、转、合逻辑秩序进行设计，景观简约，反映出清晰的观念和思考。

Overview 项目概况

Strategically located in Lishui Town, Nanshan District of Foshan City, the core area of Guangzhou-Foshan, Ecological Office District is close to Foshan First Ring Road in the west and Jinshazhou, Guangzhou with a gross land area of approx. 300,000 m² and floor area of 500,000 m². It is the first ecological commercial park of Pearl River Delta with independent property right as well as independent office buildings.

中企绿色总部·广佛基地，位于广佛同城核心区域——佛山市南海区里水镇，西侧紧邻城市主干线佛山一环，毗邻广州市金沙洲，总占地面积约30万m²，建筑面积约50万m²。项目是珠三角首个独立产权和独立办公的绿色商务园区。

Site Plan 总平面图

Positioning Strategy 定位策略

Following the management concept of "adopting featured office district to promote development of industries, using gorgeous environment to attract advanced technology and to create industrial cluster with outstanding enterprises", the project collects multi-functions including research & development, education, demonstration, display, production and marketing with the ideas of low-carbon commerce, management and services, dedicating to be the first ecological and low-carbon office district demonstration area of Guangzhou-Foshan economic circle.

整个项目秉承"以特色园区带动产业发展、以良好环境吸引先进技术，以优势企业形成集群效应"的经营理念，从低碳商务、低碳运营以及低碳服务三个方面，集研发、教育、示范、展示、生产、销售等功能于一体，致力打造成为广佛经济圈首个绿色低碳产业商务总部集群示范区。

Planning Layout 规划布局

Project is comprised of independent office buildings and eight top-level supporting facilities including International Convention Centre, a business hotel, serviced apartment, exchange center, central park, featured commercial street and club.

项目由企业独栋花园总部商务区和八大顶级服务配套：国际会议会展中心、商务酒店、酒店式公寓、交流中心、中央公园、风情商业街、会所等组成。

Design Concept 设计理念

Inspired by Chinese spatial concept of "entering the hall before walking into the room", landscape design of the project is simple and concise, following the order of "introduction, elucidation, transition and summing up" expressing distinct concept and reflection. The comparison of concise geometric lines, shapes and masses is expressed through the order to represent strict logic and fresh ideas which are restraining and quiet. The entrance space is smooth and free, hiding the exaggerated and flashy elements to reveal unique and personalized connotation.

　　景观设计以中式"登堂入室"空间概念作为景观创意蓝本，以起、承、转、合逻辑秩序进行设计，景观简约，反映出清晰的观念和思考。简洁的几何线、形、体块的对比，按照既定的原则推导演绎，表现出严格的逻辑、清新的观念，深沉、内敛、静穆。整个入口空间表达自如，平淡中见奇崛，轻落中有纵生，仿佛有一种力量在心头跳跃闪动，夸张被隐藏，浮华被抹去，体现个性化的情愫。

Landscape Design 景观设计

Entrance Space
Frontage of the entrance space is designed for displaying its image with open landscape space. While for the back entrance space, natural, semi-private and enclosed methods are adopted to create courtyard space. Grand atmosphere and iconic landscape shape is designed for entrance space, selecting featured architecture as background to attract people's eyes, besides, textured pavements have highlighted the theme one more time. Cooperations of sentimental, aesthetic and metrical elements have represented the artistic conception of the project freely.

Landscape Courtyard
Grey tone pavement and yellow tone facades are built in the courtyard, which is concise, simple grand and quiet, breaking up traditional Chinese grey and white tones to create a landscape courtyard that is fashionable and elegant.

Pattern Design
Traditional fretwork is adopted for watchhouses, feature walls and flower base, cooperating with antique lion sculptures, gate piers and bronze wall lamps to embody the elegance and dignity of the gates.

入口前后空间
入口前空间，以形象展示为主，打造开敞性景观空间。入口后空间，采用自然、半私密、围合式处理，打造庭院式空间。入口大气且具有地标性的景观造型，以特色建筑作为背景，第一时间吸引过往人群的眼球，夸张而富有质感的地面铺张，再次点亮主题。通过情感的、美学的、韵律的反复糅合，意境才能呈现，表达才能释放扩张。

景观庭院
铺地呈灰色调，立面呈黄色调。简洁、质朴、沉稳、大气、宁静且富有质感，打破传统灰搭配白的中式色调，创作出具有时尚感、尊贵气息的景观庭院。

纹样设计
岗亭、景墙、花基设置了传统的回纹元素，仿古狮子雕塑、石鼓门墩、铜雕壁灯，这些都将宅门的尊贵感体现出来。

Meiju Encyclopedia 美居小百科

Introduction, elucidation, transition and summing up are originally used for structure of Chinese regulated classic writing. While "introduction" is the causation and start of the essay; "elucidation" stands for the process of the event; "transition" means changing to another viewpoint and "summing up" is some comments regarding the event. It is being widely used in the landscape design to form a brand new design logic sequence.

起承转合，原为诗文写作结构章法方面的术语。"起"是起因，文章的开头；"承"是事件的过程；"转"是事件结果的转折；"合"是对该事件的议论，是结尾。起承转合，后来被跨界运用到景观设计领域中，成为一种设计逻辑秩序。

A French Style and Vigorous Villa Area That is Both Ecological and Well-equipped

法式风情、充满活力且兼具生态修复功能的实用别墅区

Master of The Mountain, Vanke Chengdu　成都万科五龙山公园

Location/ 项目地点：Chengdu, Sichuan, China/ 四川省成都市
Developer/ 开发商：China Vanke Co., Ltd. / 成都万科新都置业有限公司
Landscape Design/ 景观设计：UC Landscape Architecture Design Group/ 深圳源创易景观设计有限公司
Land Area/ 占地面积：1 850 000 m²
Floor Area/ 建筑面积：91 805 m²

Keywords　关键词

French Garden; Flat Independent Villa; Vertical Flow Technology
法式园林　独栋平层化　垂直流技术

 "China's Most Beautiful Residential Landscapes" Finalist
入围"中国最美人居景观"奖

Upon sufficient consideration of relationship between landscape and architecture, the designers create unique large-scale space of French garden, while assuring the landscape quality of the project. Plenty of French elements are adopted in landscape accessories, which highlight the nobility and elegance of French garden and create characteristic villa area.

设计师充分考虑到景观与建筑的关系，在确保园区景观品质的同时，塑造出法式园林特有的大尺度空间。景观小品的设计中大量运用法式元素，突显出法式园林的尊贵与典雅，营造出别具风情的别墅区。

Overview　项目概况

The project is situated in Chengdu, Sichuan, the whole project is built close to the mountain, next to the water, with altitude differences up to 80 m at the maximum. The project is in possession of a lake area of 120,000 m², with attractive view facing the water against the mountain. The design remains the original forest to the maximum extent, which creates exquisite villa area with mountain, water and forest resources close to urban center.

项目位于四川省成都市，整个园区依山而建，拥有12万 m²的湖区，场地内地势最大高差达80 m，依山面水、风景绝佳，且最大程度地保留了五龙山的原生森林，成为城市近郊独享山、水、森林的别墅区。

Site Plan 总平面图

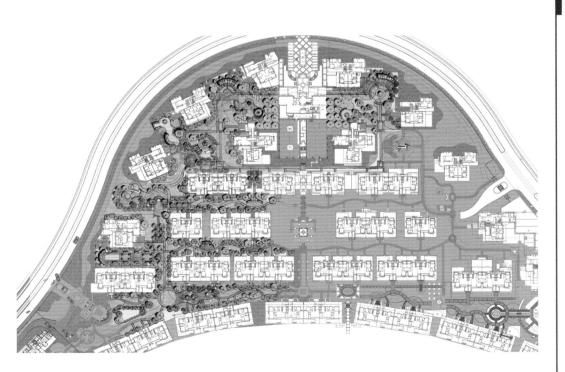

Planning Layout 规划布局

The project is aimed at creating a unique French style and vigorous villa area that is both ecological and well-equipped. Based on the China Green Building Three Star Certification Standards and respect for the nature, the designers lay emphasis on the quality of landscape design, improve the quality of feature nodes, sidewalks, terrain transformation, drainage and storage, and promote plant recover and preservation of habitat.

项目旨在打造一个具有独特的法式风情、充满活力且兼具生态修复功能的实用别墅区。项目的定位以中国绿色建筑三星认证标准为准则，以尊重自然为前提，注重景观设计的品质，提高景点、景观通道、地形改造、排水、蓄水的质量，促进植被的恢复与栖息地的保护。

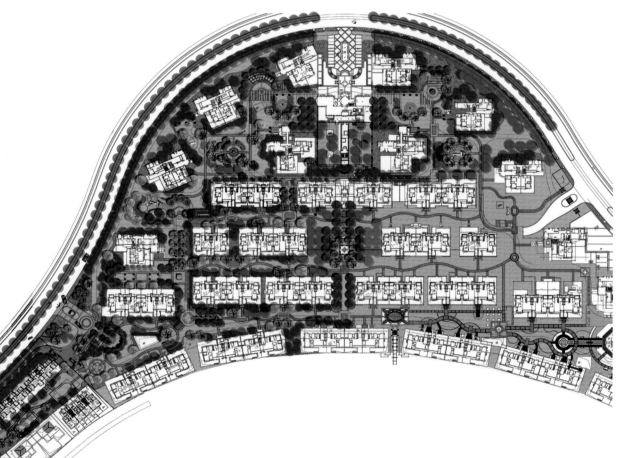

Residence Design 住宅设计

It is the first crossover project of Vanke, pioneering model of "flat independent villa · wide HOUSE", moving study room and master bedroom and washroom on the second floor to the first floor, comprising brand new flat villa with living room, dinning room and kitchen in the first floor, being a member of TOP series of Vanke. It is the first "green three star" residence project of southwest of China, equipped with floor heating, water purification and central air-conditioning, with a maximum width of 18 m, floor height of 3.3 m and a length of approx. 8m.

万科集团首次跨界作品——全国首创"独栋平层化·宽HOUSE",将独栋二层的主卧、主卫、书房平移至一层,与客厅、餐厅、厨房组成全新大平层,置入万科TOP系独栋资源之中。西南首个"绿色三星"住宅,地暖、净水、中央空调全系装修;最大约18 m面宽、约3.3 m层高、约8 m大横厅,媲美独栋的装修大平层。

Landscape Design 景观设计

The lake area is up to 120,000 m², the self-restoration system is built up to ensure the water quality. The design of three wetlands based on the vertical flow technique achieves the goal of water recycle through physical method.

Upon sufficient consideration of the relationship between landscape and architecture, the designers create unique large-scale space of French garden, while assuring the landscape quality of the project. Plenty of French elements are adopted in landscape accessories, which highlight the nobility and elegance of French garden and create characteristic villa area.

五龙山有 12 万 m² 的湖体面积，为了保证水体质量，设计师在湖泊内设置了自我修复系统，并采用生态湿地垂直流技术修建了三个湿地，通过物理方式达到循环净化水质的目的。

设计师充分考虑到景观与建筑的关系，在确保园区景观品质的同时，塑造出法式园林特有的大尺度空间。景观小品的设计中大量运用法式元素，突显出法式园林的尊贵与典雅，营造出别具风情的别墅区。

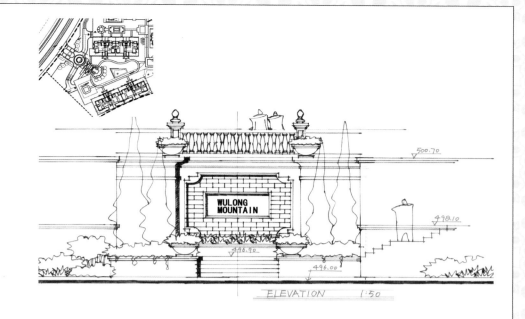

WULONG MOUNTAIN

ELEVATION 1:50

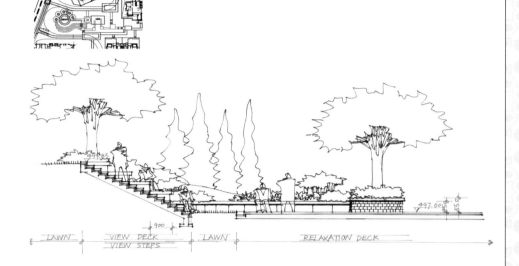

| LAWN | VIEW DECK / VIEW STEPS | LAWN | RELAXATION DECK |

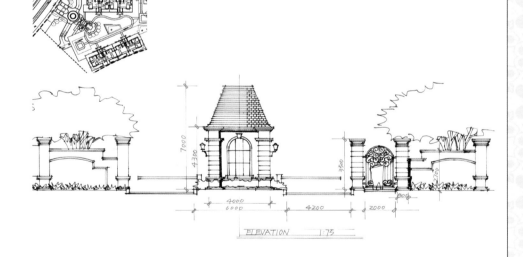

ELEVATION 1:75

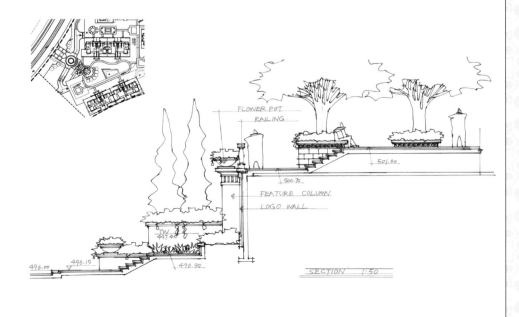

FLOWER POT RAILING

FEATURE COLUMN
LOGO WALL

SECTION 1:50

Model of Modern Landscape

曲岸观澜现代自然山水之典范

Showcase of Thai Hot High-rises 泰禾海沧高层展示区

Location/ 项目地点：Xiamen, Fujian, China/ 福建省厦门市
Client/ 业主：Thai Hot (Xiamen) Real Estate Co., Ltd. / 厦门泰禾房地产开发有限公司
Landscape Design/ 景观设计：L&A Design Group/ 奥雅设计集团
Area/ 项目规模：9 000 m²

Keywords 关键词

Artistic Conception of Landscape; Sense of Spirituality and Refinement; Changeable
山水意境 灵动细腻 富有变化

 "China's Most Beautiful Residential Landscapes" Finalist
入围"中国最美人居景观"奖

The water features around the showcase are placed on both sides of the road and enclosed by plants to improve the visual effect and listening pleasure.

项目样板房附近的水系则通过绿化的围合，让水系若有若无的显现于道路两侧，增加了视觉的变化和悦耳的流水声。

Overview 项目概况

The showcase is situated on Xinyang North Road, Haicang District, Xiamen City, west of Changgeng Hospital and close to Cathay Courtyard which is an outstanding project of L&A Design Group. It is designed as a modern and high-end residential community enjoying sophsicated supporting facilities and transportation.

泰禾海沧高层展示区位于厦门市海沧区新阳北路，长庚医院西侧，紧挨奥雅优秀作品厦门院子，周边配套设施齐全，区位优势明显，项目整体定位为现代自然高档住宅区。

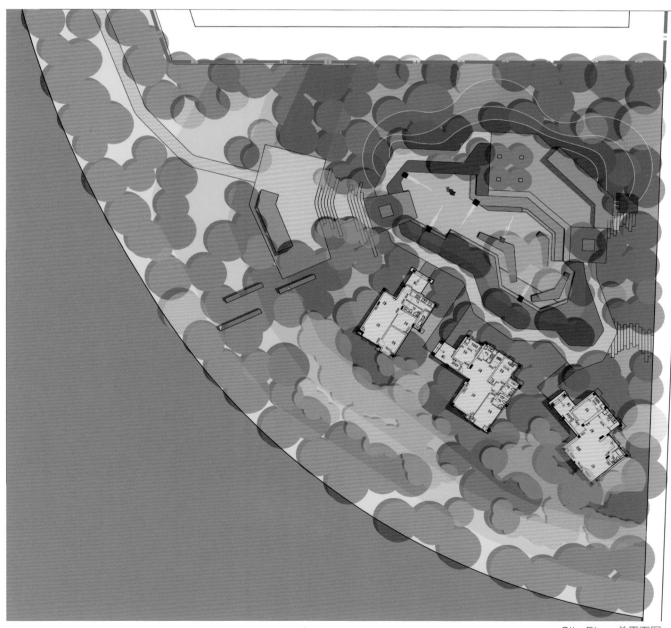

Site Plan 总平面图

Design Concept 设计理念

Its design concept is inspired by the river of the site, and it represents mountains, banks and river and cooperates with the terrain height difference to make the overall landscape space be smooth and changeable, to create an artistic conception of modern landscape. The landscape spatial layout is being well-proportioned, while the trees and flowers planted on both sides are adding a sense of spirituality and refinement to the project.

泰禾海沧高层展示区的设计理念源于对基地河流的感受，通过对山峦、曲岸及水脉的现代演绎，利用整体和两侧的落差处理，使得整体景观空间流畅而富有变化，营造出曲岸观澜的现代自然山水意境。整体景观空间布局的错落有致，两侧的树阵和花境，更添灵动细腻的感觉。

Landscape Design 景观设计

Multi-layer water features and semi-enclosed platforms are set at the entrance of the showcase which have made the entrance more welcoming. The water features around the showcase are placed on both sides of the road and enclosed by plants to improve the visual effect and listening pleasure. For the greening design, L&A Design Group adds the amusement and relaxing atmosphere of public parks into the project to provide an interesting experience for visitors.

展示区入口处设计了多层次的观水平台和半围合的平台，使得入口的仪式感得到了很大提升。样板房附近的水系则通过绿化的围合，让水系若有若无的显现于道路两侧，增加了视觉的变化和悦耳的流水声。在绿化设计上，奥雅将公园的休闲性和轻松氛围也融入其中，营造了更加愉悦的参观体验。

China's Most Beautiful Villas
中国最美别墅

Aesthetics	美 学 性
Livability	宜 居 性
Commercial Value	商 业 价 值 性
Humanity	人 文 性
Sustainability	可 持 续 性

Eco Friendly Modern Quadrangle
生态环保的现代四合院

The Lake Dragon, Guangzhou (Zone F) | 广州玖珑湖小区发展项目（F区）

Location / 项目地点：Huadu District, Guangzhou, Guangdong / 广东省广州市花都区
Developer / 开发商：Sun Hung Kai Real Estate Agency Co. Ltd. / 新鸿基地产代理有限公司
Architecture /Landscape/Interior Design / 建筑 / 景观 / 室内设计：Ronald Lu & Partners / 吕元祥建筑师事务所
Land Area / 占地面积：385 326.3 m²
Floor Area / 建筑面积：194 652 m²
Plot Ratio / 容积率：0.40
Green Coverage Rate/ 绿化率：56.7 %
Category / 项目类型：Residential / 住宅
Status / 项目状态：Built / 已建成

Keywords 关键词

Stylish Mansion; Modern Appearance; Eco Friendly
风范大宅 现代外观 生态环保

"China's Most Beautiful Villas" Award
荣获"中国最美别墅"奖

The architects have skillfully introduced surrounding views into the interior by designing large rooms and large glass doors, and adjusted the height and orientation of each villa to get natural ventilation and unobstructed views.

在别墅设计上，建筑设计师巧妙地"引景入室"，大量采用大开间、大幅玻璃门元素，以此调整每一栋别墅的高度与方位，形成户户南北对流且不被遮挡的景观优势，力求令建筑与自然融为一体。

Overview 项目概况

As a low-density luxurious residential development, the Lake Dragon offers 245 villas, houses and clubs. Phase I "Origins" locates on a gentle slope of the hills by the Dragon Lake, adjacent to the golf course designed for 2010 Guangzhou Asian Games. Ronald Lu & Partners have combined the Western and traditional Chinese architectural elements, and created a modern architectural style according to the natural topography and environment.

玖珑湖为低密度豪华住宅发展项目，当中包括245栋别墅洋房和会所。第一期"悦源"位处九龙湖一片连绵群山的微坡上，毗邻为2010年广州亚运会设计的高尔夫球场。吕元祥建筑师事务所在设计上，将西方和中国传统建筑元素揉合，再配合天然地理及环境，构成了现代建筑风格。

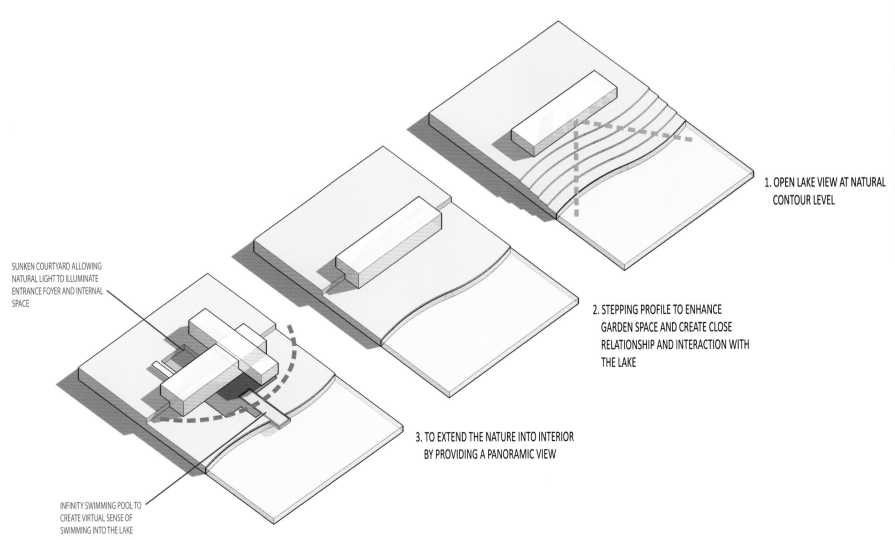

1. OPEN LAKE VIEW AT NATURAL CONTOUR LEVEL

2. STEPPING PROFILE TO ENHANCE GARDEN SPACE AND CREATE CLOSE RELATIONSHIP AND INTERACTION WITH THE LAKE

3. TO EXTEND THE NATURE INTO INTERIOR BY PROVIDING A PANORAMIC VIEW

SUNKEN COURTYARD ALLOWING NATURAL LIGHT TO ILLUMINATE ENTRANCE FOYER AND INTERNAL SPACE

INFINITY SWIMMING POOL TO CREATE VIRTUAL SENSE OF SWIMMING INTO THE LAKE

Modern Quadrangle 现代四合院

The villas feature modern appearance, square spaces, carefully designed landscapes and world-class living atmosphere. The biggest one among these villas keeps the idea of "modern quadrangle" to provide wonderful views with the front and back yards as well as the wing rooms. The combination of the spirit of Chinese traditional quadrangle and the modern aesthetics has created a "modern quadrangle" mansion for the Lake Dragon, which will bring the high-end residential market with unique product and innovative design ideas.

Buildings of the Lake Dragon are both modern and traditional, western and oriental. The modern quadrangle mansion keeps the spirit of the traditional quadrangle and meets modern living requirements at the same time.

别墅外型现代，布局方正实用，配以精致用心的园林，营造了国际级居住氛围。其中一幢超大别墅引入"现代四合院"理念，运用前庭后院及厢房设计，预留观景位置，让自然景色一览无遗，将中国传统四合院建筑精髓与现代时尚美学相融合，雕琢出玖珑湖式"现代四合院"府邸，为高端住宅市场提供了独特的产品与创新的设计理念。

玖珑湖的建筑既是现代的，也是传统的；既是西方的，也是东方的。其现代四合院设计，取传统四合院之精髓，却兼有现代人生活的各类元素。

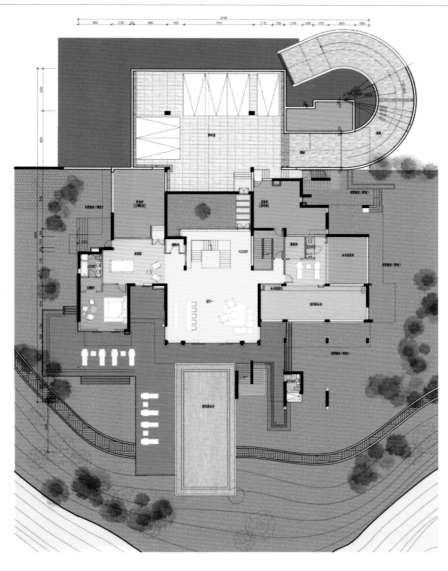

First Floor Plan 一层平面图

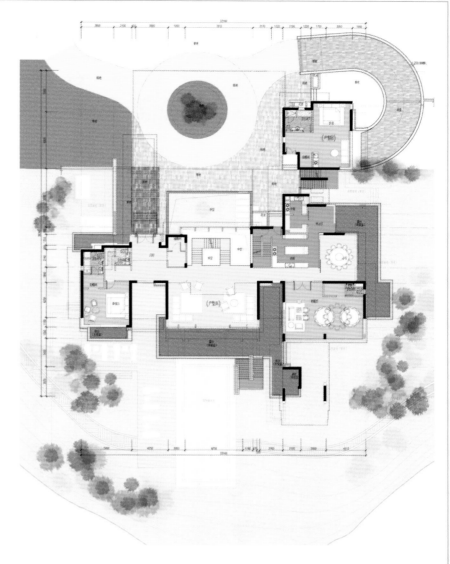

Second Floor Plan 二层平面图

Design Idea 设计理念

Inspired by traditional Chinese culture and adopting international architectural design ideas, the project has successfully met the real needs of the residents. By introducing the concept of "home", it has combined traditional Chinese wisdom with the essence of modern architecture perfectly. Currently when Chinese trend becomes more and more popular around the world, the stylish mansions of Lake Dragon style have been highly regarded in the international market.

项目既保留传统文化精髓,又结合国际化建筑设计理念,成功做到用心体会居住者的真正需要。当中以"建筑"发扬"家的概念",通过"家"将中国传统智慧与现代建筑精髓完美糅合。在时下中国思潮日益引领世界的全新格局下,玖珑湖式风范大宅已备受国际市场推崇。

Feeling of "Home" "家"的感觉

Quadrangle will remind people of the strong feeling of "home", for example, with an atrium. The architects have introduced the concept of "home" into modern life by using elements like glass windows and landscapes. In addition with the atrium for parties and communications, big families can therefore stay together, and yet enjoy the privacy of their own space.

四合院关于"家"的感觉是很强烈的,比如中庭的设计。建筑设计师将四合院"家"的模式融入现代生活中,运用玻璃窗、景观等这些元素,令现代四合院吸收了传统建筑中对"家"的理解。此外,中庭的设计既提供家族聚会、共享交流、几代同堂的场所,又兼顾了现代家庭的独立、隐私、个性化空间。

Bringing Views into the Interior 引景入室

Villas of different types, ranging from 343 m² to 875 m² in size, are designed with the idea to "bring views into the interior", which allows all units to have great views of the gardens, golf course and the mountains. Chinese traditional quadrangle combines with modern design to create a landscape system that connects the central, front and back yard.

面积为 343～875 m² 的不同户型别墅，均以"引景入室"为设计理念，每户都可以欣赏到花园、高尔夫球场和群山的秀丽景色。将中国传统四合院建筑融合到现代设计之中，以内庭园为基础概念，宽大的前园饱览户外秀丽的景色，兼能与后花园之景一脉相连。

Humane Care 人性化关怀

The atrium will encourage family communication, while the beautiful landscapes and the outdoor spaces will bring people close to nature. Three-dimensional layout provides more spacious and comfortable living spaces; multi living rooms as well as the western-style and Chinese-style kitchen enable two generations to have their own space. Modern quadrangle has combined tradition with modernity, showing great care to modern families. It is introverted yet outward.

中庭空间促进了亲情的沟通，而开杨的景观、室外空间的延伸也令住客享受大自然的乐趣。其立体式布局使生活空间更大更舒适，多厅设计和中、西厨为两代人提供了各自独立的私人活动空间。现代四合院设计融合传统与现代，令现代家庭生活增添和谐及关怀的道路，彰显着家庭伦理。它既是内向的，也是对外延伸的。

Eco Friendly 生态环保

As an eco residential development among mountains and waters, the Lake Dragon has greatly applied the latest environment-friendly ideas. The wastewater of the community will be collected and transferred to the sewage treatment station, and then used for irrigation, flushing toilet and waterscapes. At the same time, the rainwater will also be collected, treated and used for water features.

Besides, spaces are reserved on the roof of all villas for future installation of solar collectors. All these eco-friendly ideas and equipment will minimize the impact upon the surrounding environment and help to establish the harmony between man and nature.

作为一个依山傍水的生态住宅项目，玖珑湖大量运用国际上最新的环保理念。区内所有污废水均经收集后装送至污水处理站处理至灌溉级别后，再送回原区作绿化灌溉、冲厕及园林水景补水之用。同时设有雨水回收系统，有效地将雨水收集后经沉淀及除油污后装送至园林水景系统，经过滤及消毒后作园林水景用。

另外，所有别墅屋顶都预留位置供业主自行安装太阳能收集器，以减少能源消耗。所有这些环保理念和设备的应用，都保障了小区周围的生态环境受到最小限度的影响，为玖珑湖营造了人与自然和谐相处的氛围。

High-end and Low-density Livable Community
高档次低密度的宜居社区

Shanghai Yango Balmy Garden　　上海阳光城花满墅

Location / 项目地点：Luodian New Town, Baoshan District, Shanghai / 上海市宝山区罗店新镇
Developer/ 开发商：Yango Group / 阳光城集团
Architectural Design / 设计单位：Shanghai Hyp-arch Architectural Design Consultant Inc. / 上海霍普建筑设计事务所有限公司
Land Area / 占地面积：34 557.7 m²
Floor Area / 建筑面积：34 903.28 m²
Main Unit Areas / 主力户型面积：Superposed Villa / 叠加别墅：180 m², Single-family Villa / 类独栋别墅：260 m²
Plot Ratio / 容积率：1.01
Green Coverage Rate / 绿化率：35%

Keywords　　关键词
Safe and Comfortable; Elegant and Dignified; Living Environment
舒适安心　典雅高贵　人居环境

"China's Most Beautiful Villas" Award
荣获"中国最美别墅"奖

The architectural design focuses on all details and shows respect to the natural texture and color of new materials. Besides precise proportions and exquisite details enable people to experience the tradition, the history and the culture.

建筑从整体到局部，精雕细琢、镶花刻金都给人一丝不苟的印象，一方面尊重和保留新材质和色彩的自然风格。另一方面，通过准确的比例调整和精致的细节设计展现出浑厚的文化底蕴。

Location　　区位概况

Located in Luodian New Town of Baoshan District, Shanghai, the development extends eastward to Fuyuan Road, westward to Luoying Road and northward to Meilan Lake Avenue. It is about 16 km away from the civic square.

项目地处上海市宝山区罗店新镇，东至抚远路，西至罗迎路，北至美兰湖大道，距离人民广场约 16 km。

Regional Value　　版块价值

The site is surrounded by beautiful views: Meilan Lake Scenic Spot is 2 km away to the west and the golf resort is closely located to the east. Meilan Lake project is developed and built as an important town in north Shanghai. Since it is adjacent to Luodian ancient town of Baoshan District, it is also called Luodian New Town.

地块周边景观资源丰富，西侧 2 km 处为美兰湖景区，东侧紧邻高尔夫景区，美兰湖是上海北部地区重点开发建设的小城镇。因其毗邻宝山区罗店古镇，也被称为"罗店新镇"。

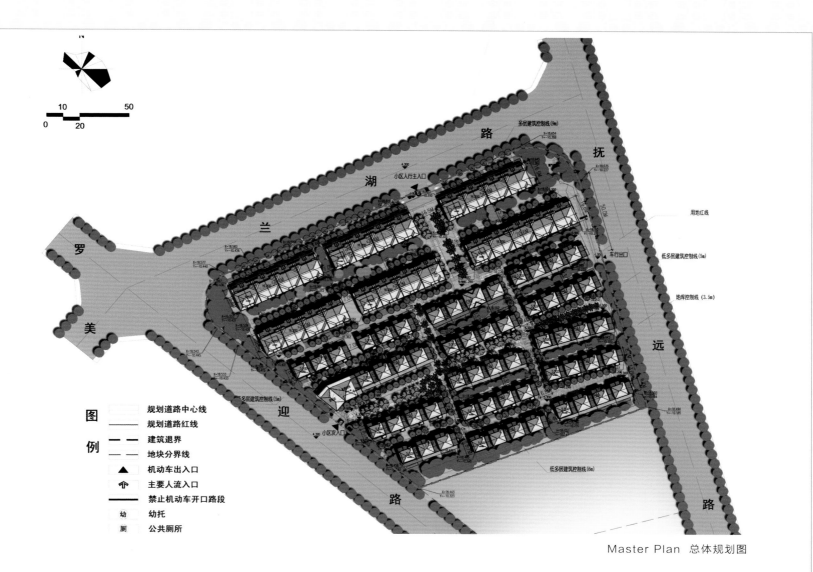

Master Plan 总体规划图

Site 项目基地

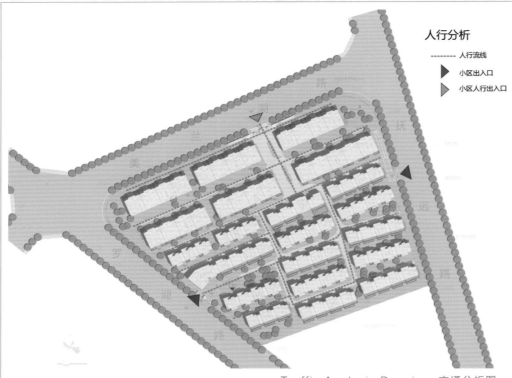

Traffic Analysis Drawing 交通分析图

Positioning 项目定位

Yango Balmy Garden is positioned as a low-density, high-end community which boasts only superposed houses and single-family like villas. Large width and multi-level gardens will give residents the experience of living in the villas.

项目定位为高端低密度社区，规划有叠拼别墅和类独栋别墅两种业态，超大面宽设计，多重花园享受，尊享别墅生活新体验。

South Elevation of Superposed Villa 叠拼南立面

North Elevation of Building 9# 9#号楼北立面

Planning and Design 规划设计

Overall Plan

There are superposed villas standing in two rows in the north of the site and single-family like villas in five rows sitting in the south. Arranged in this way, multi-storey buildings in the north will not block the sun of the lower villas in the south. Taking advantage of the inverted trapezoidal site, the architectural arrangement follows the topography, landscape and orientation to create interesting spatial forms. Along with the optimized landscape, it builds a natural and harmonious residential environment which features rich space levels and changing views.

Traffic System

There are two driveway entrances on the east and west, and an entrance for walkers on Meilan Lake Avenue in the north. Thus the pedestrian flow is separated from the vehicles, and the ramps at the entrances will ease traffic congestion, creating a safe, green and pedestrian-friendly environment throughout the whole development. Carpark is mainly underground, while a few parking spaces are set on ground for visitors' cars. Every unit is equipped with 1.1 parking spaces on the average.

总体布局

场地北面为两排叠加别墅，南面为五排类独栋别墅，北高南低，北面布置多层，南侧布置低层，对日照间距影响最小，本地块呈倒梯形结构，建筑排布顺应地势景观及朝向的变化，形成多个意趣盎然的空间形态。在此基础上，将景观加以优化，创造具有更加丰富的空间层次，步移景异，形成自然的人居环境，为居住者赢得空间、建筑、景观与生活四者充分交融的宜人场所。

交通系统

整个地块在东西两侧共设两个车行出入口，北侧美兰湖大道上设置一个人行主出入口，采用人车分流设计，通过在入口设置的坡道来消化所有车流，保证小区内部连续而绿意盎然的纯人行空间，为居住者提供舒适安心的居住安全感和自然惬意的宜居生活空间。机动车停车以地下停车为主，地面设置少量的访客停车位，机动车停车位配置为每户1.1个。

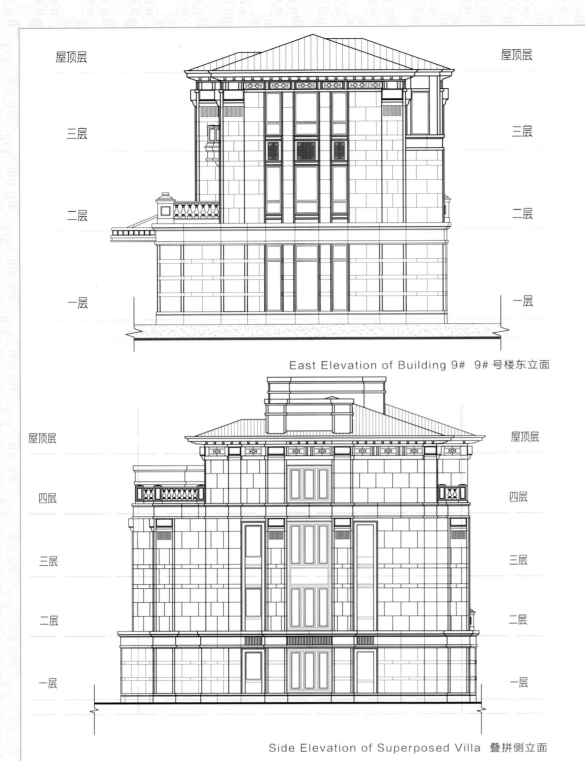

East Elevation of Building 9#　9#号楼东立面

Side Elevation of Superposed Villa　叠拼侧立面

Architectural Design 建筑设计

Unit Types
Catering to the local market's demand for a high-end and low-density community, Yango Balmy Garden offers two types of units: superposed villas and single-family like villas. The villa design is people oriented to provide an elegant and reasonable layout, maximizing the utilization rate of the floor area and adding comfort for living.

Style Analysis
Designed in a mixed style of neoclassical style and traditional Chinese style, it highlights the elegance, dignity and value of the buildings and shows respect to traditional culture. The architectural design focuses on every detail, keeps the natural texture and color of new materials, abandons redundant fabrics and decorations, and simplifies the lines to form a modern and elegant appearance. Besides precise proportions and exquisite details enable people to experience the tradition, the history and the culture.

产品类型
开发商结合当地市场需求，打造一个高档次低密度的宜居社区，产品有叠拼别墅和类独栋别墅两种业态，室内设计遵循"以人为本"的宗旨，内部空间动静分区，居寝分离，洁污分离，室内布局大气、合理，提高面积的使用率和居住的舒适度。

风格分析
项目采用新古典与传统中式结合的风格，体现建筑典雅高贵有价值感的同时，也体现对传统文化的尊重。建筑从整体到局部，精雕细琢、镶花刻金都给人一丝不苟的印象，一方面尊重和保留新材质和色彩的自然风格，摈弃过于复杂的肌理和装饰，简化线条，保持现代和简洁的审美倾向。另一方面，仍然通过准确的比例调整和精致的细节设计很强的感受传统的历史痕迹与浑厚的文化底蕴，精致、端庄、对称、典雅。

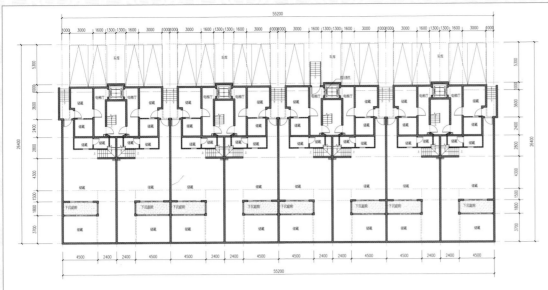

Plan for Basement Floor 地下一层平面图

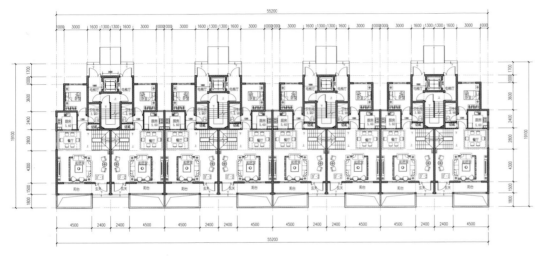

First Floor Plan 一层平面图

Second Floor Plan 二层平面图

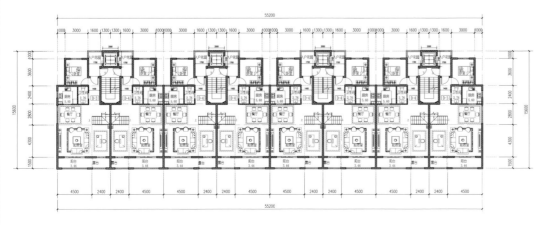

Third Floor Plan 三层平面图

Unit Design 户型设计

The innovative single-family like villa is about 260 m², boasting a 12 m big width to get more sunlight and natural landscape views, a 5.1 m high basement to bring great enjoyment, a 6.5m double-height living room and a presidential-level master bedroom with spacious study. The independent doorway and the front and back yards will enable people to enjoy great privacy as well as the urban landscapes. The superposed villa is about 180 m². The lower part is accessed through the southern courtyard, while the upper part through the northern courtyard or by elevator. The sunning courtyards increase the levels of villa space and the 3.9 m high basement provides perfect functional spaces, well interpreting the luxury lifestyle together with the 6.9 m big width and the grand master suite.

项目原创类独栋别墅户型面积约为 260 m²，拥有 12 m 极致面宽，能更好接纳阳光和自然景观，地下 5.1 m 层高设计，给地下活动空间带来超值享受，6.5 m 的客厅挑空，总统级主卧套房兼阔达书房，彰显大宅风范，独立入户，前庭后院，别墅的私家领地，尽享都市园林生活。叠拼别墅户型约为 180 m²，下叠别墅南向庭院独立入户，上叠别墅由北侧庭院及电梯独立入户，阳光庭院增加别墅生活空间的层次感，3.9 m 的地下层高打造完美的功能空间，6.9 m 超大面宽和豪华主卧套布局演绎超豪华居住意境。

Multiple Levels, Flexible Spaces
视觉层次丰富，空间变化多样

Konka Moon River, Jiangsu　江苏康佳水月周庄

Location / 项目地点：Kunshan, Jiangsu / 江苏省昆山市
Developer / 开发商：Kunshan Konka Investment and Development Co., Ltd. / 昆山康盛投资发展有限公司
Architectural Design / 建筑设计：Cube Architects / 深圳市库博建筑设计事务所有限公司
Land Area / 占地面积：334 000 m²
Floor Area / 建筑面积：320 210 m²
Plot Ratio / 容积率：0.96
Programs / 项目类型：Convention Center, Hotel Villas, Commercial Center / 会议中心、酒店别墅、商业中心
Status / 项目状态：Built / 已建成

Keywords　关键词
Private Space; Private Garden; Courtyard Experience
隐私空间　私家花园　体院感

"China's Most Beautiful Villas" Award
荣获"中国最美别墅"奖

Villas are built near to water, and the boundary of the island features a density that's similar to a classic water town. A palette of vegetation sets a partition between the large water surface and the small water courses, which also provides great privacy for the villas. There are three different courtyard spaces far away from or near to water, namely, public courtyard, private gardens for the guestrooms and the landscape platforms along the water courses.

别墅临水而设，在岛的边界上创造出在与古典水城类似的密度。丰富的植被为大水体和小水道之间提供间隔，也为别墅创造隐私空间。别墅单元与水面或远或近，形成了公共庭院、私家花园和沿水景观大平台。

Visual Level Design　视觉层次设计

Vertical elements such as the observation deck, bridges, water courses and the straight buildings play important roles in offering perfect functions and architectural experience. With the building volumes growing or shrinking; with the small nodes set along the footpath and the water channel, it enables visitors to have a rich space experience of the big square and small intimate courtyards, of the lake views and the water courses.

The entrance yard of the hotel and the big square of the convention center are located on the same level. Two sides of the square are defined by walls of the meeting room and the banquet room, another side opens to the river, and the last one is defined by the hotel tower. The buildings of modern style stand along the square, allowing guests to experience the changes of the topography.

Villas are built near to water, and the boundary of the island features a density that's similar to a classic water town. A palette of vegetation sets a partition between the

large water surface and the small water courses, which also provides great privacy for the villas. There are three different courtyard spaces: public courtyard, private gardens for the guestrooms and the landscape platforms along the water courses. The driveway runs through the island and connects with the bridges, complemented by landscape walls and gates along two sides. The platform in the front yard with screen will help to protect the privacy of the owners, which looks like lantern to decorate the island at night.

嘹望台、桥、水道以及建筑的竖向感觉，无论从功能上还是建筑体验上，都是很重要的。让建筑的体量生长或者收缩；在步行道和水渠中制造一些缩小的节点，将让参观者得到从大的广场到小的亲密空间庭院、从湖泊景色到水道的广泛空间变化体验。

酒店入口前庭和会议中心的大广场，在同一层面上。这个广场两边由会议和宴会厅的房间的墙面限定，一端面对河流开放，另外一端由酒店的塔楼限定。建筑为现代风格，沿着广场的边界，酒店客人可以体验到地形上的变化。

别墅临水而设，在岛子的边界制造出一个在尺度上与古典水城类似的密度。一组一组的丰富的植被，为大的水体和小的水道之间提供隔离，也为别墅提供隐私。别墅单元远离或者贴近水面，形成3种庭院空间：公共的到达庭院、客房的私家花园以及大的沿水景观平台。小车环路穿过岛屿，由桥连接，沿途都是景观植物墙和大门。带屏障的前庭平台为业主特别是面对面的别墅业主，提供隐私。到了晚上，这些带屏障的前庭平台，成为点缀别墅岛屿的灯笼。

Site Plan 总平面图

Legend 图例

1. 万国旗广场 Flag Plaza
2. 主入口 Main Entrance
3. 车行次入口 Secondary Entrance
4. 酒店 Hotel
5. 会议中心 Conference Center
6. 酒店泳池休闲区 Hotel Amenity Pool Deck
7. 别墅式酒店 Hotel Villa
8. 服务通道 Seveice Road
9. 商业中心 Commercial Center
10. 美术馆 Art Gallery
11. 湿地公园 Wetland Park
12. 主要公共开敞空间 Main Public Openspace
13. 绿色空间联系 Greenspace Connection
14. 人行连接 Pedestrian Connection
15. 高层或多层 Multi-Story & Highrise Condo
16. 叠拼 Stacked House
17. 联排 Townhome
18. 院墅 Courtyard Villa
19. 急水港 Jishui River
20. 全旺路，通往周庄 Quanwang Road, to Zhouzhuang
21. 锦州路，通往苏州 Jinzhou Road, to Suzhou
22. 秀海路 Xiuhai Road
23. 湖面 Lake

Commercial Village 商业村

The commercial village is carefully designed to create a relaxing and unforgettable environment. The space experience will be exciting and inviting, giving both the visual impact and the breathing space. Being attractive like the historical canal town — Jinxi Town and Shanghai Xintiandi, the commercial village will be a unique place to create unforgettable memories.

商业村被细致地打造成为一个轻松的和令人难忘的环境。这里所创造的体验感，将是令人兴奋的，让人不自觉参与进来，是视觉上有冲击度却又有喘息的机会的一种旅途体验。同历史运河城金溪镇和上海新天地一样具有较强的吸引力，商业村将成为一个独特的场所，为该区域周围的人留下难忘的记忆。

Traffic Organization 交通组织

The hotel will draw clear routes for the guests and visitors of the hotel and convention center. Visitors of the commercial village will have many chances to shop in retail stores. A series of interlaced routes including the driveways, water courses and walkways will enable this new district to be operated as the nearby town which was established several centuries ago.

　　酒店将为会议中心的客人、酒店客人以及其他来访者提供明显且容易到达的交通路线。商业村的来访者将有很多机会在零售商店中往来。一系列的交织在一起的循环路径，结合了车辆、水上及行人交通，将让此新区，与几个世纪前就形成的相邻市镇一样运作。

Landmark and Iconic Museum 地标和标志性博物馆

The hotel tower, together with the performance center, will become the landmark on the south bank. Modern footbridge serves as the connection between the hotel and the commercial village. In the center of the commercial zone there will be an iconic building — the Museum which can be seen from most platforms of Phase I. With flexible traffic and visual design, tourists will have many chances to take a view of the museum; they can also access to the building directly by water or by walk.

台阶式的酒店主体，与演艺中心一起，将成为湖南岸主要的一个地标。现代的行人桥提供从酒店到商业村的连接。在商业发展的中心点，将是一座标志性建筑——博物馆。从第一期的大多数平台上，都能看到这座建筑。以多种模式的交通和视线为重点，游客将有很多不同的机会，看到这栋楼，或者通过水上的上车或步行，到达这座建筑。

French-style Community of Nobility, Elegance and Romance
尊贵、典雅、浪漫的法式建筑居住社区

Yongding River Peacock Lake 永定河孔雀城大湖

Location/ 项目地点：Langfang, Hebei, China/ 中国河北省廊坊市
Developer/ 开发商：CFLD/ 华夏幸福基业
Architectural Design/ 建筑设计：CDG (Concord Design Group) / 西迪国际（CDG 国际设计机构）
Size/ 建筑规模：440 000 m²
Overground Floor Area/ 地上面积：880 000 m²
Plot Ratio/ 容积率：2.0
Building Coverage Rate/ 建筑覆盖率：22%
Green Coverage Rate/ 绿化率：35%

Keywords 关键词

Large Scale; French Style; Close to Nature
尺度开阔　法式气度　融于自然

 "China's Most Beautiful Villas" Award
荣获"中国最美别墅"奖

Classic slope roof, together with multi-level lines, has skillfully integrated French-style colonnades, carvings and lines to highlight the elegance of the French-style residences.

经典坡屋顶的设计加以层次丰富的线条，巧妙将法式柱廊、雕花、线条等经典元素一一归位美化，法式大宅的气度在层叠的工巧形式中得以尽显。

Overview 项目概况

The development is situated within Gu'an Peacock Ring Road of Langfang, Hebei, connecting with the Peacock Lake, adjacent to the second airport of Beijing, and attaching itself to the only 130,000 m² park in south Beijing — Peacock Lake Park.

项目位于河北省廊坊市固安孔雀环路内侧，与孔雀大湖相接。毗邻北京第二机场，依附京南唯一的 13 万 m² 孔雀大湖公园。

Project Positioning 项目定位

Taking advantage of the lake park's natural environment, the development is positioned to be a noble, elegant and romantic high-end residential community that provides boutique lakeview residences and low-density townhouses.

为顺应大湖公园的自然环境，项目定位于低密度的沿湖观景精品住宅和联排低密住宅，创造一片尊贵、典雅、浪漫的滨湖高档居住社区。

Site Plan 总平面图

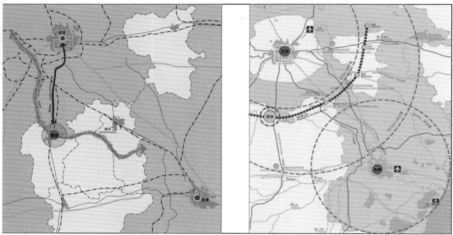

Location Diagram 区位关系图

Current Traffic Condition 区位交通现状图

Design Idea 设计理念

Yongding River Peacock Lake has made reference to the Burke Lake Park in Washington, trying to integrate art into life, making art part of life, and characterizing itself by the unique lakeside lifestyle. To cater to the international life experience and meet family requirements, it has created a 130,000 m² lake park. Based on the idea of "building an intimate relationship between human beings and nature", 9 activity areas of different themes are set around the central lake, turning the boring park-style spaces into venues for communication and gathering. It aims to get back the intimacy between human and nature.

永定河孔雀城大湖，以华盛顿伯克湖公园度假地为蓝本，力图巧妙连接艺术和生活，让艺术成为生活触手可及的一部分，塑造其独有的湖居魅力。为契合国际生活感受，满足全家庭所需，首度开创13万 m² 大湖公园，这里以"建立人与自然的亲密关系"为理念，围绕中心水域逐级向外设计9大湖岸主题活动区，将原本单调的公园式风景变成了可用来交流情感、开放的生活聚会场所，致力于还原人与自然的亲密无间。

West Elevation 西立面图

East Elevation 东立面图

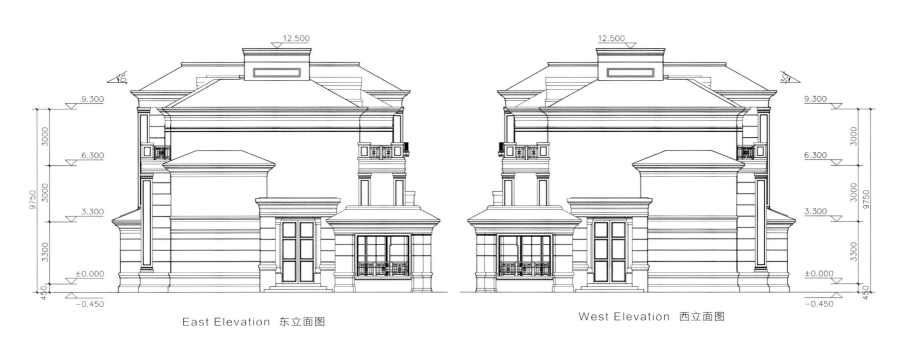

East Elevation 东立面图

West Elevation 西立面图

Design Concept 建筑设计构思

The buildings are designed with the architectural elements of classical French palace. Full stone facade enriches the levels of the whole building and makes it feature the romance and unpretentiousness of French-style architecture. The architraves are carefully designed, the interior walls are dominated by warm yellow, and the exquisite dormers in the roof look like eyes of the building, showing the dignity and mystery of the building. This kind of tone has defined the elegant architectural style and unique sense of rhythm for the development. And the design of the courtyards has followed the style of the French-style buildings.

永定河孔雀城大湖借鉴法式古典宫廷建筑元素，采用全石材立面，使整体建筑线条层次丰富趣味十足并散发着法式建筑的浪漫和不张扬的气度。精工细作的细腻线脚、暖黄的色调、屋顶上精致的老虎窗好似建筑的眼睛，开合之间透出建筑的高贵与神秘，整体建筑厚重稳健的基调形成了大湖典雅的建筑气质与独特的韵律感。大湖每一幢法式风情庭院的细节刻画，皆是法式建筑精粹的秉持。

Plan for Roof Floor 屋顶层平面图

Second Floor Plan 二层平面图

Architectural Design 建筑设计

Characterized by graceful scales as well as the classical French-style architectural elements, the buildings well interpret the elegance and magnificence of the neoclassical-style architecture.
Classical slope roof, together with multi-level lines, combine with the French-style columns, carvings, and lines skillfully, highlighting the elegance of French-style residence. The exterior wall is decorated all by durable stones, giving a noble and dignified expression.
Large space scales and exquisite details are defined by graceful lines, showing the noble quality that dialogues with the grand lake. All details of the buildings have reflected the great charm of the French architectural style.

建筑以雅致与恢弘大气的尺度，汲取法式建筑的古典元素，演绎一曲辉煌大气的法国新古典建筑诗篇。

经典坡屋顶的设计加以层次丰富的线条，巧妙将法式柱廊、雕花、线条等经典元素一一归位美化，法式大宅的气度在层叠的工巧形式中得以尽显。外墙以纯正耐久的高贵石材装饰，洋溢着体面而尊贵的建筑表情。

开阔的空间尺度、精致的建筑细节在优雅的线条勾勒下散发出与大湖环境相衬的尊贵典雅的气质，每一栋建筑细节的刻画，皆是法式建筑的秉持。

Meiju Encyclopedia 美居小百科

Dormer window, also called dormer skylight in Shanghai, is a kind of window set into the roof. Due to the cold and snowy climate in the UK, the residences there are mostly designed with steep slope roof to reduce the pressure of snow. And, to increase daylight and natural wind, windows will be set in the roof. The pronunciation of the English word "roof" is similar to "Laohu"(tiger) in Shanghai dialect, thus this kind of window was pronounced as "Laohu Chuang" in pigeon-English. After 1920s, to solve the problem of housing shortage, Shanghai citizens begun to make full use of Shikumen Residence which features high second floors and slope roofs, and build duplexes between them. To get more daylight and natural ventilation, windows were opened in the roof. Therefore, today's "dormer window" usually refers to the roof window in Shanghai Shikumen Residences.

老虎窗，又称老虎天窗，上海俗语，指一种开在屋顶上的天窗。而地处北欧的英国气候寒冷多雪，为了避免积雪对建筑物的压力，他们的房屋大多为高坡度、尖顶样式，为了增加采光和通风，又在屋顶上开设了许多屋顶窗。英文屋顶为"Roof"，其音近沪语"老虎"，于是，这种开在屋顶的窗就被洋泾浜英语读做"老虎窗"。20世纪20年代后，随着上海住房困难的加剧，上海人就利用石库门住宅的二楼空间较高及有斜屋顶的特点，在二层与屋顶之间加建阁楼，为了增加阁楼的采光和通风，也在屋顶上开窗，这种窗也被称之为"老虎窗"。因此，现在"老虎窗"一般多指上海石库门住宅中的屋顶窗。

Top Single-family House Community
顶级纯独栋别墅社区

Kingdom Park 金臣别墅

Location / 项目地点 : Jinguang Road, Minhang District, Shanghai / 上海市闵行区金光路
Developer / 开发商 : Sunan Longfor Properties Co., Ltd. / 苏南龙湖地产
Landscape Design / 景观设计 : EGS Landscape & Urban Design Architects / 上海易境景观规划设计有限公司
Land Area / 占地面积 : 111 549 m²
Floor Area / 建筑面积 : 44 809 m²
Plot Ratio / 容积率 : 0.35

Keywords 关键词

Return to Nature; People Oriented; Ecological and Natural
返璞归真 以人为本 生态自然

"China's Most Beautiful Villas" Finalist
入围"中国最美别墅"奖

The planning design is inspired by nature and takes account of the comfortableness in modern western architecture to avoid barrack-style layout and well arrange those villas among hills, forests and lakes. It reproduces a natural environment within the city and provides semi-open space that is independent from the private spaces.

在规划设计上,以自然为蓝本,萃取西方现代化建筑对于舒适度上的考量,突破了传统兵营式排布,巧妙地将山地、森林、湖泊这3大自然界元素融入别墅群落,在城市中再造自然,营造出独立于私属空间的半开放式空间布局。

Regional Value 板块价值

Located in the core of the world-class Hongqiao CBD, Kingdom Park is the only and top newly built residential community that provides single-family houses. The development is only 10-minute's drive to the biggest transportation hub of Asia — Hongqiao Hub, and it is to the north of the world's biggest exhibition and convention center — National Exhibition and Convention Center (Shanghai), to the south of the biggest medical service institution — Shanghai International Medical Center, and to the west of the future CBD of Yangtze River Delta — Hongqiao CBD. The superior location is unique and can't be duplicated.

Kingdom Park is sitting in the center of Jinfeng international community, which, as one of the oldest international communities in Shanghai, boasts the best education resources in Yangtze River Delta: there are seven international schools within the community. "Villa lifestyle and downtown facilities" make it the top priority of the Asia-Pacific CEOs from the Fortune Global 500 companies. Due to land limitation, newly-built single-family houses are out of supply for years within this area, so the birth of Kingdom Park will target this niche market.

金臣别墅是世界级CBD——虹桥商务区核心唯一新建的顶级纯独栋别墅社区。项目距亚洲最大的交通枢纽——虹桥枢纽仅需驱车10分钟；南邻世界最大的会展中心——国家会展中心；北邻全国最大的医疗服务机构——国际医学中心；东邻未来长三角中央商务区——虹桥核心商务区，铂金地段不可复制。

金臣别墅位于虹桥核心金丰国际社区腹地，作为上海老牌涉外国际社区，金丰国际社区集结七所国际学校，拥有长三角最完善的国际教育资源。"别墅型生活、市区型配套"的优势，使其成为世界500强亚太区CEO的首选住区。限于土地供应，板块内已多年未有新建独栋别墅房源上市，项目的亮相，将弥补市场的翘首以盼。

Positioning Strategy 定位策略

With the planning and development of Hongqiao CBD, the land price here is increasingly rising. With a global perspective, Kingdom Park, sitting right in the center of Hongqiao CBD, has become the first private garden villa development in Shanghai that integrates hills, forests and lakes. Its unique resources and large scale are incomparable. It shows great respect to the land with innovative ideas and high quality, and creates a high-end circle especially for the high-net-worth individuals.

随着虹桥商务区的规划发展，虹桥的土地价值日趋上升，在如此寸土寸金的虹桥中心，项目以前瞻性的全球眼光，打造出上海城市别墅中首个集山、林、湖于一体的私家园林别墅，其臻稀资源与超大手笔非寻常高端住宅可比拟，项目以其创新的理念与追求品质的高度表达了对土地的崇高敬意，更为全球高净值人群打造匹配的高端生活圈层。

Planning and Design 规划设计

The planning design is inspired by nature and takes account of the comfortableness in modern western architecture to avoid barrack-style layout and well arrange those villas among hills, forests and lakes. It reproduces a natural environment within the city and provides semi-open space that is independent from the private spaces, showing the balance between buildings and nature as well as the harmony among city, human beings and natural environment.

在规划设计上，以自然为蓝本，萃取西方现代化建筑对于舒适度上的考量，突破了传统兵营式排布，巧妙地将山地、森林、湖泊这3大自然界元素融入别墅群落，在城市中再造自然，营造出独立于私属空间的半开放式空间布局，充分展现了建筑与自然，城市、人与环境的平衡理念。

Private Ecosystem 私家生态圈

With emphasis on "user-friendly functionalism", it's achieved perfect harmony between human, nature and architecture. The architects, based on the idea of "people orientation", have taken advantages of the natural hills, forests and lakes to create a 16,000 m² forest park. Landscape elements of British style courtyard were used to form quintuple vertical gardens, while shrubs and trees of different heights were well arranged to create a private ecosystem for Kingdom Park.

项目对于"人性化功能主义"的突出，使人与自然、建筑达到和谐相融的至高境界。设计师秉承以人为本的理念，利用山林湖的自然天赋，独辟 16 000 m² 的私家森林公园，融合英伦庭院造景元素设置出五重立体院落，用不同层次的植灌编排，围合出一个金臣专享的"私家生态圈"。

Living Space 居住空间

The design of living spaces also shows the people-oriented planning idea. 6,600 m² water channels around the periphery of every building provide natural screens for the houses. Vivid buildings, poetic spaces, organic life...... from every corner of Kingdom Park, one can experience the romance of the walls, the warmth of the green lands and pleasantness of the legendary residences.

By designing the semi-open community space, it provides a most comfortable space for neighborhood communication, which turns the villa community into a dynamic and lifeful social space. Meantime, this semi-open spatial layout also well connects the owners' life, sports and social activities together. One can have a walk in this forest park, play tennis in the standard hard court and hold outdoor parties on the spacious lawn. And children can enjoy themselves in the British-style maze.

在居住空间的打造上，更是体现了以人为本的规划理念，打造6 600 m²的自然水景环绕穿引在建筑之间，为每栋别墅提供天然屏障，鲜活的建筑、诗意的空间、有机的生活……金臣别墅从各个角度营造出一个围墙浪漫、绿地温暖、心情灿烂的传奇居所。

通过半开放式社区空间的设计，营造一个邻里之间最舒适的社交距离，让别墅社区变成一个充满活力的社交空间。与此同时，半开放式的涉外空间布局也将业主们的生活、运动、社交有机地结合起来。平日不用出社区就可以在森林公园里散步、在标准硬地网球场尽兴比赛、在英式迷宫让孩子们尽情嬉戏、在无边的香槟草坪上举办户外派对。

Architectural and Landscape Style 建筑与景观风格

The best reason why modernism architectural style is so popular in the world is that, it advocates nature-oriented planning as well as simple and elegant style. The idea to remove meaningless artistic items and make all things natural and organic caters to the psychological needs of the rich in modern society. The landscape design has abandoned traditional manmade landscape items like the flower beds and fountains, and added some British-style tree groups, maze, lawns, bridges, sculptures and putting green instead. More than one thousand of plants will provide the garden colorful views all year round. Luxury and nobility can be experienced under this kind of simplicity. All these combine together to create a private, harmonious and pleasant living environment for the owners and their families.

探索现代主义风格之所以成为全球风靡的建筑风格的原因，最贴合的是以自然为本的规划主张，简约的风格主张，在艺术上消除无意义的东西而使一切事物变得十分地自然有机，返璞归真的理念契合现代富豪们简约而回归心灵本质的需求。项目在景观的设计规划上，摈弃人工痕迹浓重的传统花坛、喷泉之类常见的景观小品，而是融入了英式园林中的树阵、迷宫、草坪、廊桥、雕塑、趣味果岭等精髓，千余种植物让园林四季都有缤纷的自然色彩，将所有的奢华与高贵都隐藏在了简约的外衣之下，为高净值人士的家庭带来私密融洽的惬意生活氛围。

High-quality Buildings Combined with Golf Course

建筑布局与高尔夫球场有机结合的高品质建筑

Meilan Lake Silicon Valley Center, Shanghai　上海美兰湖硅谷中心

Location / 项目地点：Baoshan District, Shanghai　/ 上海市宝山区
Developer / 开发商：Shanghai Gold Luodian Development Co., Ltd.　/ 上海金罗店开发有限公司
Architectural Design / 建筑设计：Shanghai Sunyat Architectural Design Co., Ltd.　/ 上海三益建筑设计有限公司
Land Area / 占地面积：40 520 m²
Floor Area / 建筑面积：106 223.71 m²
Plot Ratio / 容积率：0.4

Keywords　关键词

Natural Ventilation; Elegant Shape; Eco-friendly
自然通风　造型简约　生态环保

"China's Most Beautiful Villas" Finalist
入围"中国最美别墅"奖

The facade is designed in neoclassicism style, with special emphasis on classic proportion and details. Ternary form, elegant colonnade and the exquisitely designed details highlight the high quality of the buildings.

建筑立面采用文艺复兴的新古典主义风格，注重古典的比例和细部，通过三段式的划分，大气的柱廊、细部的刻画突显建筑的高贵品质。

Overview　项目概况

Located in Baoshan District of Shanghai, Meilan Lake Silicon Valley Center sits on the bank of the 300 mu Meilan Lake, enjoying 3,000 mu unparallel golf views.

上海美兰湖硅谷中心位于上海市宝山区，依托300亩美兰湖水景，坐拥3 000余亩高尔夫超一流生态景观资源。

Planning　项目布局

The site is divided into two zones: the south and the north, totally providing 247 low-storey buildings. There are mainly ring roads that divide those buildings into different groups. The south and north entrances are displaced to create diversified external spaces.

地块分为南北2个区，共247栋低层建筑。以环状道路为主，形成不同组团，项目以两种办公平面布局为主，按照南北入口错落分布，营造了丰富的外部空间。

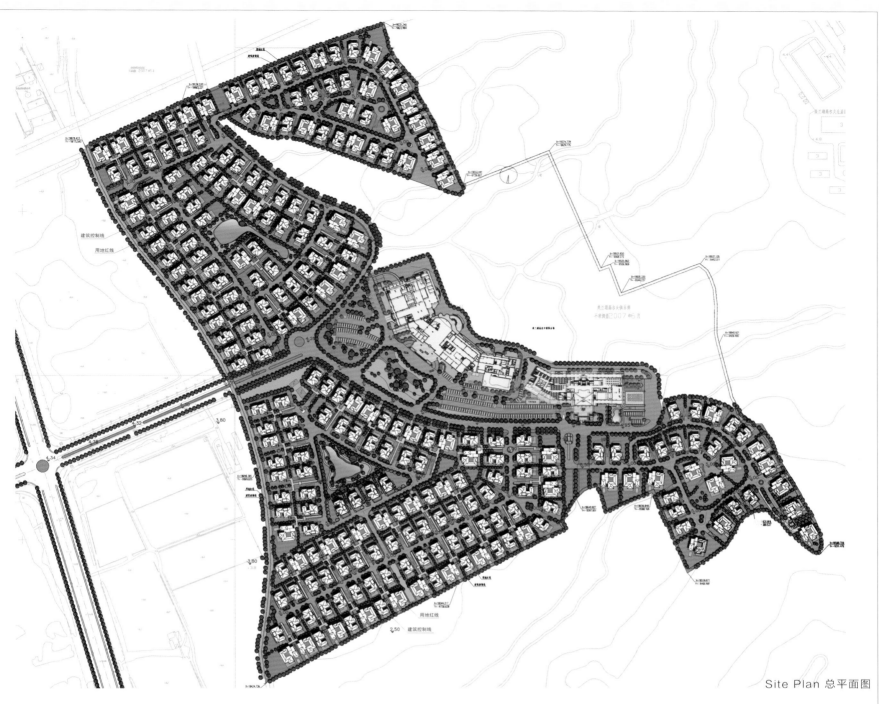

Site Plan 总平面图

Aerial View 鸟瞰图

Side Elevation 1 轴立面图 1

Characteristics of the Buildings 建筑特色

One of the biggest characteristics of the development is that the buildings are planned with the golf course. According to the topography, the headquarters buildings are staggered, allowing all offices to have great golf views. What's more, to ensure an excellent landscape environment, the newly-built headquarters base is designed with a multi-level landscape system which includes entrance landscape, main road landscape, internal public landscape, building entry landscape and courtyard landscape.

这个项目的最大特色之一就是建筑布局与高尔夫球场有机结合，企业总部建筑错落排布，结合地形，使每栋办公楼都有能看到高尔夫球场的视角，创造了高景观价值的总部基地环境。

此外，为保证良好的景观空间，新建企业总部地块拥有包括入口景观、主干道公共景观及区域内部公共景观、建筑入户景观及院落景观等在内的多层次景观。

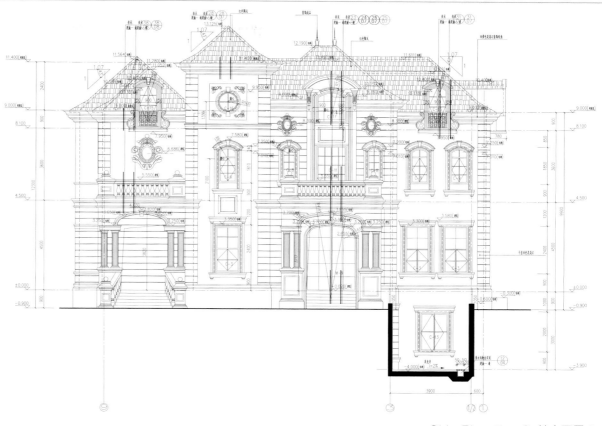

Side Elevation 2 轴立面图 2

Side Elevation 3 轴立面图 3

Building Facade 建筑立面

The facade is designed in neoclassicism style, with special emphasis on classic proportion and details. Ternary form, elegant colonnade and the exquisitely designed details highlight the dignity of the buildings. The most noteworthy are the single-family landscape buildings within the site that surround the golf course. The building facades are modern and elegant, while the interior spaces are flexible. Simple appearance forms sharp contrast with the complicated space structure. By using the fifth facade (folded plates) and the special interior space structure, it introduces natural light and landscape from the outside and realizes the natural ventilation within the buildings. The design idea echos the development's future position as a sci-tech and innovation center for Shanghai enterprises.

建筑立面采用文艺复兴的新古典主义风格，注重古典的比例和细部，通过三段式的划分、大气的柱廊、细部的刻画突显建筑的高贵品质。

值得一提的是，该项目地块内含环高尔夫球场独栋景观建筑。建筑立面风格现代简约，建筑内部空间灵动，简约的造型与其复杂的内部空间结构形成鲜明反差。建筑通过第五立面的丰富折板与室内异形结构空间的运用，引入室外光线与景观，采用自然通风打造生态环保的流动空间。其设计理念与该项目未来打造上海企业的科技中心、创意中心的定位相得益彰。

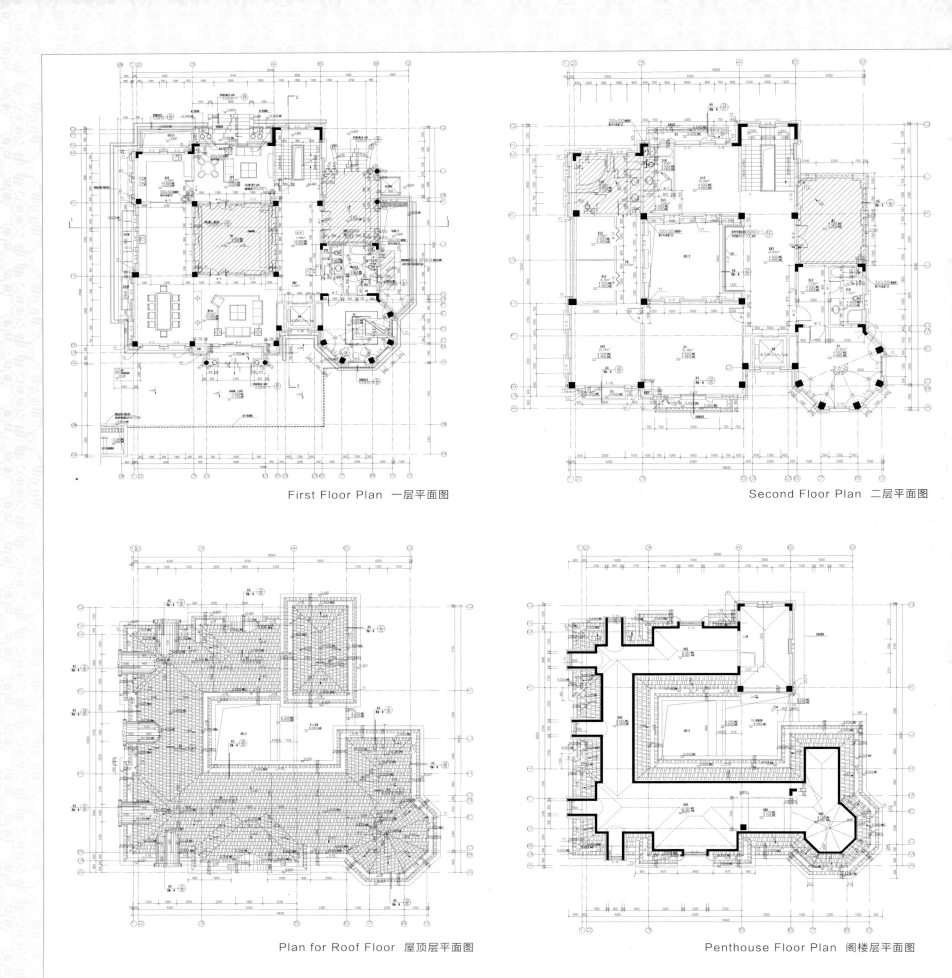

First Floor Plan 一层平面图

Second Floor Plan 二层平面图

Plan for Roof Floor 屋顶层平面图

Penthouse Floor Plan 阁楼层平面图

Meiju Encyclopedia 美居小百科

Neoclassicism Style, improved from classical style, has kept the general style of materials and colors, and abandoned excessive textures and decorations, which enables people to have strong experience of the tradition and the profound culture. The lamp of neoclassical style has been simplified and has presented new appearance by using modern materials. With this kind of diversified thinking, nostalgia is combined with modern living to present designs that are dignified, elegant, fashionable and modern. It reflects the aesthetic ideas and cultural taste of the post-industrial age.

新古典主义的设计风格,其实就是经过改良的古典主义风格。一方面保留了材质、色彩的大致风格,仍然可以很强烈地感受传统的历史痕迹与浑厚的文化底蕴,同时又摒弃了过于复杂的肌理和装饰,简化新古典主义建筑风格的启示——建筑历史与设计简化了线条。新古典主义的灯具则将古典的繁杂雕饰经过简化,并与现代的材质相结合,呈现出古典而简约的新风貌,是一种多元化的思考方式。将怀古的浪漫情怀与现代人对生活的需求相结合,兼容华贵典雅与时尚现代,反映出后工业时代个性化的美学观念和文化品位。

Chinese Traditional Living Ideal
中国传统居住理想

Carec Sanctuary　中航云玺大宅

Location / 项目地点：Kunming, Yunnan / 云南省昆明市
Owner / 业主：CAREC (Kunming) Co., Ltd. / 昆明中航房地产有限公司
Architectural Design / 建筑设计：Shenzhen Huahui Design Co., Ltd. / 深圳华汇设计有限公司
Land Area / 用地面积：600 000 m²
Floor Area / 建筑面积：600 000 m²
Plot Ratio / 容积率：1.0

Keywords　关键词

Traditional Space; Prairie Style; Refined Internally and Externally
传统空间　草原风情　内外兼修

"China's Most Beautiful Villas" Finalist
入围"中国最美别墅"奖

Based on Wright's prairie style which is characterized by its oriental complex, it introduces more Chinese cultural elements to interpret traditional Chinese living ideal from space, architecture and details. It makes the project significant to our times.

建筑风格在赖特颇具东方情结的草原风格基础上赋予更多中国文化的元素，在空间、建筑、细部几个层面诠释中国传统居住理想，并赋予一定的时代意义。

Overview　项目概况

Located by the Dian Lake, SANCTUARY is adjacent to the Wujiatang Wetland Park, set in favorable natural environment.

中航云玺地处滇池之畔，紧邻五甲塘湿地公园，自然环境优越。

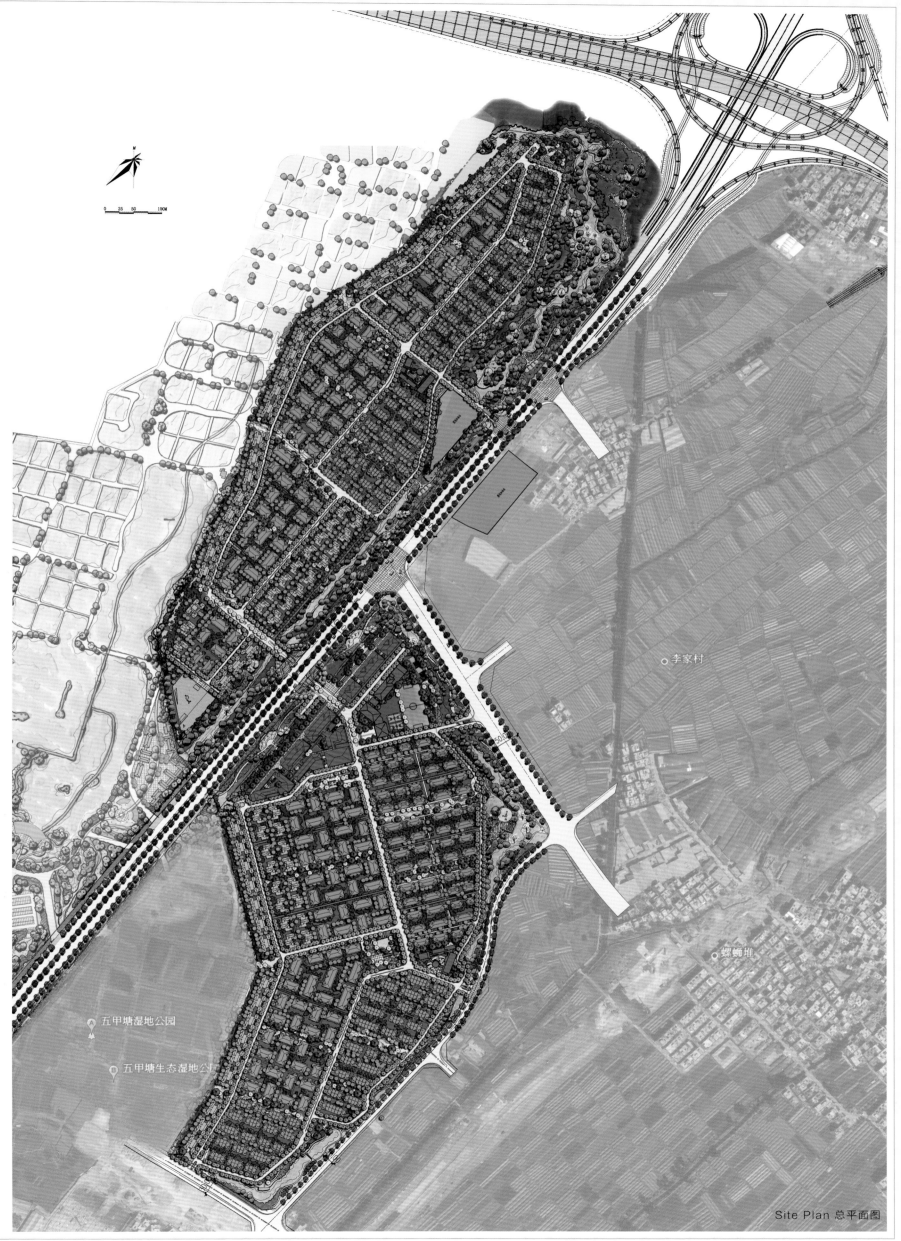
Site Plan 总平面图

Aerial View of the North Plot 北地块鸟瞰效果图

Design Concept 设计构思

Due to the height limitation of 12 m and the requirements for low-density villa layout, the vast area still has limited space. In the traditional Chinese living space, the residence is as important as the courtyard with the former representing the internal and the latter representing extension, which is the traditional Chinese understanding to the world—attaching the same importance internally and externally. In this project, the architects hope to combine this traditional Chinese idea on living with a modern idea of life. Therefore, the ground is designed to be used by people and the parking lot placed underground. The sunken garden makes full use of the underground space and presents more landscape views to create a relaxing and pleasant atmosphere throughout the community.

由于有12 m限高的要求,以及业主方对于低密度别墅总量的要求,虽然场地很大,但能发挥的空间是很有局限的。在中国传统居住空间中,宅与院同等重要,一个是内在,一个是外延,这也是中国传统对于世界的理解,讲究内外兼修。在这个项目里,设计师也希望把中国的传统居住理想与现代的生活理念相结合,为此,设计师首先把地面还给居民人行使用,所有停车在地下室解决,而下沉庭院则把地下室空间景观化、地面化,从而为社区塑造轻松怡人的环境奠定了基础。

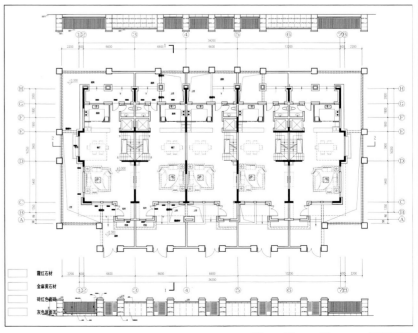

First Floor Plan 一层平面图

Section 1-1　1-1 剖面图

Architectural Style　建筑风格

Based on Wright's prairie style which is characterized by its oriental complex, it introduces more Chinese cultural elements to interpret traditional Chinese living ideal from space, architecture and details. It makes the project significant to our times.

建筑风格在赖特颇具东方情结的草原风格基础上赋予更多中国文化的元素，在空间、建筑、细部几个层面诠释中国传统居住理想，并赋予一定的时代意义。

Creating French-style Living Experience
创造法式生活体验的主题

Royal Mansion, Suzhou 苏州中吴红玺

Location / 项目地点：Xushuguan Town, Suzhou, Jiangsu / 江苏省苏州市浒墅关镇
Developer / 开发商：Jiangsu Zhongwu Properties Co., Ltd. / 江苏中吴置业有限公司
Architectural Design / 建筑设计：HIC / 上海翰创规划建筑设计有限公司
Land Area / 占地面积：58 749 m²
Floor Area / 建筑面积：82 248 m²
Plot Ratio / 容积率：1.4
Green Coverage Rate / 绿化率：35.4%

Keywords 关键词
Open View; Exotic Style; Theme of Experiencing
开阔视野 异国情调 体验主题

 "China's Most Beautiful Villas" Finalist
入围"中国最美别墅"奖

The facade is designed in French style with classic proportion, which looks elegant and dignified to enable people to experience the exotic style.

建筑在立面处理上采用了法式建筑的手法，古典的比例创造了优雅尊贵的感受，为客户带来异国情调的体验感受。

Overview 项目概况

The project is located in Xuguan Town of Suzhou High-tech Zone, Jiangsu, to the east of Beijing-Hangzhou Grand Canal and to the north of Shanghai-Nanjing Railway. Occupying a flat site, the development enjoys convenient transportation and complete supporting facilities.

项目位于江苏省苏州高新区浒关镇，京杭大运河以东，沪宁铁路以北，地势平坦，交通便利，各类生活盛业配套设施完善。

Regional Analysis 地形分析

Xushuguan Town sits in the northwest of Suzhou ancient city, which is known as the "heaven on earth". It is only 10 km away from downtown Suzhou, 20 km from the center of Wuxi and 100 km from Shanghai center. Surrounded by Suzhou Industrial Park on the east, downtown Suzhou on the south, Taihu Lake to the west and Wuxi High-tech Zone on the north, it has won a reputation as the "Top 100 Towns of Jiangsu and Top One Town of Xushuguan" (Inscription by General Hong Xuezhi).

浒墅关镇位于素有"人间天堂"之称的苏州古城西北，距苏州市中心仅10 km，与无锡市中心和上海市中心分别相距20 km和100 km。东依苏州工业园区，南接苏州市区，西临太湖，北靠无锡高新区。盛享"江苏百家名镇、浒墅关第一镇"（洪学智题）的美誉。

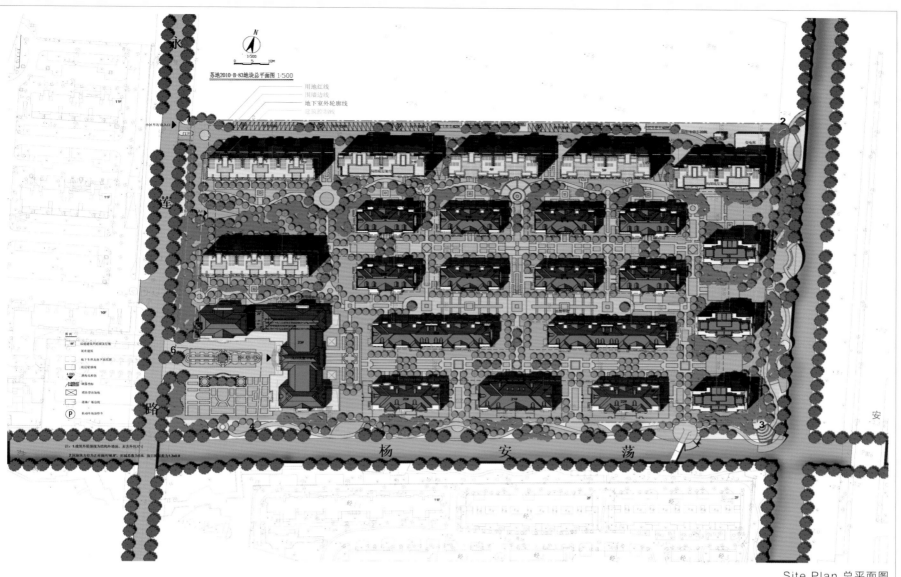

Site Plan 总平面图

Aerial View 鸟瞰图

Planning and Layout 规划布局

Responding to the high-end market position of the development, the design has focused on providing French-style living experience. Axisymmetric layout has been introduced which combines with the open green floors to create a natural and fresh atmosphere. Meanwhile, this kind of layout helps to form natural landscape levels. Along the open French-style landscape axis, pleasant green belts, interesting green lands and the semi-outdoor courtyard that extends to every residential unit are arranged in order, keeping in harmony with the surroundings.

设计结合高端的市场定位，利用创造法式生活体验的主题，规划在引入轴线对称布局的同时结合内部架空层的绿化渗透，与自然清新的气息融为一体。同时中轴对称的规划逻辑，也使得景观自然而然分级有序。在中心开阔视野的法式景观轴线统帅下，清新宜人的绿化带、富有情趣的小区组团绿地以及进入每个单元的半室外庭院分合有序，浑然天成。

Facade Design 建筑立面

The facade is designed in French style with classic proportion, which looks elegant and dignified to enable people to experience the exotic style. Stone and stone paint are used alternately to well control the cost and ensure a caring living environment at the same time.

建筑在立面处理上采用了法式建筑的手法，古典的比例创造了优雅尊贵的感受，为客户带来异国情调的体验感受。同时在材料上石材与真石漆结合有致，充分考虑到成本对客户的影响，为更多的人享受更好的生活提供关爱。

Unit Design 户型设计

All residential units can be accessed directly with an independent lift. With a total area no more than 270 m², each unit is designed with an 8 m wide living room and a terrace-style balcony to meet the requirements for gathering, meeting, health care, etc. All bedrooms are designed in suite. In this way, the public spaces are separated from the private spaces perfectly. With emphasis on the nature of family life, it has created a place where the heart belongs without losing comfort and dignity.

作为大平层的单体建筑，在户型的设计上，所有的住户均采用了电梯单独入户的方式，在控制整体单户面积不超过 270 m² 的基础上，公共空间考虑到未来客户聚会、会客、养生等功能，均采用 8 m 面宽的大厅结合露台化的情景阳台，私密空间则做到所有卧室均为套房的生活标准，做到动则灿若星河，静则润物无声。户型设计概念上强调家庭生活的本质，在创造心灵回归场所的同时，不失生活的舒适感与尊贵感。

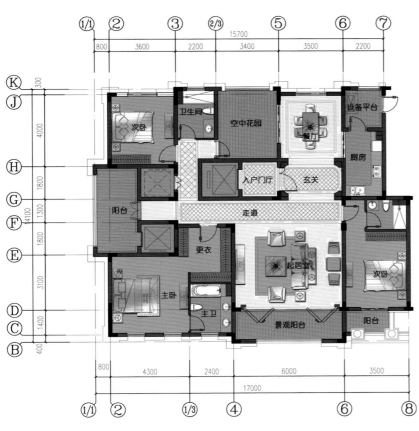

Layout 1 户型图 1

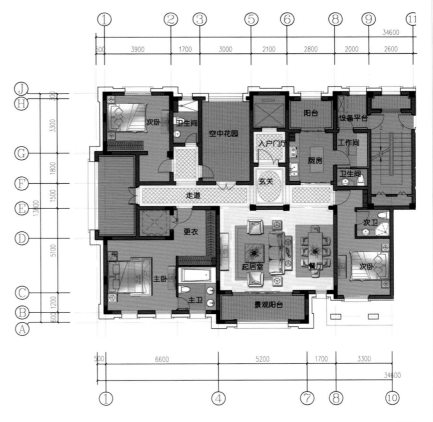

Layout 2 户型图 2

Meiju Encyclopedia 美居小百科

French-style architectural elements: building of French style usually features a mansard gambrel-style roof which is punctured by dormer windows of different shape. The exterior wall is usually decorated by stone or archaistic stone. And the details such as the French-style columns, carvings and lines will highlight the elegant and romantic style. The whole building will be symmetrical and look magnificent.

　　法式风格建筑元素：屋顶多采用孟莎式，坡度有转折；屋顶上多有精致的老虎窗，且或圆或尖，造型各异；外墙多用石材或仿古石材装饰；细节处理上运用了法式廊柱、雕花、线条，呈现出浪漫典雅风格。整个建筑多采用对称造型，气势恢宏。

Amenities 高端配套

Royal Mansion is a self-contained development offering a complete suite of facilities. The clubhouse, which also serves as the entrance lobby for pedestrians, accommodates functions such as restaurant, SPA, fitness center, entertainment space, etc. These facilities not only highlight the quality of the community but also provide multiple options for the residents.

高端的定位势必结合高端的配套设施，兼为小区人行入口大堂的会所，为小区住户提供各类配套公共活动设施，从高端餐饮到养生 SPA，健身娱乐一应俱全，增强小区高端气氛的同时，也为未来客户的生活情趣平添了一份色彩。

Family Mansion with Multi-Style Facade Organically Integrated
多风格立面有机共融的家族大宅

Dragon Palace, Phase I 博恩·御山水一期

Location / 项目地点：Dandong, Liaoning / 辽宁省丹东市
Developer / 开发商：Dandong Bo'en Real Estate Development Co., Ltd. / 丹东博恩房地产开发有限公司
Architectural Design / 建筑设计：CCDI / 悉地国际（CCDI）
Land Area / 占地面积：1 055 900 m²
Floor Area / 建筑面积：340 000 m²

Keywords 关键词

Big and Orderly; Well Proportioned; Dignified and Elegant
大而有序 近人尺度 高贵典雅

"China's Most Beautiful Villas" Finalist
入围"中国最美别墅"奖

It tries to create a symbol for high-end residences in China's market and a landmark of family mansion. The architectural form follows the example of European palace to present the luxury of Mediterranean Style, the pureness of British Style, the dignity of French Style and the elegance of Neoclassical Style.

规划力求打造中国高端住宅符号，一个里程碑式的家族大宅，建筑形式效仿欧洲王庭：地中海建筑的宫廷华贵、英式建筑的原乡质朴、法式建筑的端庄宏伟、新古典主义建筑的高贵典雅在社区建筑上很好的得到了体现。

Overview 项目概况

Located at the foot of Jinshan Development Zone, boasting a total floor area of 350,000 m², the project is close to the express landscape road from Dandong to Wulongbei. Being 8 km away from downtown Dandong, Dragon Palace is surrounded by Jinshan Mountain, Wulong Mountain and Hutou Mountain on three sides. With the Jinshui River that runs through the site, it enjoys advantaged landscape resources. As a landmark villa project in Dandong City, it offers villas of British style, French style and Tuscan style, trying to integrate different facade styles within one development by skillful planning and design.

项目位于金山开发区脚下，总建筑面积约35万 m²，紧邻丹东至五龙背快速景观路，距离丹东市区8 km，金山、五龙山、虎头山三山环绕，金水河穿流而过，景观资源丰富。百万平米稀缺别墅大盘，是辽宁省丹东市标杆性项目，多风格立面有机共融在一个大盘内，融汇英式、法式、托斯卡纳等多风格别墅产品，并具有高水准的设计实现度。

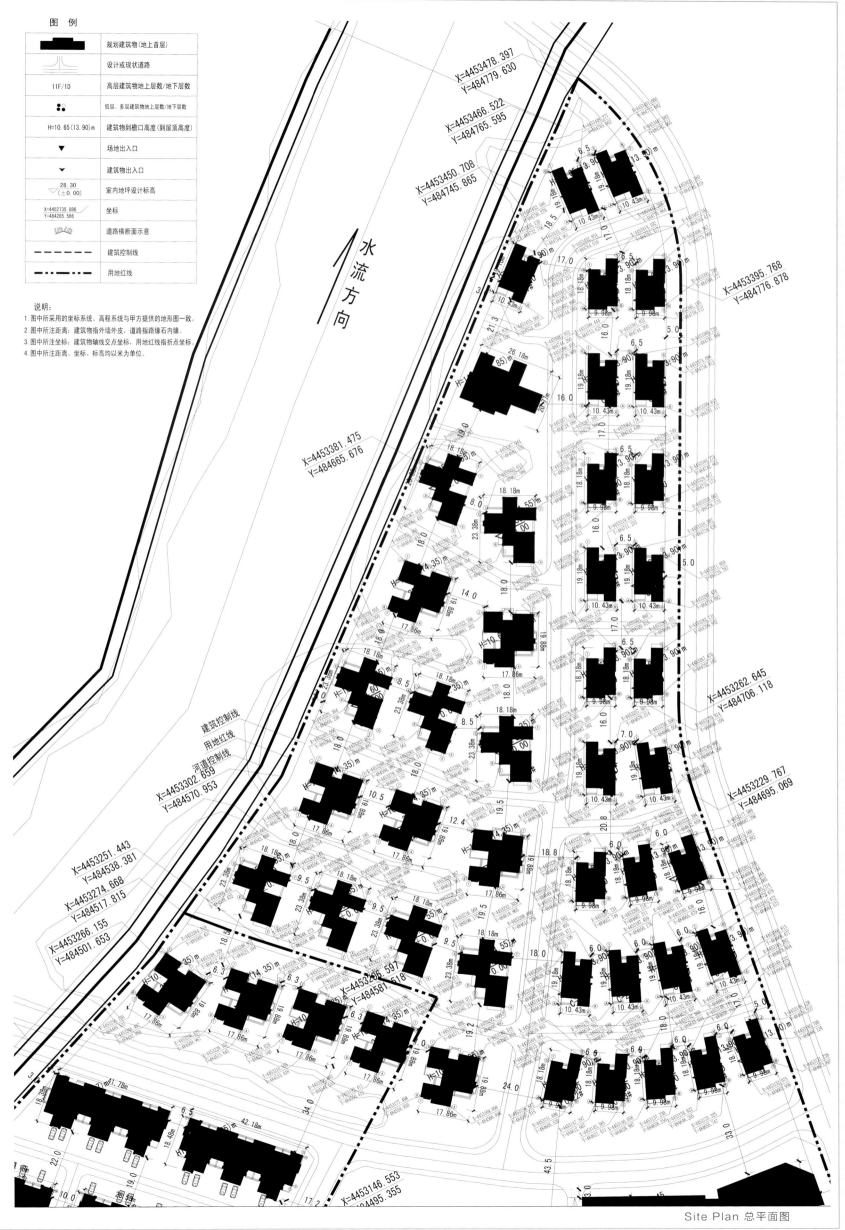

Site Plan 总平面图

Planning 项目规划

According to the planning, it tries to create a symbol for high-end residences in China's market and a landmark of family mansion. The architectural form follows the example of European palace to present the luxury of Mediterranean Style, the pureness of British Style, the dignity of French Style and the elegance of Neoclassical Style. While the architectural arrangement, the form of the courtyard and the division of space follow the rules of oriental residence to be "big and orderly".

规划力求打造中国高端住宅符号，一个里程碑式的家族大宅，建筑形式效仿欧洲王庭：地中海建筑的宫廷华贵、英式建筑的原乡质朴、法式建筑的端庄宏伟、新古典主义建筑的高贵典雅在社区建筑上很好的得到了体现；建筑格局和院落形式、空间分割上遵从"大而有序"的东方住宅法则。

Unit Types 产品类型

The development offers single-family villas, single-family like villas, townhouses, courtyard villas, garden apartments, high- and medium-rise apartments. According to different categories and values, these units are well arranged together with the roads and landscape system, creating a relatively independent and harmonious community. Different residential groups feature their own architectural style, landscapes, facilities, signage system and pleasant proportion.

产品类型包括独栋别墅、类独栋别墅、联排别墅、合院别墅、花园洋房、高层和小高层等，结合道路和景观系统，按照产品价值分布规则，形成相对独立又和谐有序的社区组团，拥有各自的建筑风格、景观园林和场地器械、标识系统和近人尺度。

Construction 项目施工

It is a joint effort between the developer, the architects and the contractor from getting approval, conceptual design, preliminary design, construction documents design and on-site construction. To finalize the plan, the architects have carefully studied the current conditions of the site and the local dwellings, fully understood the suggestions of the developer, defined and realized the unique architectural style in this region — a mixed European style of Tuscan and French style. The development is recognized and highly praised by the developer and the local residents. Now the built showflats, single-family villas, townhouses, garden apartments and high-rise apartments have fully realized the design concept. The architects have also been keeping close eyes on the construction process and visited the site regularly to solve different problems, trying to realize the whole plan perfectly.

项目从立项，到方案初稿、初步设计、施工图设计、现场建设，无不凝聚建设方、设计方、施工方的辛勤汗水。方案确立过程中设计师紧密结合现场及当地住宅的现状，充分理解建设方的意见建议，完成并实现了在当地独一无二的建筑风格，即西班牙托斯卡纳结合法式的欧派建筑风格，得到了建设方及当地居民的高度认可及赞扬。目前，在已经建成的样板区、独栋别墅区、联排别墅区、花园洋房区、高层住宅区等，近乎100%的实现了设计方案，毫不掩饰其"最佳设计实现"的特质。从开工的第一锹土开始，设计师时刻关注现场的建设进程，在有计划的现场回访中及时准确地解决建设过程中的要点、难点及重点问题，使得整个项目的建设，完全的按照设计师及建设方既定的实施方案。

Modern Residential Community Integrating Elegance, Rationality and Romance
融大气、雅致、理性与浪漫于一体的现代化居住区

Vanke Chefoo Island, Yantai 烟台万科海云台

Location/ 项目地点：Yantai, Shandong / 山东省烟台市
Owner/ 业主：Yantai Vanke Real Estate Development Co., Ltd. / 烟台万科房地产开发有限公司
Architectural Design/ 建筑设计：Concord Design Group / 西迪国际（CDG 国际设计机构）
Floor Area/ 建筑面积：311 600 m²
Plot Ratio/ 容积率：1.42
Green Coverage Rate/ 绿化率：42%

Keywords 关键词
Comfortable and Pleasant; Architectural Etiquette; Modernization
舒适宜人 建筑礼制 现代化

"China's Most Beautiful Villas" Finalist
入围"中国最美别墅"奖

The buildings are designed in New Chinese style and built with modern building materials. Design elements in traditional Chinese architecture have been abstracted and used to meet modern living requirements and show respect to the history as well.

项目建筑为新中式风格，采用现代化建筑材料，在充分吸收传统中式建筑文化内涵的基础上，进行抽象总结与概括，在保证当代人民的居住功能需求的同时也体现了对历史的尊重。

Regional Analysis 区位分析

As one of the earliest trading ports in modern China, Yantai has experienced an integration of Chinese and Western Culture since its opening, and modern architectures of different styles have emerged all around the city during that period. Vanke Chefoo Island is strategically located at the shore of the Yellow Sea, on Zhifu Island in north Yantai, enjoying beautiful surrounding views and a cultural environment. It creates a unique architectural spirit as well as cultural spirit, being a language of brick and stone flowing in the blood of architecture. The project has highlighted the essence of Western architecture, shown respect to Chinese traditional architecture, and been endowed with the quality of dignity, elegance and unpretentiousness.

烟台作为我国近代最早的通商口岸之一，在开埠以后，中西文化相互交融，绚丽多彩的近代建筑，在烟台市区遍地开花。本项目位于黄海之滨、烟台北部的芝罘岛上，地理位置优越，周围环境优美、洋溢着文化，也塑造着一种独特的建筑精神、人文精神，它是流动在建筑血液中的砖石语言。建筑整体彰显西洋建筑精粹，尊崇中国传统建筑礼制，使其拥有端庄、低调、内敛的高贵气韵。

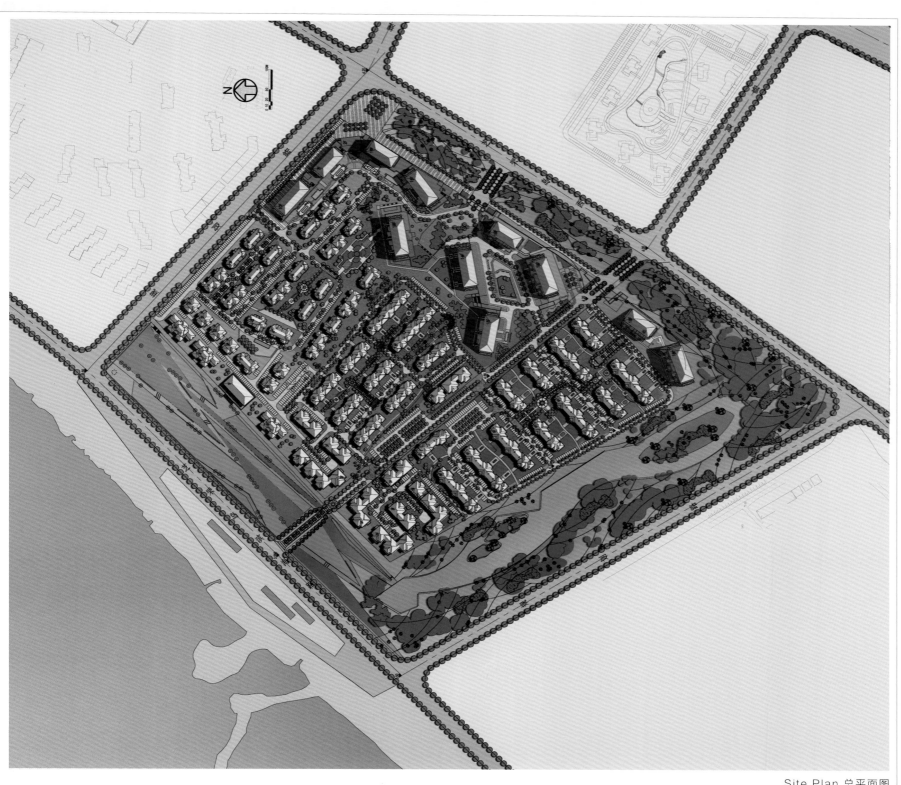

Site Plan 总平面图

Design Concept 设计构思

Taking the surrounding environment into account, the design has skillfully borrowed the natural landscape and introduced the concept of urban block, creating a new lifestyle that shows the combination between residences and natural environment. By using modern planning theories and space organization skills, it has created a modern residential community which is elegant, rational and romantic as well.

设计时充分考虑结合周边环境，巧妙的借用基地周边良好的自然景色，并将城市街区概念引入其中，创造一种全新的居住和生活模式，体现居住与环境的结合关系，使社区完全融于城市的自然环境之中，同时采用现代的规划理论与空间环境塑造手法，力求创造出融大气、雅致、理性与浪漫于一体的现代化居住区。

Location Map 区域位置图

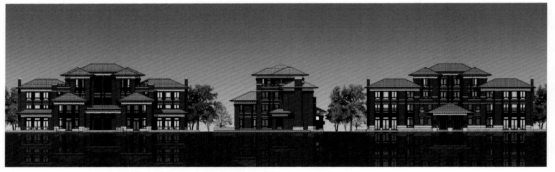

North Elevation of Large Semi-detached Villa
大双拼北立面图

Side Elevation of Large Semi-detached Villa
大双拼侧立面图

South Elevation of Large Semi-detached Villa
大双拼南立面图

South Elevation of the 6th Floor (Garden Apartments)
洋房四联拼6层南立面图

Side Elevation of the 6th Floor (Garden Apartments)
洋房四联拼6层侧立面图

North Elevation of the 6th Floor (Garden Apartments)
洋房四联拼6层北立面图

South Elevation of the 4th Floor (Garden Apartments)	Side Elevation of the 4th Floor (Garden Apartments)	North Elevation of the 4th Floor (Garden Apartments)
洋房四联拼4层南立面图	洋房四联拼4层侧立面图	洋房四联拼4层北立面图

North Elevation of Small Semi-detached Villa	Side Elevation of Small Semi-detached Villa	South Elevation of Small Semi-detached Villa
小双拼北立面图	小双拼侧立面图	小双拼南立面图

Architectural Design 建筑设计

The buildings are designed in New Chinese style and built with modern building materials. Design elements in traditional Chinese architecture have been abstracted and used to meet modern living requirements and show the respect to the history as well. The design is inspired by Chinese traditional architecture but pays more attention to modern living values. Different from buildings that simply intimate the ancient architecture, buildings here are dedicated to enhance the comfortableness by providing great privacy and increasing sunlight and natural ventilation. It also attaches great importance to the bathroom and kitchen, and well organizes and separates the bedrooms for the elderly, the children and the masters. In addition, Chinese traditional architectural elements such as the courtyard, the sunken yard and the veranda are upgraded by innovate design to better cater to modern life.

项目建筑为新中式风格，采用现代化建筑材料，在充分吸收传统中式建筑文化内涵的基础上，进行抽象总结与概括，在保证当代人民的居住功能需求的同时也体现了对历史的尊重。建筑在沿袭中国传统建筑精粹的同时，更注重对现代生活价值的精雕细刻。同单纯的仿古建筑不同，该建筑着力提高居住的舒适度，比如在设计中更多考虑私密性，增强采光通风，更有效地提高卫生间、厨房在居室中的地位，更好地使老人、孩子、夫妇间的居室环境合理分隔与有机协调等。另外，在庭院、地下室的处理中，也吸纳了更多现代生活流线的创新之笔，比如庭院、下沉庭院、内游廊等，让中式建筑以一种更自然、更现代、更具生命力的品相出现。

Landscape Design 景观设计

Landscape design of the development is inspired by traditional gardening culture and establishes a landscape system that is suitable to the residents. It creates a people-oriented residential environment based on the society, economy, culture, environment and technology of the 21st century.

It has abandoned the conventional way to arrange buildings around a large area green center. In stead, it has divided the green center into small gardens between buildings which enables each building to share a garden. What's more, a small square is set in the public area of the fusiform site for gatherings and activities.

小区的景观设计汲取中国传统造园文化的精髓，化整为零，探索适合本小区安置对象的景观模式，体现21世纪以社会、经济、文化、环境和技术为基点的"人为本"、"人性化"的人居环境。

景观设计摒弃了常用的大面积中心绿地四周布置建筑，化整为零，把景观庭院化，充分利用宅间空地把内容做丰富，把原来一家一院的小家小院化为一楼一院的大家大院。另外，在梭形地块的公共区内辟出一个小广场，来满足人数比较多的活动需求。

Multiple Skills to Realize Harmonious Coexistence Between Architecture and Nature

多技法营造建筑与自然和谐共生的意境

Emerald Lemmon Lake, Jinan 济南翡翠莱蒙湖

Location/ 项目地点: Zhangqiu, Jinan, Shandong/ 山东省济南市章丘市
Developer/ 开发单位: CNHTC Real Estate Development Co., Ltd. / 中国重型汽车集团房地产开发公司
Landscape Design/ 景观设计: Botao Landscape (Australia)/ 澳大利亚·柏涛景观
Land Area/ 用地面积: 300 900 m²
Floor Area/ 建筑面积: 55 000 m²
Landscape Area/ 景观面积: 220 900 m²
Plot Ratio/ 容积率: 1.6
Green Coverage Rate/ 绿化率: 41.3%

Keywords 关键词

Spatial Sequence; Harmonious Coexistence; Eco-friendly
空间序列 和谐共生 生态环保

 "China's Most Beautiful Villas" Finalist
入围"中国最美别墅"奖

The overall planning starts from creating urban interface and internal core landscape, making the inner and outer ring and the core water feature as the main theme. Low-rise residences within the inner ring are divided into three groups and organized in courtyard layout to form enclosed neighborhood spaces, which presents a spatial sequence by the the public spaces of the community, the public spaces of the residential groups, the public courtyard spaces and the private garden spaces.

整体规划以城市界面和内部景观核心的打造为主要出发点，以内外两环及核心水景为构思的主要中心。内环住宅分为三个组团，采用大合院的布置方式，形成围合式的邻里空间，完成小区公共空间、组团公共空间、合院公共空间、私家庭院的层层深入的空间序列。

Overview 项目概况

The project is located in Zhangqiu city, which is the hinterland of Shandong Province with beautiful mountains and rivers and known as the "Small Springs City" because of groups of spewing springs. The site is flat and the surrounding environment is good. 30-minutes ride to Jinan downtown on the west and 15 km away from Zhangqiu downtown on the east, it enjoys highly convenient transportation. On the south of the site, there is a mountain called Huangqi Mountain which is covered by lush green trees. Thus, the project enjoys unparallel green views of the surroundings and the air quality is pretty good here.

项目位于章丘市，章丘市地处齐鲁腹地，山明水秀，群泉喷涌，素有"小泉城"的美誉。基地平整，周边自然环境良好，西距济南市区30分钟的车程，东距章丘市区15 km，交通十分便利。在基地南侧有黄旗山，山上绿树成荫。整个基地及周边地区绿化条件及空气质量较好。

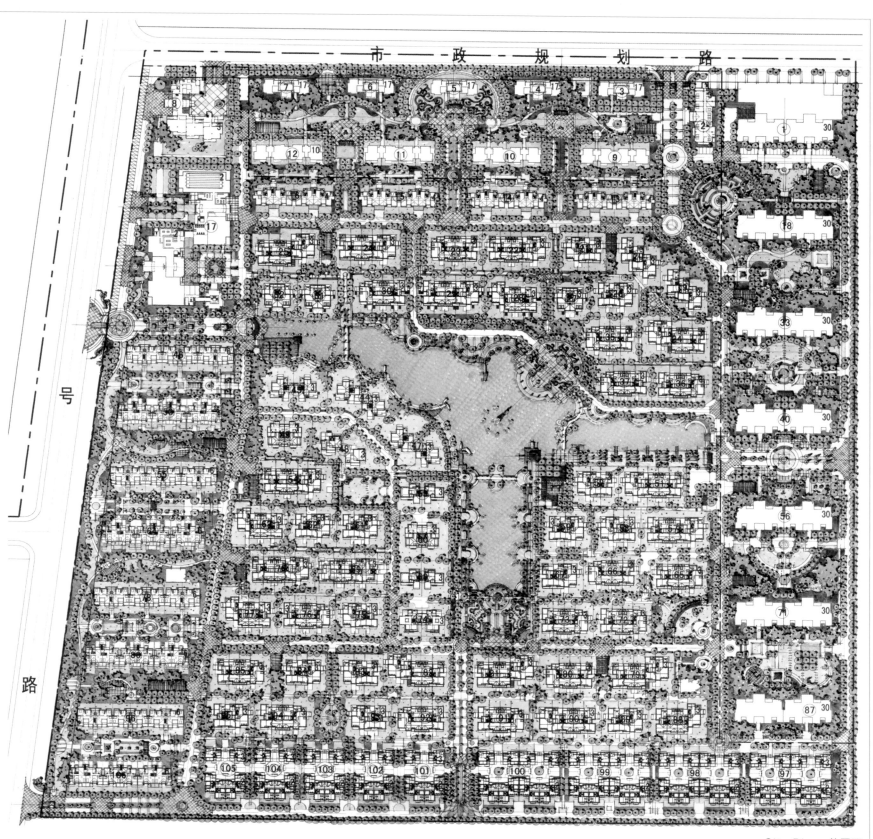

Site Plan 总平面

Planning Concepts 规划构思

The overall planning starts from creating urban interface and internal core landscape, making the inner and outer ring and the core water feature as the main theme. The annular zone between the inner and outer ring is set with high-rise buildings. Considering the urban interface and the adjacent properties, the height of the high-rise buildings decreases gradually from the east to the north, then to the west and south, just like a ring-shaped stream flowing down to the inner ring road, reaching the central water feature, and forming the core landscape area of the community. Meanwhile, the high-rise area also looks like a cornucopia which gathers energy from the outer ring into the center of the community. Within the inner ring are mainly high-end low-rise residences which are arranged around the central landscape belt. Rich in form and low in space, these residences combine together with the waterscape to form the landscape center of the community, providing a beautiful fifth facade for the high-rise buildings outside the inner ring road. These low-rise residences are divided into three groups and organized in courtyard layout to form enclosed neighborhood spaces. In this way, it forms a spatial sequence by the the public spaces of the community, the public spaces of the residential groups, the public courtyard spaces and the private garden spaces.

整体规划以城市界面和内部景观核心的打造为主要出发点，以内外两环及核心水景为构思的主要中心。外环内环之间的环形地带主要布置高层区，考虑城市界面和相邻物业的形态，高层区沿东—北—西—南逆时针降低层数，如环状的溪流，逐级缓缓落下，最后从低层区流入内环，汇聚于中心的景观水体，形成整个小区核心景观区，也如聚宝盆一般，将沿外环注入的能量集中于中心，形成小区的精髓。内环以内为高端低层产品为主，围绕核心景观带呈岛状布局，由于其产品自身丰富的形态和低矮的空间关系，也和景观水体一起共同成为小区的景观中心，为内环以外的高层产品提供了优美的第五立面。内环住宅分为三个组团，采用大合院的布置方式，形成围合式的邻里空间，完成小区公共空间、组团公共空间、合院公共空间、私家庭院的层层深入的空间序列。

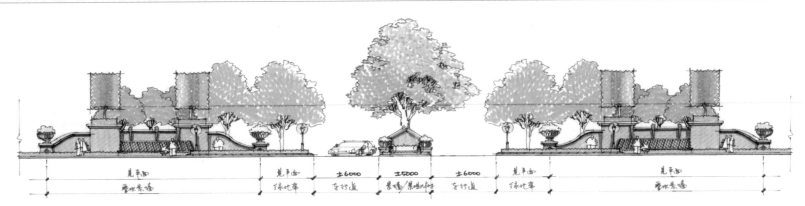

Elevation 1 立面图 1

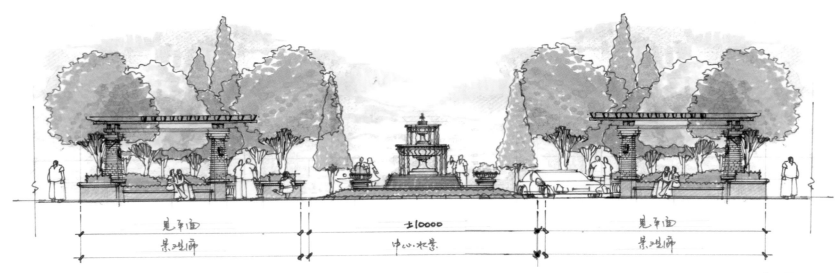

Elevation 2 立面图 2

Landscape Design 景观设计

Water is the main theme when arranging and building the low-rise residential groups. All residences are built around the lake or embrace the public gardens to provide enclosed neighborhood spaces and realize the harmony between architecture and nature. Along the south boundary of the site are courtyard-style residences which consist of several individual buildings and a central courtyard, presenting a compact and cohesive space structure. With emphasis on "green" and "landscape", it integrates all landscapes spaces by skills like penetrating, borrowing, echoing, staggering and division as well as the structures like axisymmetry and vestibule, combining the natural landscape with the cultural landscape perfectly. Therefore, living culture is integrated into green landscape, and the courtyard-style landscape spaces are distinctive and pleasant.

低层住宅组团以水为媒，自然形成组团式布局，住宅或环水分布，或围绕组团公共花园形成围合式邻里空间，营造建筑与自然和谐共生的意境。南侧沿用地边界布置合院住宅，其布局是由若干栋单体建筑和内聚型的中庭院组成，体现紧凑的、内聚的空间特点。通过景观空间的渗透、借景、对景、错景、轴线、通廊、分割等手法的运用将居住区景观融为一体，突出"绿"，强调"景"，将自然景观和人文景观有机的结合在一起。景观设计使居住文化融入绿化之中，全区景观体现院落空间的特征，达到景观特色鲜明的效果。

Private Island Villas of Italian Style
以意式风情为依托的私岛别墅

Purple Shore   紫岸

Location/ 项目地点：Handan, Hebei/ 河北省邯郸市
Developer/ 建设单位：LONGJITAIHE Industrial Development Co., Ltd./ 隆基泰和实业有限公司
Landscape Design/ 景观设计：America Leedscape Planning and Design Co., Ltd./ 美国俪禾景观规划设计有限公司
Design Team/ 设计团队：Chen Juncheng, Ma Yanjie, Li Chunsheng, Liu Ning, Wang Hongrui, Xu Qing, Zhang Rui/ 陈君成 马燕杰 李椿生 刘宁 王红蕊 徐庆 张睿
Landscape Area/ 景观面积：75 000 m²

Keywords   关键词
Cultural Atmosphere; Noble Quality; Green and Ecological
人文氛围 高贵特质 绿色生态

"China's Most Beautiful Villas" Finalist
入围"中国最美别墅"奖

Villas on this private island are noble and elegant, while the landscape is designed in Italian style, presenting the natural and cultural beauty of Tuscany. Buildings and landscape architectures echo each other, and natural landscape integrates with the local culture, creating a unique high-quality community for Handan.

紫岸项目整体建筑突显私岛别墅的高贵特质，景观设计以意式风情为依托，还原托斯卡纳自然风光及历史人文之美，建筑与景观交相辉映，风景及文化锦上添花，成为邯郸独具特色的高品质社区。

Overview   项目概况

Located in Handan City of Hebei Province, to the south of Zhizhang River, to the north of Zuoxi No.1 Road and to the west of Nanbao Street, the Purple Shore is 3.3 km away from Handan South Exit of Beijing-Zhuhai Expressway and 7.3 km from Handan Airport, enjoying convenient transportation. Adjacent to three high-education schools of Handan City, it boasts strong cultural atmosphere. Villas on this private island are noble and elegant, while the landscape is designed in Italian style, presenting the natural and cultural beauty of Tuscany. Buildings and landscape architectures echo each other, and natural landscape integrates with the local culture, creating a unique high-quality community for Handan.

紫岸项目位于河北省邯郸市支漳河以南、左西一路以北、南堡大街以西，距离京珠高速邯郸南出口 3.3 km，距邯郸市机场 7.3 km，出行交通便捷。周围毗邻河北邯郸三所高校，人文氛围浓厚。紫岸项目整体建筑突显私岛别墅的高贵特质，景观设计以意式风情为依托，还原托斯卡纳自然风光及历史人文之美，建筑与景观交相辉映，风景及文化锦上添花，成为邯郸独具特色的高品质社区。

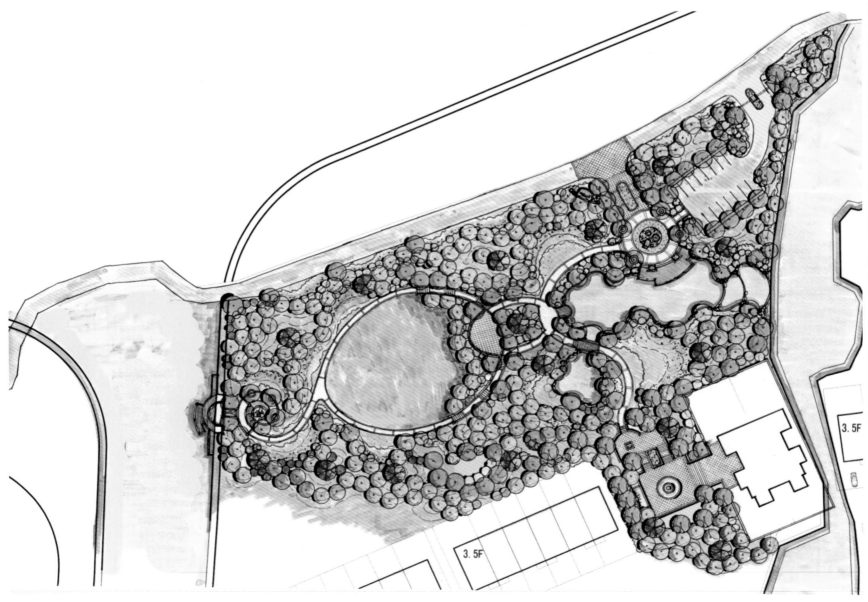

Site Plan 总平面

Aerial View 鸟瞰图

Design Idea 设计理念

Tradition— the designers have drawn inspirations from Tuscan traditional landscape and rich cultural heritage, and explored leisureliness, romance and warmness in living.

Nature—another idea comes from the colorful European natural landscape: big trees and shrubs are randomly arranged to soften the building volumes and enclose the spaces, reminding people of the natural forest.

Integration—Tuscan noble culture is introduced to complement the natural landscape, presenting wonderful views. Exquisite colonnade, pavilions, dancing fountains, pergola climbed with plants, large lawns, closed woods and graceful cambered surfaces, all combine together to create a high-quality living environment of Italian style. Landscape elements such as the pavement, landscape wall and ironwork are exquisitely designed and used to follow the architectural style. Meanwhile, textures of the natural materials, namely, wood and stone, enable people to experience the rich and peaceful country life with sunshine, hillside, farm and vineyard.

传统——设计方案从传统托斯卡纳的美丽风景以及丰富的艺术遗产中摄取灵感，在居住中寻求悠然、浪漫、温暖的情怀。

自然——方案另一方面源于欧洲色彩艳丽的自然风光，唤起自然森林感觉的大树及灌木以一种"随意"的方式分布，软化建筑体量风格并围合空间。

融合——设计过程中引入托斯卡纳贵族文化，融合自然风光元素，营造丰富精彩的景观。精致的柱廊、景亭、潺潺涌动的喷泉，爬满植物的藤架，大片的休闲草坪、围合密林以及优雅的弧面，共同打造出意式风情浓郁的高品质人居环境。设计师将建筑元素体现在铺地、坐墙和铁艺上，使景观与建筑风格协调一致，同时运用天然的材料如木头、石头等丰富材质的肌理，让沐浴其中的人在阳光、山坡、农庄、葡萄园中能享受到朴实富足的田园生活。

Landscape Design 景观设计

Watchtower at the Entrance: evergreen cedars stand straightly to form a rhythmic and heavy background for the tower. Behind the cedars are common species such as the ash trees and ginkgo trees, forming a vast and dense forest. In front, single-colored seasonal flowers are planted under the LOGO; beautiful cambered lines add a breath of romance for this space.

Water Garden: the existing wooden trestle is useful to provide a sightseeing spot for the visitors. The designers have transformed the water area into a water garden, trying to bring as much fun as possible.

Flower Sea: weeping willows gesture gracefully and stand elegantly along the road. Wild flowers of red, yellow, white are mixed and planted to highlight the sunshine and warmth of the country life.

入口区岗楼：由竖线条的常绿乔木雪松构成塔楼背景，营造一种强烈的阵列感，凝重的色彩也增添了背景的厚重感。雪松背后选择普通绿化树种白蜡、银杏等，形成大量的密林区域。前方Logo下栽植单一色彩的时令花卉，优美的弧线轮廓充满了浪漫气息。

水生花园：现有的木栈道是该场地的有益因素，为游客提供游赏空间。设计师将此地的水面改造为水生花园，尽可能提供较多的乐趣。

花海：乔木选择姿形姣好的垂柳列植于道路两旁。地被以野花组合的方式栽植，色彩以红色、黄色、白色系为主，体现乡野的阳光与热情。

01	圆冠阔叶大乔木
02	高冠阔叶大乔木
03	高塔型常绿乔木
04	低矮塔型常绿乔木
05	圆冠型常绿乔木
06	球类常绿灌木
07	修剪色带
08	小乔木
09	柱形灌木
10	团形灌木
11	可密植成片的灌木
12	普通花卉型地被
13	长叶型地被

Vegetation Analysis 区域种植分析图

High-quality Residential Courtyard Promoting Chinese Culture
弘扬中式文化的高品质居住宅院

Cathay Courtyard　泰禾·北京院子

Location/ 项目地点：Sunhe Village, Chaoyang District, Beijing / 北京市朝阳区孙河乡
Developer/ 开发商：Beijing Zhongwei Taihot Real Estate Development Co., Ltd. / 北京中维泰禾房地产开发有限公司
Architectural Design/ 建筑设计：Architectural Design: A&S International Design Co., Ltd. / 北京翰时国际建筑设计咨询有限公司
Floor Area/ 建筑面积：106 168.9 m²

Keywords　关键词
Innovative Form; Chinese Culture; High Quality
形式新颖　中式文化　品质高端

"China's Most Beautiful Villas" Finalist
入围"中国最美别墅"奖

Buildings are designed mainly in new Chinese style with elegant appearance and exquisite details. The high quality and innovative form have distinguished the development itself in the low-density residential market which is dominated by European-style architectures. It has greatly highlighted the profound Chinese culture.

建筑风格以新中式为主导，整体设计简洁大气又不失细节，使整个居住区形象品质高端、形式新颖，在以西方欧式建筑为主体形态的低密度住宅市场中更加别具一格，将中式文化体现得淋漓尽致。

Overview　项目概况

The site locates at plot W of Beidian West Village in Sunhe Town, Chaoyang District, belonging to Beijing's Central Villa District (CVD). It is a residential project offering a total floor area of 106,168.9 m².

项目位于北京中央别墅区，朝阳区孙河乡北甸西村 W 地块，总建筑面积 106 168.9 m²。

Design Inspiration　设计灵感

The architects have drawn inspirations from the New Chinese-style living experience and absorbed the quintessence of traditional culture to promote Chinese culture with courtyard-style residences.

项目单体以新中式生活感受为设计灵感，提取传统中式文化的精髓，使人们赖以生活的居住宅院成为弘扬中国文化的一部分。

Aerial View 鸟瞰图

Planning and Design 规划设计

The development is planned with the idea of Chinese traditional neighborhood. To establish a Chinese-style layout, it makes full use of the underground space and creates an underground traffic system which separates pedestrians from vehicles completely in the residential area.

规划设计秉承中式街坊的理念。为体现更纯粹的中式布局模式，规划充分利用地下空间，形成地下交通体系，使整个居住区流线实现完全人车分离。

Architectural Style 建筑风格

Buildings are designed mainly in new Chinese style with elegant appearance and exquisite details. The high quality and innovative form have distinguished the development itself in the low-density residential market which is dominated by European-style architectures. It has greatly highlighted the profound Chinese culture.

项目建筑风格以新中式为主导，整体设计简洁大气又不失细节，使整个居住区形象品质高端、形式新颖，在以西方欧式建筑为主体形态的低密度住宅市场中更加别具一格，将中式文化体现得淋漓尽致。

Unit Types 户型设计

There are five unit types, including single-family villas, courtyard villas, townhouses, superposed villas and garden houses. Multi-storey residences in the north are all designed with two units on each floor to share one staircase. There are both flats and duplexes. The interior spaces are square and well organized to separate activity spaces from private spaces, separate dry from wet, and provide clear functions. Low-storey courtyard residences open to the south or north and are designed in U shape with private outdoor spaces in the center and exterior spaces in the south and north. The interior and exterior spaces can be connected. This kind of design will not only save the land but also cater to low-density lifestyle.

Based on the principle of providing comfortable living and reasonable layout, all units are designed with large width and small depth to get enough sunlight, cross ventilation, clear routes as well as square and practical rooms. Villas are characterized by emphasizing the integration of interior spaces and courtyard spaces. All rooms enjoy sufficient daylight and natural ventilation. The courtyard is square and enclosed by buildings and walls to provide great privacy.

住宅产品分成五类：大独栋别墅、合院别墅、联排别墅、叠墅及洋房。北侧多层住宅组团住宅均为一梯两户单元式户型，分为平层户型和跃层户型，户型内部空间设计遵循动静分区、洁污分区、功能空间细化的原则，房间形状方整好用。低层合院住宅分为南北户型，其基本构成为U型，中心为私密室外空间，南北侧为南北户型的外部空间，内外部空间可以连通。此种户型既节省土地又具备基本的低密度住宅品质。

户型设计以居住舒适、布局合理为原则。所有户型均为大面宽小进深设计，采光充足，南北通透，交通流线清晰，分区合理且房间方整实用。别墅产品强调室内房间和室外庭院的空间交融。所有房间采光及通风充分。庭院方整连续，利用建筑围合和院墙体系保证每个别墅户型的私密性。

Small Town of European Style
欧洲风情小镇

Sunny Melody, Yunfu 云浮远大美域

Location/ 项目地点：Yunfu, Guangdong/ 广东省云浮市
Developer/ 开发商：Yunfu Guangyuan Properties Co., Ltd. / 云浮远广置业有限公司
Architectural Design/ 建筑设计：Chen Shi Min Architectural Design Consultant Ltd. / 深圳市陈世民建筑设计事务所有限公司
Land Area/ 占地面积：333 900.26 m²
Floor Area/ 建筑面积：662 292.65 m²
Plot Ratio/ 容积率：1.5
Green Coverage Rate/ 绿化率：35%

Keywords 关键词

Artistic Beauty; Return of Human Nature; Green and Ecological
艺术美感 人性回归 绿色生态

"China's Most Beautiful Villas" Finalist
入围"中国最美别墅"奖

The project is positioned as a small town of European style. The building facade is dominated by Italian style with classical architectural elements of the Renaissance, the Gothic, the Roman times, the Tuscany, the Baroque, etc. Colors, materials and details are skillfully integrated to create an exotic style European town.

项目定位为欧洲风情小镇，立面采用以意大利为主的欧式风格，摒弃国内传统对意大利风格的狭隘认知，糅合了意大利文艺复兴、哥特、罗马、托斯卡纳、巴洛克等不同时期的建筑元素，在色彩、材料、细节等巧妙搭配，形成了有强烈异域风情的欧洲小镇。

Overview 项目概况

The development is located in Xijiang New Town of Yunfu City, Guangdong Province, sitting among mountain and water and overlooking the whole Xijiang River. The surrounding environment is highly protected, and the site leans to the mountain and extends along the mountain ridge and the riverbank. The development consists of a high-quality Italian-style building that accommodates a five-star hotel, a stylish villa area, a commercial street, high-rise residences and commercial buildings. The biggest challenge for the project is how to fully respect the existing eco environment and create a development rich in culture connotation, human care and artistic beauty by innovative design at the same time.

项目地处广东省云浮市西江新城，坐落在层峦起伏的山水间，可以俯瞰整个西江。周边自然环境保护度很高，且依山伴水沿着山脊和湖边水岸自然的延伸展开。该项目由一栋背山面水的高品质意式五星级酒店、风情别墅区及风情商业街和高层住宅及商业组合而成，最大的挑战在于如何在充分尊重保护现有生态环境的大前提下，通过设计上的创新来营造一个富有文化内涵、突出人性回归、且具有新型艺术美感的建筑产品。

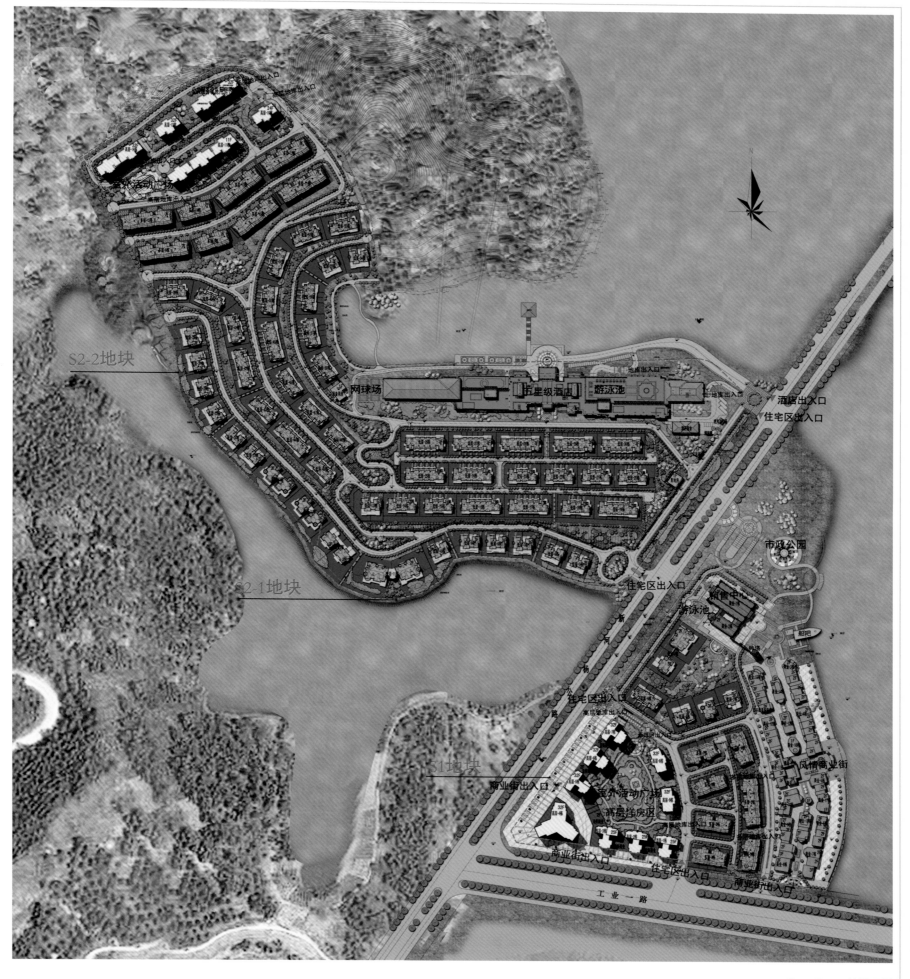

Site Plan 总平面图

Design Idea 规划设计和理念

The team has abided strictly by the planning and design idea of "environment, space, culture and effectiveness" by the master Chen Shi Min, and started the all-round thinking and study on the project. It proposed to take advantage of Yunfu's natural landscape resources and create a stylish villa area by borrowing surrounding views, which will be people-oriented, green and ecological, characterized by its exotic style and culture.

设计团队认真遵循着陈世民设计大师提出的"环境、空间、文化、效益"规划设计理念，对项目展开了全方位的思考与研究。通过分析提出了"发挥云浮自然山水资源优势，通过"借景入区"的方式打造"以人为本，以绿色生态"为核心，具有异域风情和文化为特色的时尚风情别墅区"的设计目标。

Architectural Style 建筑风格

The project is positioned as a small town of European style. The building facade is dominated by Italian style with classical architectural elements of the Renaissance, the Gothic, the Roman times, the Tuscany, the Baroque, etc. Colors, materials and details are skillfully integrated to create an exotic style European town. It is the first project of this style in domestic market, distinguishing itself by the innovative designs.

项目定位为欧洲风情小镇，立面采用以意大利为主的欧式风格，摒弃国内传统对意大利风格的狭隘认知，糅合了意大利文艺复兴、哥特、罗马、托斯卡纳、巴洛克等不同时期的建筑元素，在色彩、材料、细节等巧妙搭配，形成了有强烈异域风情的欧洲小镇。整体风格特点为国内首创，推陈出新，自成一体，与众不同。

Planning 规划布局

In terms of overall planning, the architects have focused on the lifestyle and demands of modern people, arranged the building functions and living spaces according to the terrain and landscapes, and created an urban living quarter by the water. Taking advantage of the elevation difference, the high-end villas are built along the mountain ridge, providing more private and quiet spaces as well as more open views.

在总体规划布局上,设计师根据现代人们更为关注的生活方式及需求,结合地形特点和景观资源优势划分建筑区功能,组织社区居民生活,并打造滨水城市生活带。设计师利用山地自然高差将高端别墅区,错落有序地沿山脊而建,这提供不仅更为私密和安静的居住空间生态,同时使视野更加开阔。

Aerial View 鸟瞰图

Facade Design 建筑立面设计

When designing the facade, the architects have drawn inspiration from the architecture in Siena, abstracted and simplified the architectural elements, kept the cultural connotation and luxurious decorations, and created colorful villa facades by using red and beige bricks. Warm colored roof tiles are used to create a warm and harmonious atmosphere. The warm red and yellow bricks as well as the bright and dark red tiles on the slope roofs make the whole residential area, the surrounding green mountain and lake bathe in the sunshine of Yunfu. Again, it enables people to experience the romance of classical European paintings.

在云浮远大美域建筑立面设计上，设计师汲取锡耶纳建筑的精华，并加以提炼和简化，保留了其厚重的历史文化感和华丽的装饰构件，在材料和颜色上引用深浅不一的红色和米黄色面砖，从而使别墅显得丰富多彩。为了让建筑统一在一片温暖和谐的气氛中，屋面则选用不同深浅的暖色调屋面瓦。暖色调的红砖、黄砖以及暗红色的亮红色的坡屋面使整个居住区和周围绿色的山峦、碧蓝的湖面共同沐浴在云浮的阳光中，使人们又一次呈现出欧洲古典油画一样的诗意当中。

Architectural Design in Demonstration Area 示范区建筑设计

Villa design for the demonstration area has paid attention to the location of the buildings and the relationship between residences, landscapes and natural environment, trying to maximize the villas' quality and the landscape views. The architects have made use of the low-rise terraces and introduced green spaces into every unit and every courtyard. And the open floors, entry platforms and roof gardens of the high-rise residences are created to provide more public green spaces.

在示范区别墅户型产品设计时，设计师综合考虑建筑所在位置，住宅与园林和自然景观之间产生互动关系，力求户型产品的品质和景观最大化。设计师通过低层住宅设置退台，将绿化空间引入每户每院，利用住宅高层的架空层、入户平台、天台花园等营造公共及私人绿化空间及公共空间。

Meiju Encyclopedia 美居小百科

Buildings in Siena are dense and highly integrated. Reddish bricks keep harmonious with the dark blueish gray hills around. Beside Roman style and Renaissance style, Gothic style is another important architectural style in the city which was influenced by the Orient during Crusades and presented with the form of narrow arcade. A great number of Gothic-style fountains which were mostly built in the 13th century have highlighted the magnificence of Siena. Siena Cathedral is one of the most famous examples of this style.

锡耶纳建筑，密集并且具有高度的建筑统一性。淡红色调子的砖块和与周围暗蓝灰色的丘陵相协调。除了罗马风格和文艺复兴风格的因素以外，意大利哥特式风格占重要地位，它受到了十字军东征时期东方的影响并且以狭窄的拱廊的形式表现出来。大部分建于13世纪的数量众多的哥特式喷泉增添了锡耶纳的华丽与协调。锡耶纳大教堂就是其代表之一。

Riverside Luxury Residence of Mediterranean Style
地中海风格的江景豪宅

Riverside Houses | 江畔豪庭

Location/ 项目地点：Dongguan, Guangdong / 广东省东莞市
Client/ 业主：Dongguan Gaowei Real Estate Development Co., Ltd. / 东莞市高威房地产开发有限公司
Architectural Design/ 建筑设计：AIM Group International / 加拿大 AIM 国际设计集团
Total Floor Area/ 总建筑面积：80 000 m²

Keywords 关键词

Bold Colors; Well Proportioned; Private and Transparent
色彩大胆 错落有致 私密通透

"China's Most Beautiful Villas" Finalist
入围"中国最美别墅"奖

The residential facade is designed in Mediterranean style with red tiles and slope roofs to dialogue with the green mountain and forest. The application of bold colors, concave-convex forms, staggered structures as well as the changes of building heights, has presented a beautiful image.

住宅的立面设计为地中海风格，红色瓦坡屋顶的建筑与葱郁山林相映成趣。色彩的大胆应用和形体上的凹凸变化、错合、弧型拼接及建筑体量的高低错落，形成丰富的建筑景观。

Overview 项目概况

Riverside Houses is situated at Zhongtang, in the northwest of Dongguan. The Water Towns District is the future subcenter of Dongguan, while Zhongtang is an important platform for the corporation between Guangzhou and Dongguan. Thus, the development is envisioned to be the future new town in north Dongguan and the first villa group in the high-end living quarter, enjoying a good environment. The south of project occupies the advantaged river views and it is the only pure villa project in Zhongtang.

江畔豪庭位于东莞西北部的中堂，水乡片区是东莞城市发展的次中心，中堂是穗莞合作的重要平台。项目是东莞北部未来的现代滨江新城，同时也是高尚生活区的首期别墅群，居住环境优越。项目南面既是中堂唯一拥有得天独厚的江景景观资源的项目，也是中堂唯一的纯别墅项目。

Aerial View 鸟瞰图

Elevation 立面图

Section 1-1 1-1 剖面图

Location Analysis 区位分析

Located in the center of Zhongtang, Riverside Houses are highly accessible and well served by main connective routes to the surroundings. Following Binjiang Avenue to the west, it is the adjacent Mayong Town; to the east it is the IT Electronic Industrial Park of Zhongtang. It takes only 10 minutes to drive northward to the center of Guangzhou Xintang through 107 National Road, 40 minutes to downtown Guangzhou and 20 munites to the downtown of Dongguan. It is near to Gaungzhou-Shenzhen Expressway within 3 minutes' walking distance, and the Zhongtang Light Rail Station is nearby within 5 minutes. It enjoys convenient traffic network which can reach Dongguan and Guangzhou within 30 minutes.

项目位于中堂中心的位置，交通便利，拥有二纵三横无缝对接的交通网络，从滨江大道西行可以前往邻镇麻涌镇，东行进入中堂IT电子产业区，经过107国道北行到广州新塘中心区只需10分钟，到广州市区40分钟，南行直走到东莞市区只要20分钟，向南右走三分钟接驳广深快速；5分钟前往轻轨中堂站，30分钟即可享受莞穗生活圈。

Basement Floor Plan
负一层平面布置图

Ground Floor Plan
首层平面图

Architectural Design 建筑设计

The design has fully considered the local climate and regional characteristics to well combine the art of architecture with the traditional culture and the local culture. The planning is clear with two residential groups respectively on the left and right. With the theme of "nature", it has created a modern-style landscape courtyard in the dynamic space. The residential facade is designed in Mediterranean style with red tiles and slope roofs to dialogue with the green mountain and forest. The application of bold colors, concave-convex forms, staggered structures as well as the changes of building heights, has presented a beautiful image.

项目充分考虑当地气候条件及体现地方特色，力求建筑艺术及传统文化，地域文化的结合。本方案规划结构清晰明确，由左右两个组团构成，设计以"自然"为线索在流动的空间创造一个现代派的风景庭院。住宅的立面设计为地中海风格，红色瓦坡屋顶的建筑与葱郁山林相映成趣。色彩的大胆应用和形体上的凹凸变化、错合、弧型拼接及建筑体量的高低错落，形成丰富的建筑景观。

Unit Types 户型设计

The development offers 135 units, of which 47 are single-family villas and 88 are semi-detached villas, ranging from 380 m² to 680 m² in size. It is the first villa community that separates pedestrians from vehicles in Zhongtang. Those Spanish-style residences boast 1.2 km long rive views, double-storey garden-style design as well as 800 m² complimentary garden area.

The buildings are arranged in line and staggered orderly according to the natural topography, allowing all the units to have great river views and great transparency. To ensure the privacy and comfortableness of the residents, the part of the site along Binjiang Road is elevated about 2 meters high for a panoramic view of the river and a natural, healthy and comfortable living environment for all units.

江畔豪庭由47套独栋和88套双拼组成，共135套，其面积为380～680 m²。江畔豪庭是中堂首个人车分流的纯别墅小区，1.2 km最长河岸江景大宅、纯西班牙血统建筑风格、双首层花园式设计、赠送800 m²超大花园。

项目规划上充分利用天然地块并采用线形布局，户型布局错落有致，每户互不遮挡，具有极好的江景景观视野和通透性。为了最大限度地保证别墅客户追求的居住舒适度和私密度，设计师将项目地势较滨江路抬高近2 m，从而实现所有别墅产品都拥有一线开阔江景视野的同时享受自然、健康、舒适的生活环境。

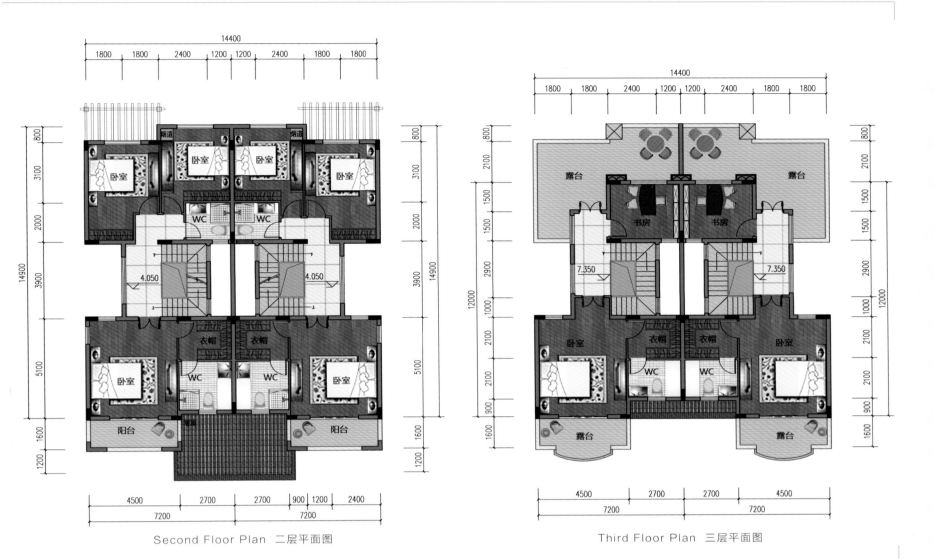

Second Floor Plan 二层平面图

Third Floor Plan 三层平面图

Boutique Resort with Mountain and Lake Views
山地湖景度假精品

Phoenix Valley, Changtai 长泰凤凰谷

Location/ 项目地点：Zhangzhou, Fujian / 福建省漳州市
Developer/ 业主：Changtai Jinhongbang Real Estate Development Co., Ltd./ 长泰金鸿邦房地产开发有限公司
Architectural Design/ 建筑设计：Shanghai ZF Architectural Design Co., Ltd./ 上海中房建筑设计有限公司
Floor Area/ 建筑面积：55 892 m²

Keywords 关键词
Natural Materials; Eco Environment; People-oriented Principle
天然材质　生态环境　人本原则

"China's Most Beautiful Villas" Finalist
入围"中国最美别墅"奖

These buildings, designed in natural and leisurely Southeast Asian style, feature hip roofs and colonnades as well as the details such as plaster, stone coping, wooden balcony and wood decorative cornice, to highlight the quality and uniqueness of the buildings.

建筑风格采取自然、休闲的东南亚设计风格，采用四坡屋顶、柱廊的形式，细部上通过局部的壁柱、石材压顶、木制阳台、木装饰檐板等做法，增加了建筑的品质感与独特性。

Positioning 规划定位

Designed with the idea of "ecology, culture, health, leisure and environment protection", the project is envisioned to be a first-class mountain and lake resort to attract both the domestic and international tourists. Located in the initiating zone of Tiantongshan Resort, plot C-01 is created as a high-quality and high-standard resort with tropical landscape, which will drive the development of the whole Tiantongshan Resort.

以"生态、文化、健康、休闲、环保"为理念打造福建区域一流的山地湖景度假精品，成为吸引国内高端游客和国外游客的度假旅游胜地。C-01地块是天铜山度假区的启动区项目，将其打造成为高品质、高水准的热带景观山水度假住宅，使其成为天铜山度假区长期良好发展的带动力量。

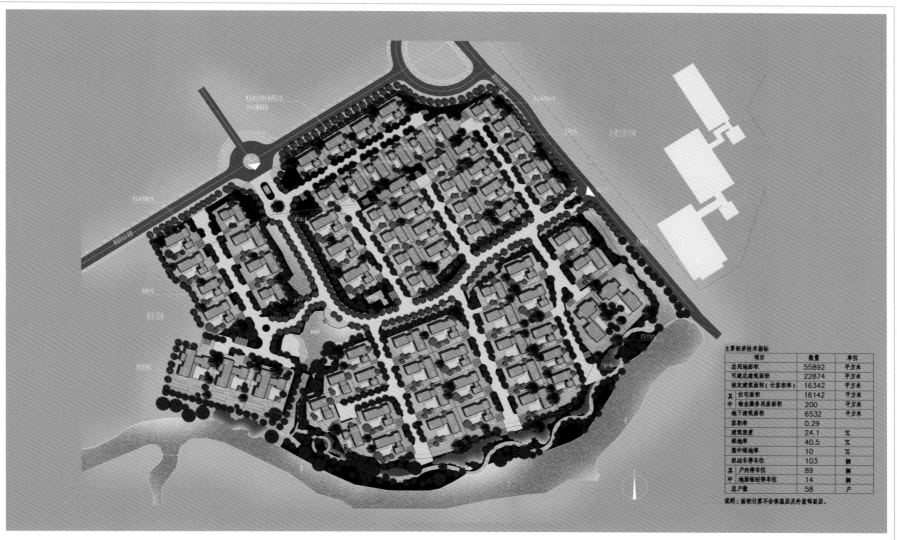

Site Plan 总平面图

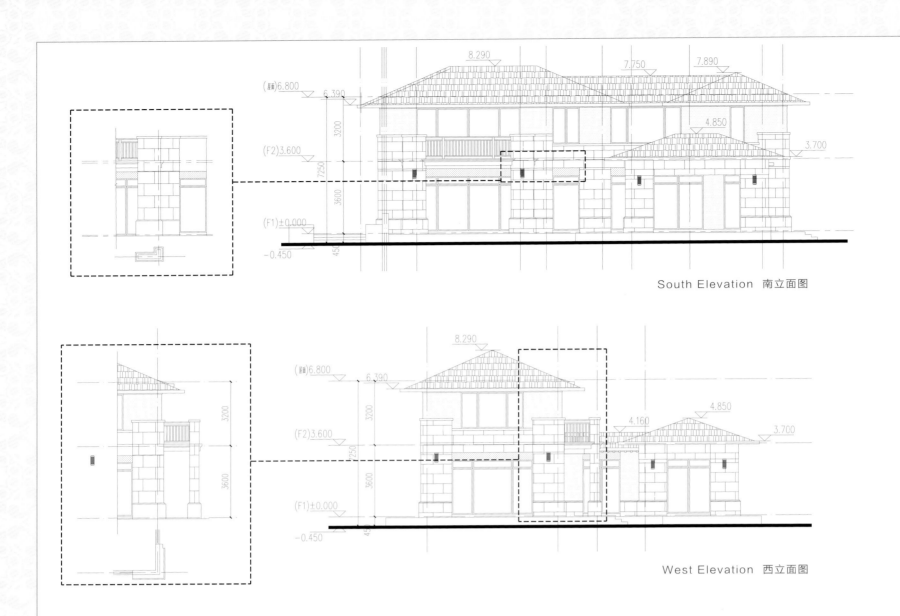

South Elevation 南立面图

West Elevation 西立面图

Planning Principles 规划原则

Ecological: to respect nature and the site and to protect the ecology; to implement the development based on effective protection and cultivating resources; to keep the integrity and original nature of the landscape resources; to maintain biodiversity and to ensure sustainable development.

Characteristic: to characterize the resort by the spirit of the times and the international perspective; to take advantage of the topography and highlight the theme of "tropic, mountain and leisure".

Integration: to abide by the overall plan of this region, to make full use of the local infrastructures, and to complement the other projects.

People Oriented: based on the principle of "people orientation and nature first" to create a comfortable and beautiful environment for the mountain resort.

生态原则：尊重自然，尊重场地，保护生态。在有效保护、培育资源的基础上实施适度开发，尽量保持景观资源的完整性、原生性及生态环境的多样性，走可持续发展之路。

特色原则：以时代精神、国际视野打造超前性的高品位旅游度假产品，根据场地的地理优势，突出"热带、山地和休闲"的主题元素。

整体原则：遵循项目区域的发展定位，充分利用当地基础设施，并与其他项目形成优势互补、整体发展的关系。

人本原则：遵循以人为本，以自然为本的宗旨，创造舒适优美的山地旅游度假环境。

South Elevation 南立面图

East Elevation 东立面图

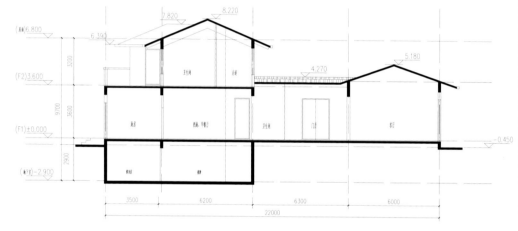

Section 1-1 1-1 剖面图

Architectural Design 造型设计

To adapt to the climate of southern Fujian, most of the buildings have two floors, and some have only one floor, which are all arranged in order and form the image of the building group. These buildings, designed in natural and leisurely Southeast Asian style, feature hip roofs and colonnades as well as the details such as plaster, stone coping, wooden balcony and wood decorative cornice, to highlight the quality and uniqueness of the buildings.

建筑造型中结合闽南气候特点，主要以二层住宅为主，局部布置一层建筑，错落有致，高低结合，成为群体的主要形象。建筑风格采取自然、休闲的东南亚设计风格，采用四坡屋顶、柱廊的形式，细部上通过局部的壁柱、石材压顶、木制阳台、木装饰檐板等做法，增加了建筑的品质感与独特性。

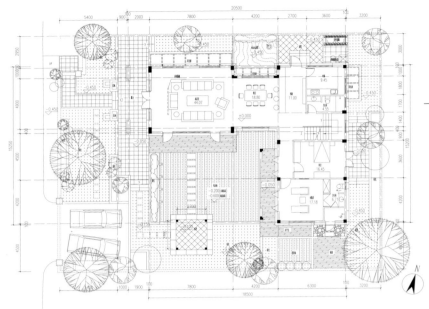

First Floor Plan 一层平面图

Floor Plan 平面设计

Taking advantage of the long site, courtyards of different sizes, on different levels are designed to organize the functional spaces. The main courtyard is the "water yard" on the ground floor that accommodates functions like the living room, dinning room and guestroom. The indoor and outdoor spaces are integrated: at the south end, a sunken area is left for outdoor barbecue to complement the indoor functions. The front yard combines with the parking space and the entryway to display the image, and two side yards are respectively for sightseeing and for enjoying hot springs. Close to the kitchen is the equipment yard which is hidden and relatively independent.

Part of the underground space is designed as the basement floor to accommodate the multi-functional hall, which complements the overground functions and allows most rooms to have natural light by installing the light well.

Flexible architectural arrangement and low-density layout are suitable for the subtropical area, which also enable people to close to nature and enjoy the beauty of life.

平面设计中充分利用基地进深较长，以不同大小，不同层次的庭院组合个功能空间。主庭院为"水院"，结合起居室、餐厅、客房等底层主要使用空间，室内外一体考虑，并在最南侧通过有意识的局部下沉形成室外烧烤区，作为室内功能的补充与延伸；前院结合停车及玄关设计，成为入口形象的主要展示；两边侧院一边为景观庭院，另一边为温泉泡池主题庭院；而以厨房紧邻的地方则作为设备庭院及后勤庭院，通过合理的遮挡，相对独立。

住宅局部设置地下室，设置多功能厅，作为地面主要居住功能的补充，并通过采光井使大部分房间能够自然采光。

丰富、灵动、错落有致的建筑布置以及较分散的建筑布局符合亚热带的气候特征，使回归自然、返璞归真的生活情趣得到充分体现。

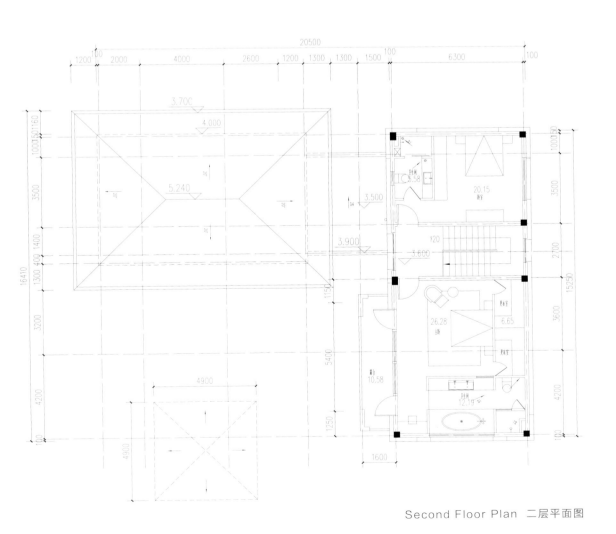

Second Floor Plan 二层平面图

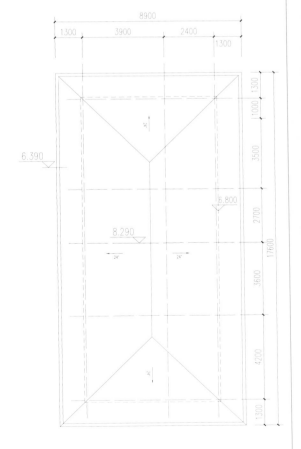

Plan for Roof Floor 屋顶平面图

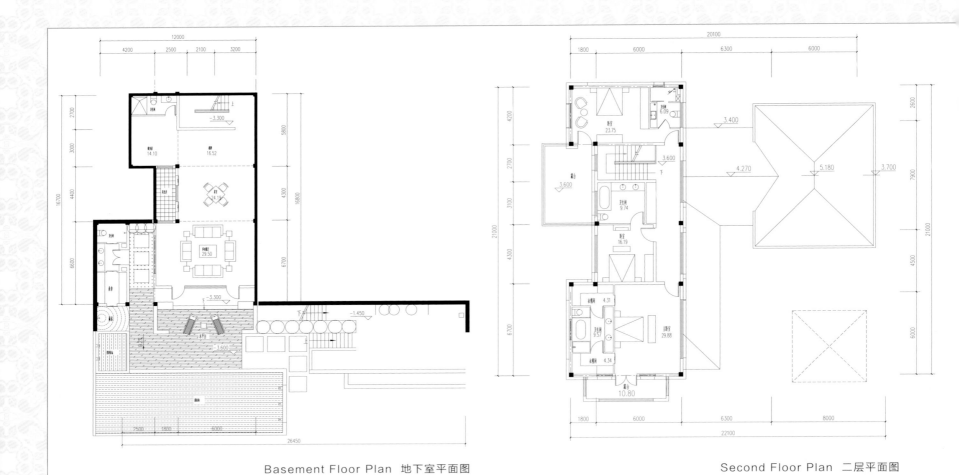

Basement Floor Plan 地下室平面图

Second Floor Plan 二层平面图

Choosing Materials 材料选择

Natural materials such as wood and terra-cotta tiles are chosen to reflect the characteristics of the mountain residences in the subtropical region— Changtai. And the utilization of stone, wood, glass, coating, plane tiles, etc. can make the buildings integrate into the surrounding eco environment.

材料选择上则主要采用木材、陶瓦等较天然的材质，体现出长泰的亚热带气候和度假区山地住宅的特点。材质采用石材、木材、玻璃、涂料、平板瓦等材料，符合基地周边自然原始的生态环境。

Meiju Encyclopedia 美居小百科

The triangle faces of the hip roof are usually covered by plain tile or pantile, while the ridges often use decorative tile or imbrex. For this kind of roof, spaces between tiles are useful for ventilation, the tiles are waterproof and heat insulated, and sometimes the roof is complemented with dormer or small duplex. Since villas have become increasingly scarce, some newly built villas and garden houses demand more on the design of roof and encaustic tiles: some follow the mountain ridge to present a multi-slope roof which echoes the beautiful terrain outline; some take advantage of the small hill and the slope landform to create an exotic-style slope roof; still others use soaring cupola to show the uniqueness.

四坡屋顶，屋面瓦用小平瓦、波形瓦，屋脊线采用装饰砖或筒瓦。此类屋面多通过瓦缝通气，保湿隔热，有的还装有老虎窗或小阁楼。由于别墅越来越稀少，一些新建别墅和花园洋房对屋顶和彩瓦提出了更高的设计要求：有的配合山脊走向，要求设计为多坡屋面，使建筑高低错落的形态与地形优美的轮廓线协调；有的根据浅丘、坡地设计为异国情调的异型坡屋顶，有的为展现其标志性和独特性，其坡屋顶如圆穹直插云霄。

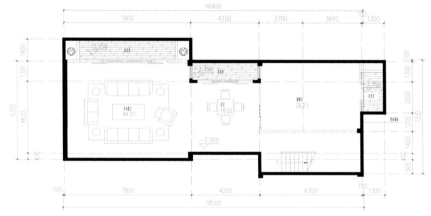

Basement Floor Plan 地下室平面图

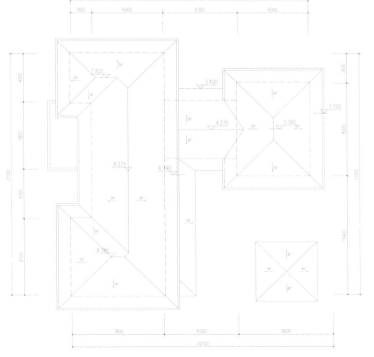

Plan for Roof Floor 屋顶平面图

Boutique Holiday Residence of Modern Southeast Asian Style
精益求精的现代东南式度假住宅

Vanke · Rancho Santa Fe, Phase II　万科·兰乔圣菲二期

Location/ 项目地点：Guangzhou, Guangdong/ 广东省广州市
Architectural Design/ 建筑设计：Guangzhou Jingsen Engineering Design and Consulting Co., Ltd./ 广州市景森工程设计顾问有限公司
Total Land Area/ 总用地面积：67 535 m²
Residential Land Area/ 居住建筑用地面积：13 500 m²
Roads and Square Area/ 道路广场用地面积：2 701 m²
Green Land Area/ 绿化用地面积：20 260 m²
Total Floor Area/ 总建筑面积：65 671 m²
Plot Ratio/ 容积率：0.95
Green Coverage Rate/ 绿化率：30%

Keywords　关键词

Elegant and Dignified; Good Daylighting; Close to Nature
庄重大气　采光良好　亲近自然

"China's Most Beautiful Villas" Finalist
入围"中国最美别墅"奖

Buildings of Phase II look elegant and dignified with simple colors. By using large-area glass windows and gray aluminum alloy frames, it creates great facade effect, provides open views and sufficient daylight, and strengthens the relationship between human beings and nature.

万科·兰乔圣菲二期形体厚实大方、色彩朴素淡雅，立面上使用的大面积灰色铝合金玻璃窗及构架，既丰富了立面造型，也使得建筑室内空间采光良好、视野开阔，加强了人与自然的联系。

Location　区位概况

Located within the grade-4A scenic spot—Huadu Lotus Peak Resort, leaning against the Prince Mountain Forest Park and facing the Lotus Peak reservoir, the project is surrounded by mountains and near a lake on three sides.

项目位于省级 4A 旅游度假区——花都芙蓉嶂度假区内，背靠王子山森林公园，面对芙蓉嶂水库，三面临湖，群山环绕。

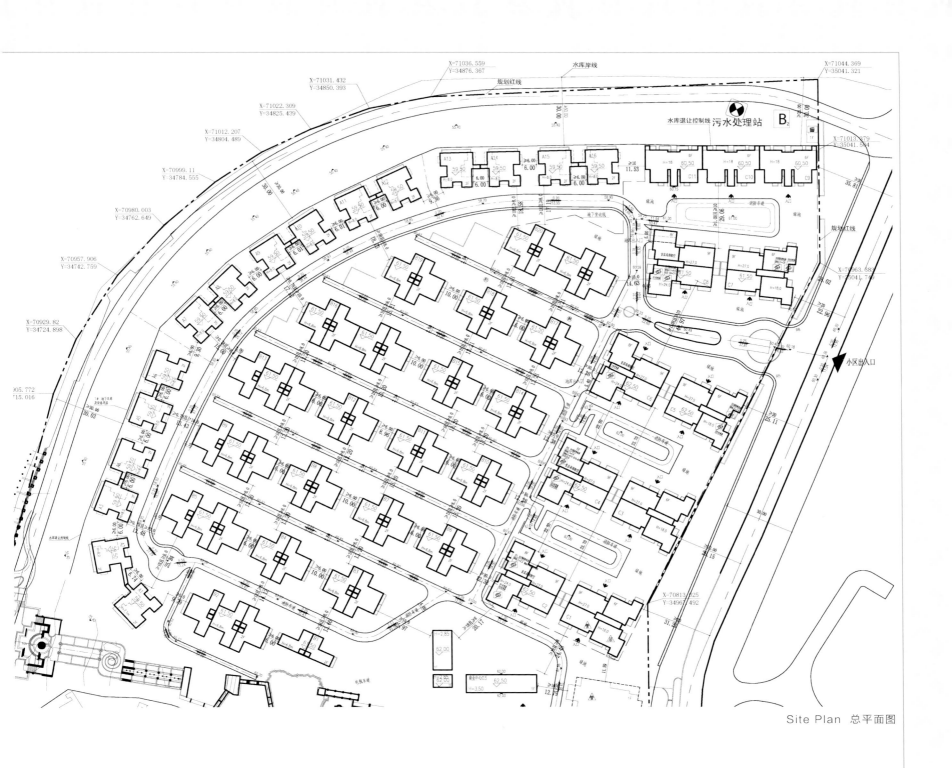

Site Plan 总平面图

Aerial View 鸟瞰图

Side Elevation 1 轴立面图 1

Side Elevation 2 轴立面图 2

Overview 项目概况

Rancho Santa Fe is a pure villa project which is rare in Guangzhou. These villas are mainly built on the slopes to make full use of the multi-spaces. The site is like a peninsula that opens to the lake on three sides. And with about 2 km long shorelines and the topography that is lower in the south, north and west and higher in the center and east, it will maximize the lake views and provide more flexible internal spaces within the community.

兰乔圣菲是广州地区罕有的纯别墅项目。项目以坡地建筑为主，既提高了空间的层次感，又可以充分的使空间最大化；地块呈半岛形状，三面望湖，有近 2 km 的岛岸线，最大限度利用沿湖景观，而且地块南、北、西三面靠湖部分地块较低，中间及东部地块较高，呈坡地形状，既能最大化的利用湖景资源又能增加社区内空间的变换。

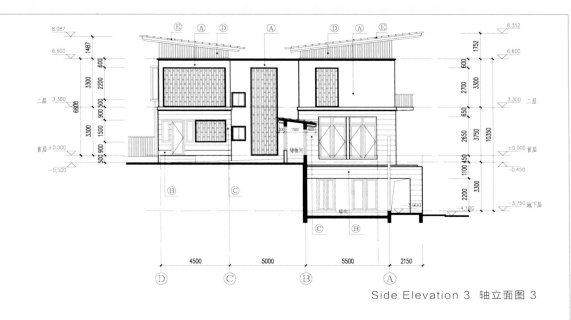

Side Elevation 3 轴立面图 3

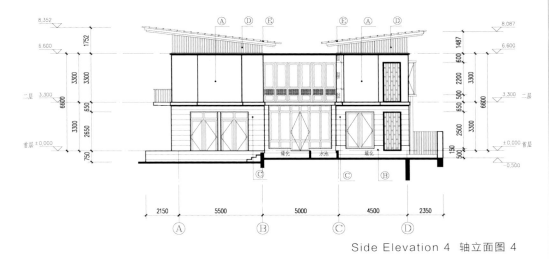

Side Elevation 4 轴立面图 4

Planning 规划布局

The development is planned with "one axis, three zones and multiple groups". "One axis" refers to the central landscape axis which runs through the site from east to west, connecting the public activity spaces like the clubhouse, the commercial street and the central square. Taking advantage of the elevation difference, the villas are arranged on "three terrace zones" of different heights to enjoy different lake views according to their distance with the lake. And for the "multiple groups", traffic system and water system have combined together to separate these villas into five peninsula-style groups which are relatively independent from each other.

兰乔圣菲采用一轴、三区、多团组规划。一轴：一是东西走向的景观主轴；串联了会所、商业街、中心广场，是住户的公共活动空间。三区：根据基地与湖水的关系划分为三线不同的景观住户，包括一线滨水呈岛状分组团布局的地台式别墅区。二线的望湖住宅，充分利用原有地形高差形成台地式建筑布局，保证每户均能不同程度的望到湖景。多团组：由交通体系和水系分割呈五个呈半岛式组团，各组团相互独立。

Architectural Design 建筑设计

Buildings of Phase II look elegant and dignified with simple colors. By using large-area glass windows and gray aluminum alloy frames, it creates great facade effect, provides open views and sufficient daylight, and strengthens the relationship between human beings and nature. It also pays attention to every detail to shape the architectural image of elegance, dignity and magnificence.

万科·兰乔圣菲二期形体厚实大方、色彩朴素淡雅，立面上使用的大面积灰色铝合金玻璃窗及构架，既丰富了立面造型，也使得建筑室内空间采光良好、视野开阔，加强了人与自然的联系。在建筑细节上精益求精，力在处理好每一个细节。建筑形象挺拔简练、庄重大方，气度不凡。

Section 1-1 1-1 剖面图

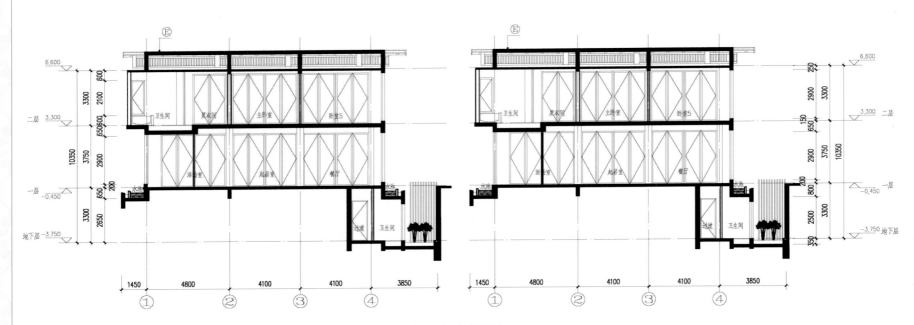

Section 2-2 2-2 剖面图

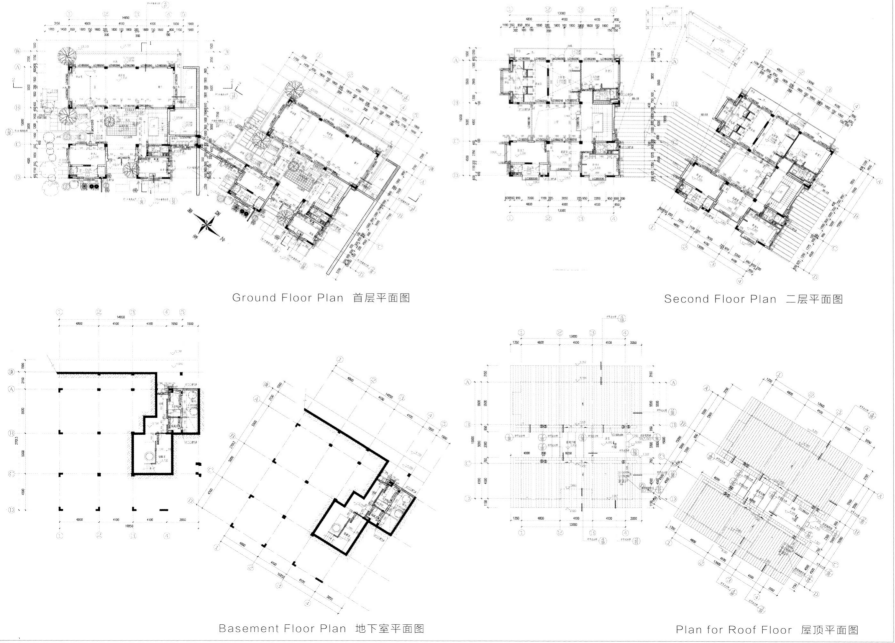

Ground Floor Plan 首层平面图

Second Floor Plan 二层平面图

Basement Floor Plan 地下室平面图

Plan for Roof Floor 屋顶平面图

Residential Design 住宅设计

Phase II of Rancho Santa Fe is designed in modern Southeast Asian style. Building A and B have two floors, while Building C has 9 floors whose ground floor is partly empty. The underground floor is used for equipment rooms and garages. Many architectural elements of modern Southeast Asia style are used, including large-area glass windows and doors, big flat cornices, aluminum alloy grilles, etc.

万科·兰乔圣菲二期是现代东南式度假住宅。其中A、B栋为两层住宅，C栋为9层住宅，底层局部架空，地下室设有设备房和地下车库。设计上使用了大量东南亚现代建筑元素，包括有落地大面积玻璃门窗、平板大挑檐、铝合金装饰格栅等。

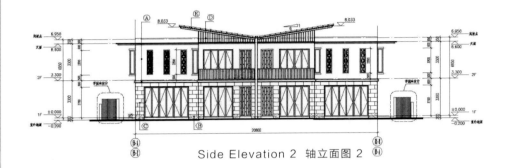

Side Elevation 1　轴立面图 1

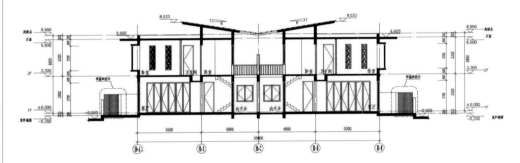

Side Elevation 2　轴立面图 2

Section 1-1　1-1 剖面图

Landscape Design 景观设计

Making full use of the site conditions and the surrounding landscape, it tries to well arrange all the buildings, allowing all units to have a great view of the Lotus Peak Reservoir and providing the lower-floor units with home gardens. And the elegant square layout enables all units to have bathrooms with SPA.

注重场地与景观方面的要素，通过分级平台使得各款住宅均能从芙蓉峰水库取得较好的景观视野，低层住宅家家户户自带花园。建筑方正简洁的布局及户户均带有 SPA 按摩池的卫生间。

China's Most Beautiful Stylish Apartments
中国最美风格楼盘

Aesthetics　　　　　美 学 性

Livability　　　　　宜 居 性

Commercial Value　　商业价值性

Humanity　　　　　人 文 性

Sustainability　　　可持续性

Building Naturally Grown Residential Groups with Unique Geographical Advantages
以得天独厚的地理优势塑造"自然生长"的居住组团

Vanke Anshan Whistler Town   鞍山万科惠斯勒小镇

Location / 项目地点：Anshan, Liaoning / 辽宁省鞍山市
Developer / 业主：Anshan Vanke Real Estate Co., Ltd. / 鞍山万科房地产有限公司
Architectural Design / 建筑设计：(CDG) Concord Design Group / 西迪国际（CDG 国际设计机构）
Building Scale / 建筑规模：606 531 m²
Overground Area / 地上面积：529 520 m²
Plot Ratio / 容积率：1.75
Green Coverage Ratio / 绿化率：26%
Building Coverage Rate / 建筑覆盖率：20.9%

Keywords   关键词

Concise & Clear; Harmonious & Serene; Various Shapes
简洁明朗 和谐安详 形体丰富

"China's Most Beautiful Stylish Apartments" Award
荣获"中国最美风格楼盘"奖

The architectural form of this project uses the model of Canadian Whistler Town, and the planning layout complies with the tendency of the mountains, which is interspersed and staggered with rich and clear levels. The space expression uses the change of the spatial levels to break the rigidity of the traditional architecture and make the rhythm, proportion and dimension comply with the geometric aesthetic principles.

项目建筑形式以加拿大惠斯勒小镇为模板，规划布局依就山势，穿插错落，层次丰富分明。空间表情通过空间层次的转变，打破传统建筑的呆板，其节奏、比例、尺度符合几何美学原则。

Overview   项目概况

The project is at the piedmont of Anshan East Mountain, where the topography is undulating mountain. The planning makes the Canadian Whistler Town as blueprint, uses the unique geographical advantages, refers to the planning design concept and practical experience of the "Naturally Grown" town and uses a main ring road which complies with the tendency of the mountain and owns rich and beautiful landscapes to connect the residential groups.

本案位于鞍山东山山麓，地形为起伏较大的山地。规划依照加拿大惠斯勒小镇为蓝本，借助得天独厚的地理条件，引用了国外"自然生长"小镇的规划设计理念和实践经验，以一条依坡就势、景观丰富优美的主干环路串联各居住组团。

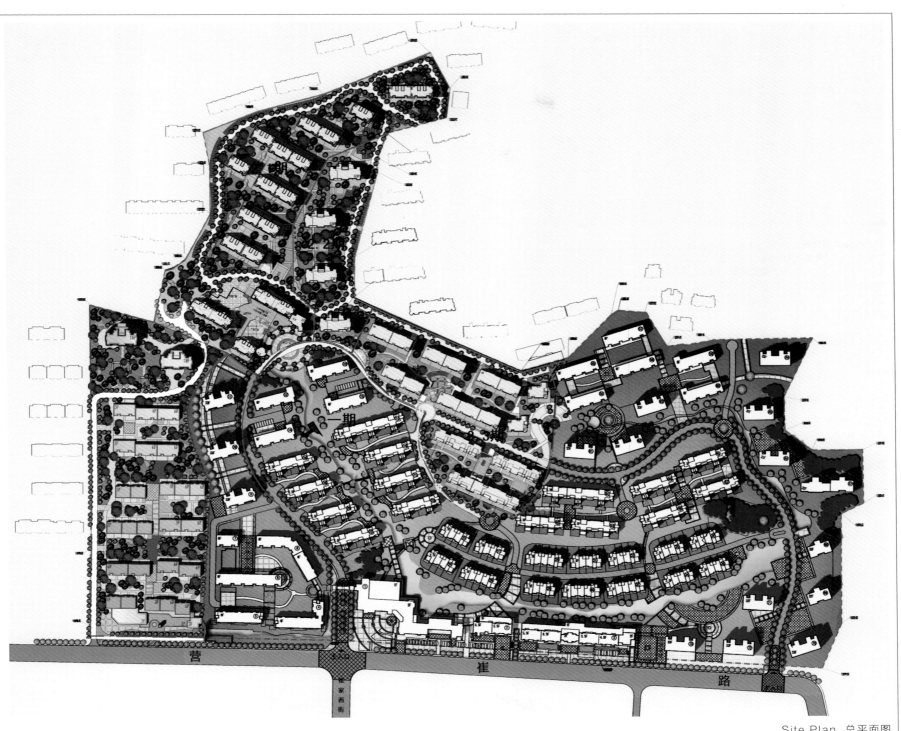

Site Plan 总平面图

Aerial View 鸟瞰图

West Elevation of 1# Building　1#楼西立面图

East Elevation of 1# Building　1#楼东立面图

North Elevation of 1# Building　1#楼北立面图

South Elevation of 1# Building　1#楼南立面图

Planning　规划布局

The planning design has fully considered the high in the north and low in the south sloping terrain. On the premise of meeting the basic living requirements such as sunshine, lighting, interval, and orientation, the designers use the approaches that emphasize order and texture, make the groups as a unit and separate the driveways and sidewalks, forming a multilayer living landscape space and providing the residents with a reliable and comfortable physical environment. Among the residential groups, the designers utilize the topographic altitude differences and the empty spaces at the bottom of the high-rise residences and design public facilities such as the small-scale neighborhood center, library, and senior center to improve the cultural connotation and sentiment index of the community.

规划设计充分考虑了北高南低的坡型地势条件，在满足日照、采光、间距、朝向等等居住基本需求前提下，采用强调序列与肌理的手法，以组团为单位，人车分流，构筑了多层次的生活景观空间，为居住者提供了一个可靠舒适的物质环境。各居住组团中，利用地势高差及高层住宅的底层架空层，设置小型邻里中心、图书馆、老年中心等公共设施，提升社区文化内涵及情感指数。

Building Arrangement 建筑布置

The designers fully consider the tendency of the current mountain, and on the building arrangement, high-rises are arranged at the highest ridge and townhouses are arranged at the central landscape belt along the valley, which highlights the change of city skyline, makes the high place higher and low place lower and intensifies the characteristics of the mountain.

The commercial street with rich humanistic sentiment, courtyard groups with continuous visions and the interpenetrative buildings and landscapes all manifest the clear streamlines of the community places in terms of space organization, and the characteristics of open and unobstructed views and landscapes with rich humanism and life sentiment.

充分考虑现状山地的地势，在建筑布置上，采用将高层布置于最高的山脊，TOWNHOUSE 布置于山谷沿中心景观带，突出了城市天界线的变化，使高处更高，低处更低，强化了山地的特点。

富有人文情调的商业街、视觉连续的组团院落，以及相互渗透的建筑与景观，无不体现着社区场所在空间组织层面上流线分明、视线开阔通畅、景致富有人性化及生活情调的特征。

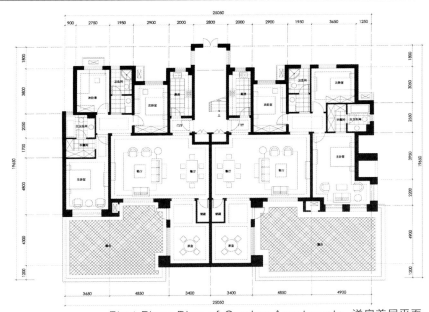
First Floor Plan of Garden Apartments 洋房首层平面

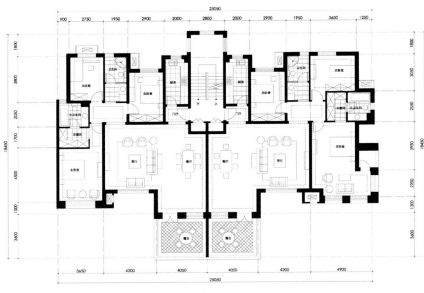
Second Floor Plan of Garden Apartments 洋房二层平面

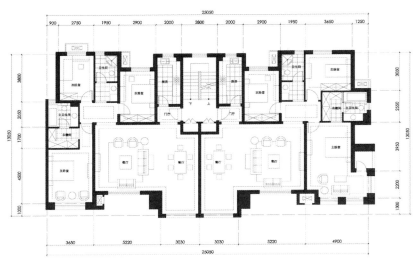
Third Floor Plan of Garden Apartments 洋房三层平面

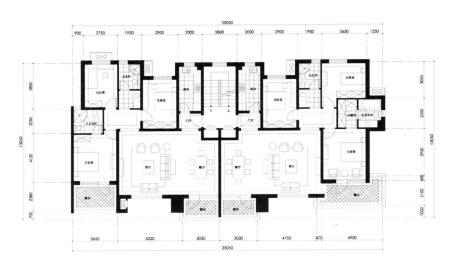
Fourth Floor Plan of Garden Apartments 洋房四层平面

Design Concept 设计理念

The architectural form of this project uses the model of Canadian Whistler Town, and the planning layout complies with the tendency of the mountains, which is interspersed and staggered with rich and clear levels. The space expression uses the change of the spatial levels to break the rigidity of the traditional architecture and make the rhythm, proportion and dimension comply with the geometric aesthetic principles.

项目建筑形式以加拿大惠斯勒小镇为模板，规划布局依就山势，穿插错落，层次丰富分明。空间表情通过空间层次的转变，打破传统建筑的呆板，其节奏、比例、尺度符合几何美学原则。

House Type Features 户型特点

1. The house type plan is reasonable and with clear zonings to ensure the living room, main bedroom and dining room with good orientation and landscape.
2. The southern space is designed for the main functional space to ensure the living quality under different climate conditions.
3. The width of the living room has been widened and connected to the dining room to make the public space feel more open, transparent and comfortable.
4. Each functional space is with comfortable dimension to try the best to meet the demand of lighting and ventilation.

  1. 户型平面布局合理、分区明确，保证客厅与主卧室、餐厅有良好的朝向与景观。
  2. 南向空间设为主要功能空间，保证不同气候条件下的生活起居质量。
  3. 客厅加大面宽设计，与餐厅连通，使公共空间感觉更加开阔、通透和舒适。
  4. 各个功能空间尺度舒适，尽可能满足采光通风要求。

Multilayer Residence 多层住宅

The multilayer residence uses the various approaches such as gable, colonnade, balcony and windows to jointly form the image of the changeable and naturally grown town. On the material components, nature imitated stones, archaized brick and wood are adopted, together with coating and siding to combine with the natural environment. The designers use exquisite detail to improve the quality of the residence and make the entire community become a high-end residential community that has various architectural forms in Anshan.

多层住宅利用多样化的山墙、柱廊、阳台、开窗等处理手法，共同形成具有丰富变化的自然生长的小镇形象；在材料部品上，采用仿自然的石材、文化砖、木材，配合涂料、挂板与自然环境相结合。利用精致的细节，提升产品的品质，使整个小区成为鞍山一个建筑形式丰富多样的高档住宅小区。

High-rise Residence 高层住宅

The high-rise uses the geometric combination on architectural form, and the forms are changeful and well-proportioned. The overall style is concise and clear. The bottom of the building uses a series of combination of multilayer architectural languages and the cooperation between the overall color and layers of the buildings to form the perfect dialogue and integration between the buildings and create a harmonious, humanized and ecological residential group.

高层在建筑形式上采用几何形体的组合，形体变化丰富，高低错落有致。整体风格简洁明朗。建筑底部利用一系列多层建筑语言的组合以及建筑整体色彩与多层的相呼应，形成两种性格建筑完美的对话与融合，创建了一组和谐安详的人文生态居住建筑组群。

Landscape Design 景观设计

The project uses the gully pool that's formed by the washing and confluence of the mountain rain as prototype and builds a waterscape theme to create a central landscape belt in the form of pool, stream and valley and to connect the activity square of each community. The project tries the best to retain the native trees and remain the original ecological topography characteristics of East Mountain.

项目利用原山体雨水冲汇形成的冲沟水潭为原形，打造一条水景主题，形成以潭、溪、谷为形式的中心景观带，串联各社区活动广场。尽最大可能保留其原生树木，保持了东山自然的原生态地形特色。

灰色金属屋面
深色木质构架
深木色窗套
深红色质感涂料
粉灰色木质挂板
浅灰色窗框
灰色文化石

South Elevation of Townhouses (entering from north)　联排北入南立面图

North Elevation of Townhouses (entering from north)　联排北入北立面图

New Landmark of Fuzhou's New Humanistic Mansion
福州新人文豪宅的新地标

Rongxin·Lan County　融信·澜郡

Location / 项目地点：Fuzhou, Fujian / 福建省福州市
Owner / 业主：Fujian Rongxin Real Estate Development Co., Ltd. / 福建省福州市融信房地产开发有限公司
Architectural Design / 建筑设计：Shenzhen Bowan Architecture Design Co., Ltd. / 深圳市博万建筑设计事务所
Land Area / 占地面积：69 618.4 m²
Total Capacity of the Construction Area / 计容建筑面积：194 931 m²
Plot Ratio / 容积率：2.8
Greening Rate / 绿地率：30%

Keywords　关键词

Modern Concept; Separation of Sidewalk and Driveway; Entire Bright Design
摩登理念　人车分流　全明设计

"China's Most Beautiful Stylish Apartments" Award
荣获"中国最美风格楼盘"奖

The design of this project is in modern minimalist style, and it brings the modern concept into the building. The wall uses glasses, which present the solemn atmosphere and also manifest the fashionable feelings incisively and vividly. The hundred-meter high-rise that stands straight from the ground with extraordinary magnificence embodies the constantly surpassing humanistic spirit and energy.

项目设计为现代简约风格，把摩登理念引入建筑，外墙采用玻璃幕墙，气质稳重的同时又把时尚感展示得淋漓尽致。拔地而起的百米高层，傲然屹立的非凡气势，表达出不断超越的人文精神和力量。

Overview　项目概况

With a gross land area of 69,618.4 m², Rongxin · Lan County is divided into three areas as A, B and C, among which the A and B areas are planned to be 12 blocks of plate-type high-rises with floors ranging from 31 to 39, and the C area is planned to be 3 blocks of plate-type high-rises with floors ranging from 14 to 31. The buildings surround the central landscape, which ensures the maximum distance between the blocks, and makes each family enjoy the beautiful scenery and maximum view out of the window.

融信·澜郡总占地69 618.4 m²，共分为A、B、C三区进行规划建设，A、B两区规划12栋31～39层的板式高层；C区规划3栋14～31层板式高层。楼体围合中央景观，既保证了楼间距的最大化，又能让每一户推窗就能见美景，视野最大化。

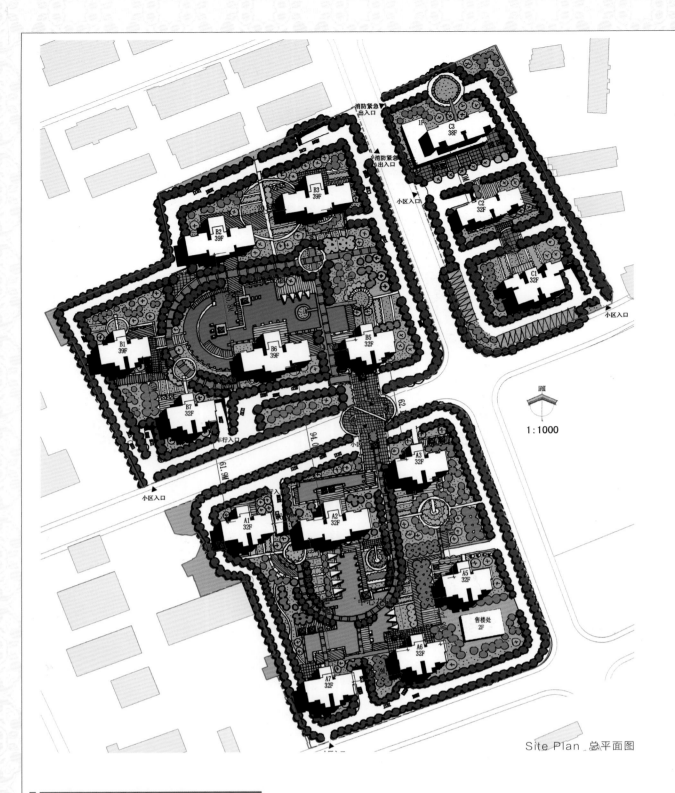

Site Plan 总平面图

Planning 规划布局

Entering from outside, the roads of this community adopts the design concept of completely separated sidewalk and driveway. The private cars enter into the underground garage directly at the entrance square instead of entering into the indoor garden of the community so that to reduce the adverse impact that the automobile exhausts and noises bring to the residents and to make the elderly and the children safely and leisurely enjoy the living environment of the community garden and the overhead pan club.

由外而入，小区道路采用完全人车分流的设计理念。私家车在入口广场直接进入地下车库，不进入小区内部花园，从而减少汽车尾气和噪音对业主的影响，使老人小孩在小区花园及架空层泛会所安全悠闲的享受居住环境。

Design Concept 设计理念

Different from the neoclassical style of the other residential projects in Fuzhou, this development is designed in modern minimalist style. It brings the modern concept into the building. The wall uses glasses, which present the solemn atmosphere and also manifest the fashionable feelings incisively and vividly. The image of these buildings is quite outstanding and distinctive. The hundred-meter high-rise that stands straight from the ground with extraordinary magnificence embodies the constantly surpassing humanistic spirit and energy.

区别于福州其它住宅项目的新古典主义风格，项目的设计风格为现代简约风格，把摩登理念引入建筑，外墙采用玻璃幕墙，气质稳重的同时又把时尚感展示得淋漓尽致；在楼盘形象上就做到出类拔萃，与众不同。拔地而起的百米高层，傲然屹立的非凡气势，表达出不断超越的人文精神和力量，也将借此打造福州新人文豪宅的新地标。

Aerial View 鸟瞰图

House Type Design 户型设计

Based on the living habits of Fuzhou citizens, the project is designed into humanized 70-160 m² units, which are diversified, ranging from 2 rooms to 4 rooms and duplex with empty spaces. This is to meet the demands of the residents with different preference and family needs. Based on the characteristics of good landscape view, light and ventilation of the high-rises, the design fully considers the wellness of the orientation and landscapes, making each unit face south and emphasizing the design principle of natural lighting and ventilation to achieve an entire bright design.

项目针对福州城市人居习惯，人性化地设计了从70～160 m²的户型面积，从二房到四房以及复式挑空的多元化设计，以适应不同心理偏好和家庭需求的业主；并根据高层景观视野好、光照及空气流通居家的特点，充分考虑朝向与景观的均好性，户户南向；强调自然采光与通风的设计原则，达到全明设计。

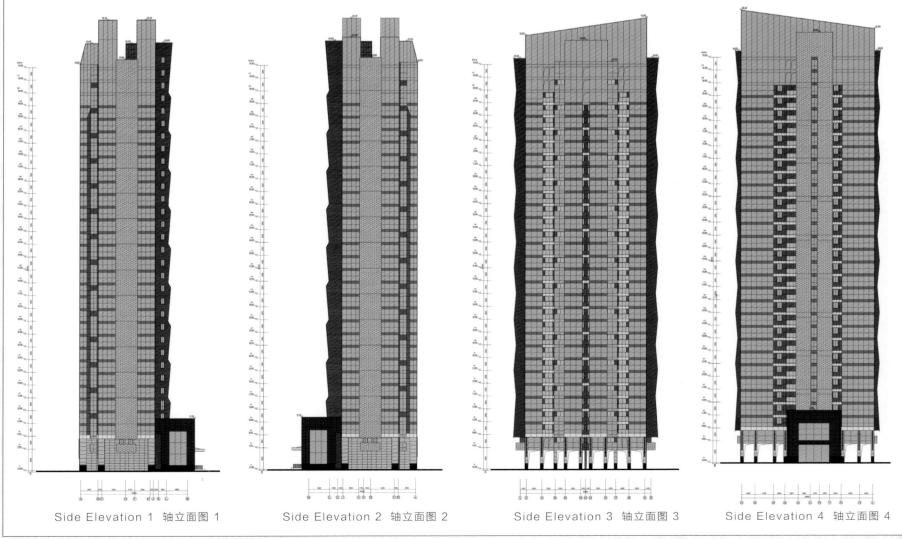

Side Elevation 1 轴立面图 1   Side Elevation 2 轴立面图 2   Side Elevation 3 轴立面图 3   Side Elevation 4 轴立面图 4

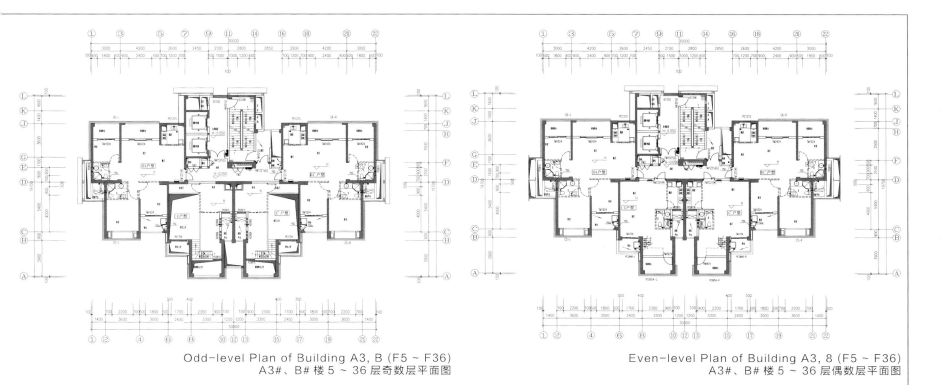

Odd-level Plan of Building A3, B (F5 ~ F36)
A3#、B# 楼 5 ~ 36 层奇数层平面图

Even-level Plan of Building A3, 8 (F5 ~ F36)
A3#、B# 楼 5 ~ 36 层偶数层平面图

Humanistic Pan Club 人文泛会所

The project has led in the first humanistic pan club of Fuzhou, which includes many high-end leisure places such as the fragrance library, humanistic cafe, yoga club, aerobic sports hall, children's talent pavilion and multi-functional tea house to meet the living demands of the residents such as leisure, fitness, entertainment and social intercourse.

项目引入了福州首个人文泛会所，包括气味图书馆、人文咖啡馆、瑜伽馆、有氧运动馆、儿童才艺馆、多功能茶艺馆等高端休闲场所，满足业主休闲健身、娱乐及社交等生活需求。

Central Park in the Hustling and Bustling City
繁华深处的中央公园

Vanke Le Bonheur | **万科·柏悦湾**

Location/ 项目地点：Zhongshan, Guangdong/ 广东省中山市
Developer/ 开发商：Vanke Real Estate Co., Ltd. /万科房地产开发公司
Architectural Design/ 建筑设计：Shenzhen Fangbiao Century Architectural Design Co., Ltd. /深圳市方标世纪建筑设计有限公司
Design Team/ 设计团队：Liu Yang, Sang Wen, Long Zhixiang, Zhang Xu / 刘洋 桑文 龙智翔 张旭
Land Area/ 占地面积：234 700 m²
Floor Area/ 建筑面积：563 000 m²
Plot Ratio/ 容积率：2.0
Greening Ratio/ 绿地率：40%

Keywords 关键词

Reasonable Layout; Elegant Proportion; Warm-tone Wall
动静分区 典雅比例 暖色调墙面

"China's Most Beautiful Stylish Apartments" Award
荣获"中国最美风格楼盘"奖

High-rise units of varied types are well organized with clear circulation and separate functions to meet different families' requirements. And the facade of the high-rises uses warm-tone coatings to create a simple and warm living atmosphere.

高层住宅户型设计功能合理，流线清晰，动静分区明确，种类多样，满足各种家庭结构需要。高层立面设计采用明快的暖色基调涂料墙面，给人以简洁和如阳光般温暖的生活气息。

Overview 项目概况

The project is located on the Torch Development Zone of Zhongshan, to the south of Bo'ai Sixth Road, to the north of the South Outer Ring Road and to the west of Beijing-Zhuhai Expressway, which is easily accessible from the whole Pearl River Delta. The site is covered by luxuriantly green and surrounded by waters, overlooking the mountain ridge in the southwest and the golf course across the river. Strategically located with convenient transportation, the development can take advantages of the built public buildings and facilities surrounding the site.

万科·柏悦弯项目位于中山市火炬开发区，北邻博爱六路，南邻中山南外环路，东面靠近京珠高速公路，辐射珠三角区域。基地植被葱郁，东西两面皆被水围绕，远眺西南方向山脊及对岸高尔夫球场。项目地理位置优越，交通便利，可充分利用周边已有的公共建筑及设施。

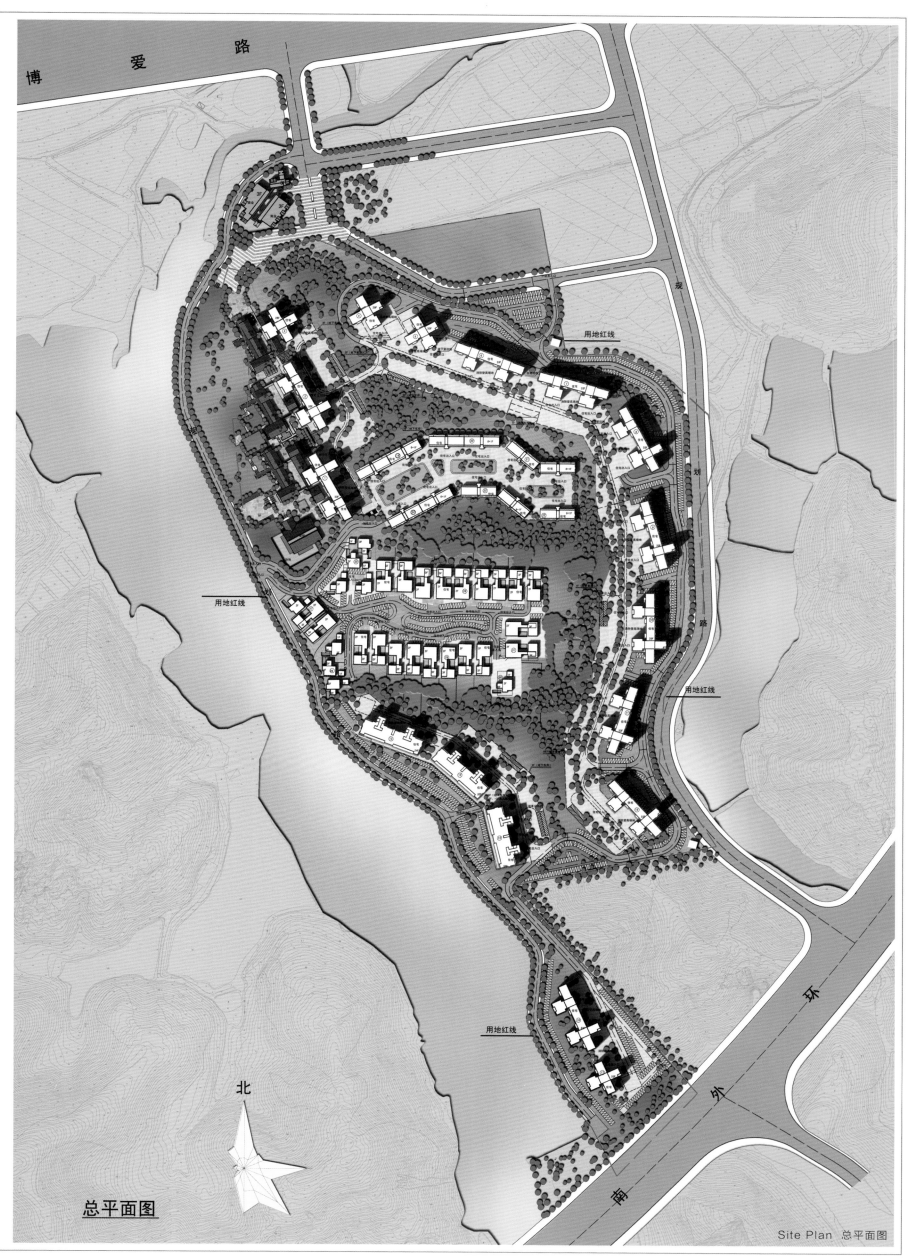

Planning 规划布局

With superior environment of water on the east and west, and mountains, it has created a layout of "water is in the site, while the site is in the mountains". By leaving the site unclosed, and building the residences according to the land – high-rise residences are built on the high level ground, and low-rise are built on the low level ground, residents are able to enjoy the view of landscape. The demonstration site is on the northwest, with low-rise commercial supporting facilities and high-rise residences, while the commercial facilities are organized in the form of "point and line" to keep perfect distance from the first-story of the residences. And the shops are connected with the club to form a linear building group over the lake, to make use of the lakeview and golf landscape maximally.

项目所在地拥有优越的生态环境，沿基地东西侧的水系及连绵的山体景观形成了"水中城，城中山"的规划布局。小区总体布局呈现围而不合的空间状态，保障小区内足够的景观视野，并采用"高区高建，低区低建"的方式实现建筑与环境的优化组合。示范区位于基地西北侧，产品以低层商业配套及高层住宅为主，商业流线通过"点—线"的方式组织，与住宅的首层空间分隔且有联系。商业与会所连成线状建筑群，沿湖边展开，将湖面及高尔夫景观最大化地利用。

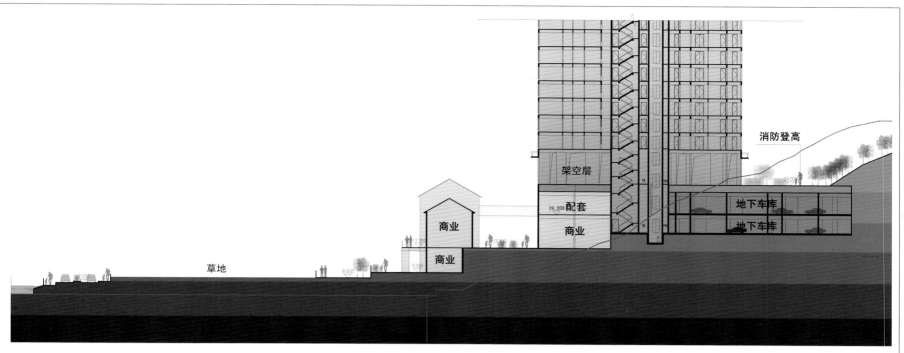

Section 剖面图

Architectural Design 建筑设计

High-rise residences feature classic proportion and match with the exotic commercial street expressing the modernity of this project. Three basic units of the shops are rebuilt. Each unit has different elevation for the outside and inside streets, while outside streets are 2- or 3-story, and 2-story for inside streets. Using brown tiles and dark grey steel roofs, they have shown the prosperous atmosphere of the commercial streets. High-rise residences feature classic proportion and activity zone with distinct streamline to meet the requirements of different families. The facades of the residential buildings are covered by bright and warm paint to create concise and warm atmosphere.

高层住宅具有典雅的比例与异域风情的商业街相搭配，不失时代感。商业部分由三组基本单元有序排列，并在相同的构架下重新塑造。每组单元分为不同标高的外街部分和内街部分，外街部分两层局部三层，内街部分为两层。建筑立面采用咖啡色砖贴面与深灰色的钢制坡顶相结合，充分展现出商业街的繁华与高贵。高层住宅户型设计功能合理，流线清晰，动静分区明确，种类多样，满足各种家庭结构需要。高层立面设计采用明快的暖色基调涂料墙面，给人以简洁和如阳光般温暖的生活气息。

Urban Space Focusing on Community Atmosphere and Neighborhood Relationship
强烈社区感与良好人际关系并重的城市空间

Baoshan Greenland Linghai Phase II & III "绿地领海"二三期

Location/ 项目地点：Baoshan District, Shanghai/ 上海市宝山区
Architectural Design/ 建筑设计：Shanghai HIC Planning Architecture and Design Co., Ltd./ 上海翰创规划建筑设计有限公司
Land Area/ 占地面积：105 696.6 m²
Floor Area/ 建筑面积：334 717.9 m²
Plot Ratio/ 容积率：2.51
Greening Ratio/ 绿地率：40.1%

Keywords 关键词
Duplex Space; Double Entry Doors; Double Balconies
复式空间 双入户门 双阳台设计

 "China's Most Beautiful Stylish Apartments" Finalist
入围"中国最美风格楼盘"奖

The development features creative unit design with "double entry doors and double balconies" which is unique in Shanghai, meeting owners' requirements to divide a duplex space into two parts. Each part will have a width reaching 7 meters and a balcony to provide enough daylight and independent functions.

创意的户型设计，上海市仅有的"双入户门 + 双阳台"设计，可以满足业主将一套房复式空间分割成两套复式空间的需求，且每套面宽均可达到 7 m 左右，并各自带一个阳台，丝毫不影响采光及平常使用功能。

Overview 项目概况

Situated in the east of the site, with Zhenchen Road on the east, residential land of phase I on the west, Chenyin Road on the south and Shitai Road on the north, phase II & III totally cover an area of 105,696.6 m², providing SOHO offices, LOFT offices, F&B and HQ business.

二、三期建设用地位于基地东侧，东至真陈路，西至一期住宅用地，南至陈银路，北至市台路，建设用地范围约 105 696.6 m²，功能构成以 SOHO 办公、LOFT 办公、配套餐饮商业以及总部商业为主。

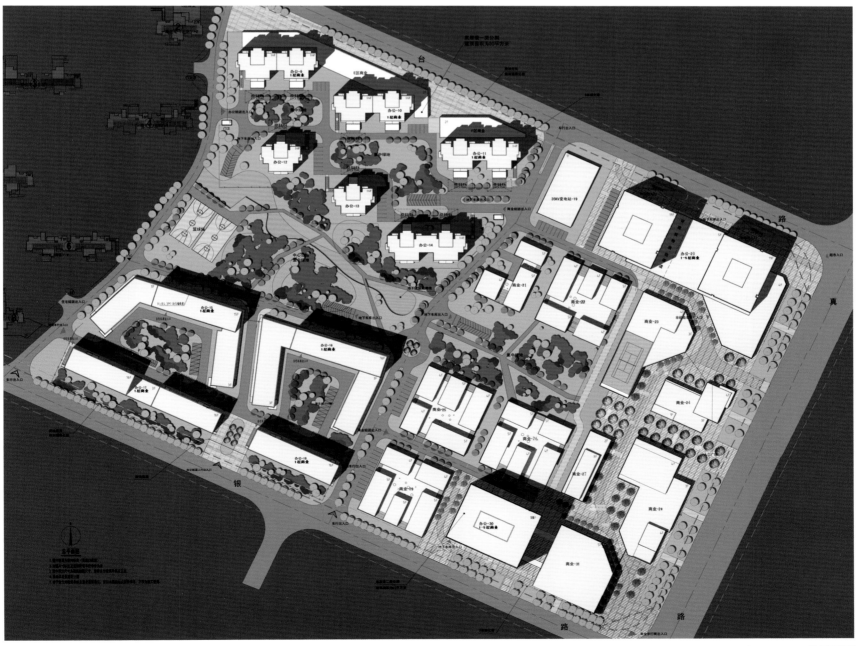

Site Plan 总平面图

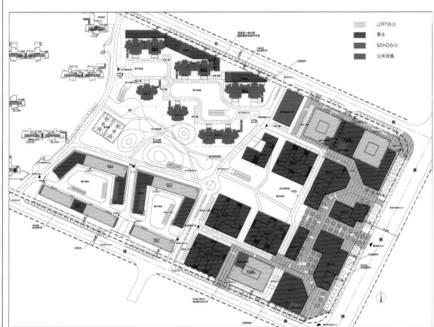
Functional Analysis Drawing 功能分析图

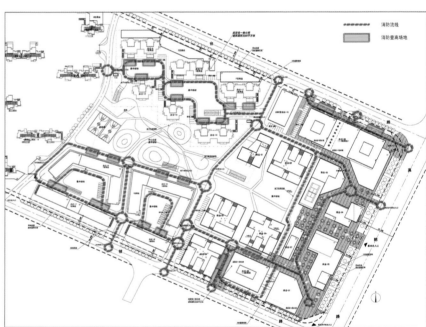
Fire Analysis Drawing 消防分析图

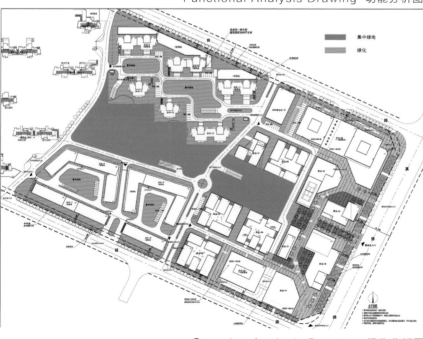
Greening Analysis Drawing 绿化分析图

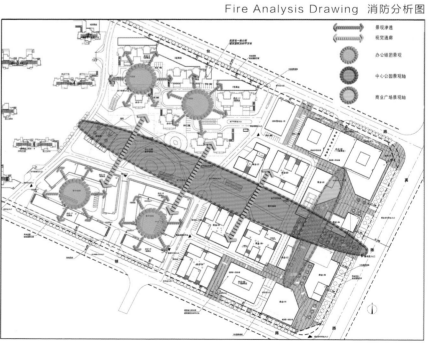
Greening Analysis Drawing 景观分析图

Planning 规划布局

Office and commercial facilities of phase II & III have followed the overall plan to be arranged in small groups, including LOFT offices, SOHO offices, F&B commerce and business supporting facilities. These groups are separated by the network of driveways and footpaths, and they are also connected with each other by a green central landscape belt that runs through from east to west.

二、三期办公商业部分在总平面布局中遵循总体规划的设计思路，依据大社区、小组团的规划原则，分为LOFT办公、SOHO办公、餐饮商业、商务配套商业等多个小组团，各组团间以格网状的车行与步行道路相分隔，同时一条中央绿色景观带至东向西贯穿其中，将各个组团联系在一起。

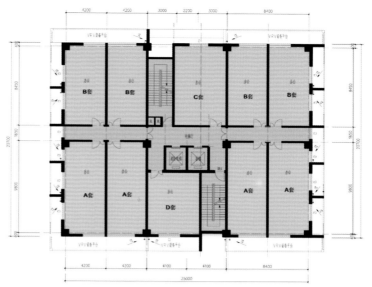

Standard Floor Plan 1 标准层平面图 1

Space Organization 空间布局

Based on the central landscape belt, it's arranged higher buildings in the south and north and lower ones in the center: the high-rise SOHO offices and LOFT offices stand along Shitai Road and Chengyin Road, the multi-storey F&B facilities are built along Chenzhen Road, and in the center there are 20,000 m² multi-storey facilities for business service. All these building groups are organized along the central green belt, allowing all buildings to enjoy the eco landscapes.

在空间布局上以中央景观带为核心，形成南北高、中间低的空间布局模式，其中高层SOHO办公与LOFT办公分别沿市台路与城银路两侧布置，东侧真陈路沿线布置多层餐饮商业，内部布置2万m²的多层配套商务服务设施。各组团均围绕中央绿化带布置。南北高、中央低的空间布局模式保证了各组团均能享受到良好的生态景观。

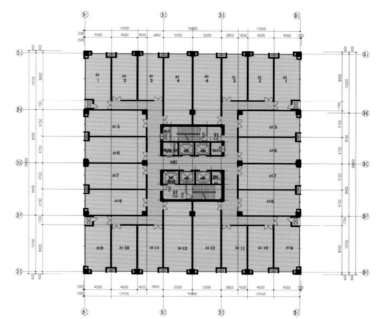

Standard Floor Plan 2 标准层平面图 2

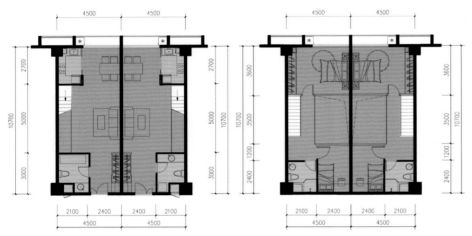

Ground Floor Plan 1 底层平面图 1 Mezzanine Plan 1 夹层平面图 1

Ground Floor Plan 2 底层平面图 2 Mezzanine Plan 2 夹层平面图 2

Design Idea 设计理念

The overall plan is based on the theme of "humane care", inspired by modern advanced urban design ideas, and insisting on the principle of "people orientation", trying to create an urban space that focuses on both the community atmosphere and the neighborhood relationship.

项目总体规划构思以"人文关怀"为设计主线，吸取了当代先进的城市设计理念，秉持以人为本的设计原则，力图打造出一个具有强烈社区感，并易于形成良好人际交往的城市空间。

Unit Types 户型设计

The development features creative unit design with "double entry doors and double balconies" which is unique in Shanghai, meeting owners' requirements to divide a duplex space into two parts. Each part will have a width reaching 7 meters and a balcony to provide enough daylight and independent functions.

创意的户型设计，上海市仅有的"双入户门＋双阳台"设计，可以满足业主将一套房复式空间分割成两套复式空间的需求，且每套面宽均可达到 7 m 左右，并各自带一个阳台，丝毫不影响采光及平常使用功能。

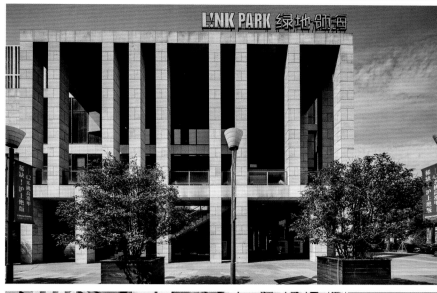

LOFT Office LOFT 办公设计

The architects have fully understood the clients' needs to design the LOFT offices with a height of 5.4 m. The single-floor spaces can be changed to two floors and reach high utilization rate of 200%. Then the clients or tenants can decorate their offices flexibly and innovatively.

绿地领海 LOFT 办公充分理解客户需求，5.4 m 层高让一层变两层，拥有 200% 超高使用率，可以利用平面二合一，空间二合一等集中组合方式打造自己的创意商务空间。

Landscape Design 景观设计

The landscape system of phase II & III is composed of the central landscape axis and the landscapes within the building groups. The central landscape axis provides views for all groups and visually connects with the central residential landscape in the west, forming an ecological "green lung" for the community. Within each group there is a landscape center which is relatively private and only opens to the buildings around. All these landscape spaces are skillfully organized and penetrate into each other to form an integrated landscape system.

项目二、三期景观体系由中心景观轴与各组团内部景观两个部分组成。其中，中心景观轴在视觉上为各组团共享，并与西侧住宅中心景观在视觉上连成一个整体，形成具有良好生态效应的中心"绿肺"，各组团内部也个分别享有组团中心景观，对组团开放，相对私密；两个部分的景观在边界上通过景观的设计，在视觉上相互渗透并相互补充，构成了一个完整的景观系统。

Pleasant Community Integrating Space, Building, Landscape and Life
空间、建筑、景观与生活充分交融的宜人社区

Fuzhou Poly Champagne International 福州保利香槟国际

Location / 项目地点：Fuzhou, Fujian / 福建省福州市
Developer / 开发单位：Poly Real Estate / 保利地产
Architectural Design / 建筑设计：Shanghai Hyp-arch Architectural Design Consultant Inc. / 上海霍普建筑设计事务所有限公司
Land Area / 占地面积：25 000 m²
Floor Area / 建筑面积：700 000 m²
Plot Ratio / 容积率：3.3
Greening Rate / 绿地率：30%
Project Type / 项目类型：Commercial Service, Residential Land / 商业服务、居住用地
Design Time / 设计时间：2012 年
Completion Time / 竣工时间：2014 年

Keywords 关键词
Close to Nature; Outdoor Greening; Diversified Units
亲近自然 户外绿化 户型多样化

"China's Most Beautiful Stylish Apartments" Finalist
入围"中国最美风格楼盘"奖

The layout of the residences emphasizes space shaping, and the designers create a humane and close-to-nature living environment with ample spaces through the creation of single house, control of space and combination of outdoor greening environment design.

住宅布局强调空间塑造，通过住宅单体的造型和空间限定，并结合户外绿化环境设计，创造空间丰富，亲近自然且具有人情味的居住环境。

Overview 项目概况

Located at the north of Huagong Road and east of Lianjiang Road, the project is near the northern plot of Helin New District Phase 9. The terrain of the development is smooth and there is no valuable building that's worth remaining. The west of the plot is 46m-wide Lianjiang Road, and the south is 45m-wide Tatou Road (planning), and both the north and east are the 24m-wide road for urban planning. The traffic of this plot is quite convenient. The north of the Sanba Road is the built affordable housing with many old buildings surrounded, which means the overall living environment remains to be improved.

项目位于化工路以北，连江路东侧，鹤林新区九期北区地块。项目地势较为平坦，用地内无具保留价值的建筑。地块西面为 46 m 宽的连江路，南面为 45 m 宽的塔头路（规划），北面、东面均为 24 m 宽的城市规划路。此地块的交通相当便利。三八路以北是建成的经济适用房，周边老旧建筑较多，整体居住环境有待改善。

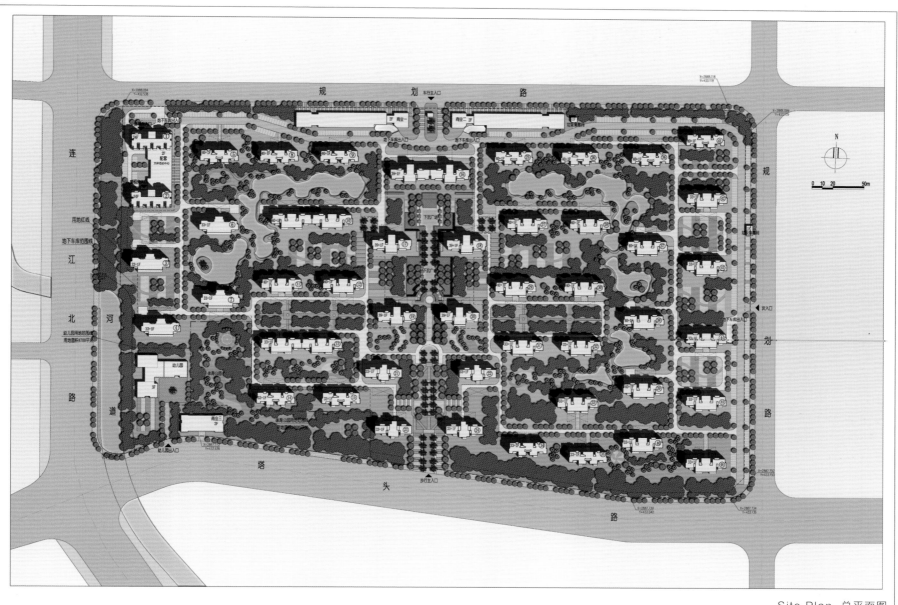

Site Plan 总平面图

Aerial View 鸟瞰图

Overall Layout 整体布局

This project has 49 blocks of buildings, including 46 blocks of residences, 2 blocks of 2-floor ancillary houses and a 3-floor kindergarten. The designers focus on the planning and design quality of the project and emphasize integrity and organic features. Based on the square characteristics of the terrain, and combined with the high-end design orientation of this community, the designers adopt the way of relatively symmetry from overall plan to single unit to present the solemn and graceful feelings. They make the plot center as axis, put the south and north entrance and the largest residential unit on this axis, and make the businesses, swimming pool and other residences symmetrical on this axis. They also arrange a sunken square at the center of the community together with rich landscapes to improve the overall quality of this community.

The 10,000 m² sports park at the southwest of the plot has provided the citizens and the residents of this community with a place for rest, entertainment and exercise and also has upgraded the landscape values of this whole community.

The business supporting service facilities are arranged along the Lianjiang Road and the northern street. It's with Art Deco and modern combined style and partially retreated disposal to be conducive to the formation of Commercial Street.

该项目共有49栋楼，其中包括46栋住宅、2栋两层的配套用房、1栋三层幼儿园。项目注重规划设计品质，强调整体性和有机性。根据本地块的地形方正的特点，结合本小区的高端设计定位，从总图到单体的设计都考虑使用相对对称的方式以体现庄重和大气的感觉。以地块中心为中轴线，将南北入口，最大户型的住宅放在中轴线上，商业、泳池和其他住宅都以此中轴线尽量对称。并在小区中央位置设计一处下沉广场结合丰富景观，提升小区整体品质。

地块西南侧设有一块用地10 000 m²的体育公园，为市民及小区居民提供一处休息、娱乐、锻炼的场所，同时也提升了整个小区的景观价值。

商业配套服务设施连江路及北侧道路沿街布置，造型上采用ART DECO风格与现代结合的方式，局部后退或退台的处理，有利于形成商业街的氛围。

Small Grouped Neighborhoods 小组团化邻里社区

Reasonable group scale is essential to the construction of a harmonious community. The site of this project is divided into four groups according to its roads and square form. The designers use the original landform texture, topography, water and roads of the site to naturally divide the plot into four areas. The building arrangement complies with the changes of terrain landscapes and the orientations, and forms many interesting spatial forms in each area. Based on this, the designers optimize the landscape to create richer spatial levels and make full use of the resources of the site to create high-quality living environment, creating a space, building, landscape and life integrated and pleasant community for the residents.

合理的组团规模是构筑和谐社区的必要条件，在项目中依据道路和基地方正形式将基地划分为四个组团，利用基地原有地貌肌理、地势、水体、道路等，将地块自然划分为四个区块，建筑排布顺应地势景观以及朝向的变化，在各个区块内形成意趣盎然的空间形态，并在此基础上，将景观加以优化，创造更加丰富的空间层次，并充分利用基地资源，创造优质人居环境，为居住者赢取空间、建筑、景观与生活四者充分交融的宜人社区。

Side Elevation 1 of Building 15#　　Side Elevation 2 of Building 15#
15# 楼轴立面图 1　　15# 楼轴立面图 2

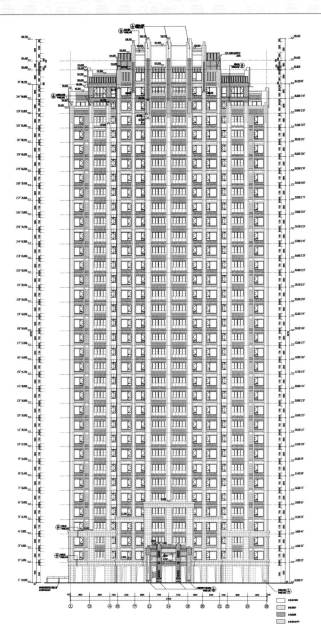

Side Elevation 1 of Building 8#
8# 楼轴立面图 1

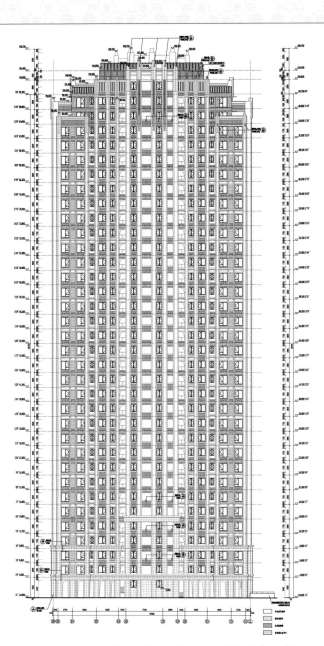

Side Elevation 2 of Building 8#
8# 楼轴立面图 2

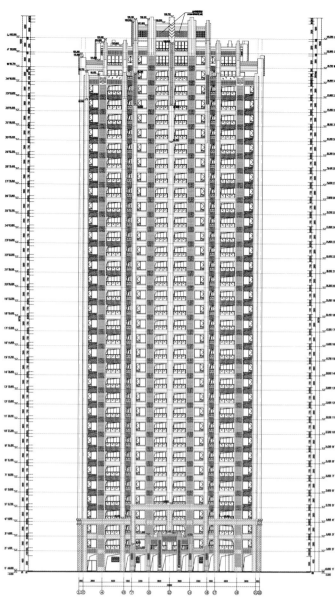

Side Elevation 1 of Building 27#
27# 楼轴立面图 1

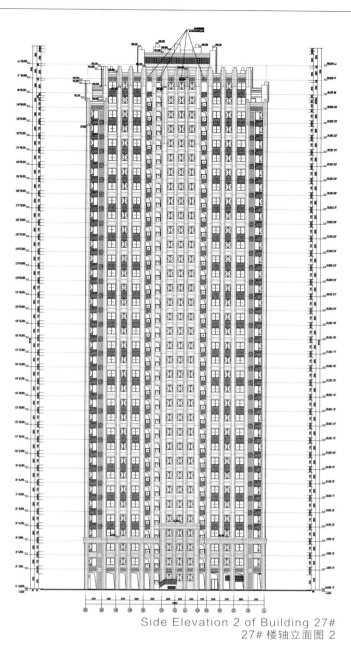

Side Elevation 2 of Building 27#
27# 楼轴立面图 2

Residential Layout 住宅布局

The community boasts 46 blocks of high-rise residences with empty spaces at the bottom to achieve the maximum landscape resources and transparent feelings, and the designers adjust measures to local conditions to integrate the residence and the environment as a whole. The layout of the residences emphasizes space shaping, and the designers create a humane and close-to-nature living environment with ample spaces through the creation of single house, control of space and combination of outdoor greening environment design.

小区共有46栋高层住宅，底层全部架空以获得最大限度的景观资源及通透感，因地制宜，使住宅与环境融为整体。住宅布局强调空间塑造，通过住宅单体的造型和空间限定，并结合户外绿化环境设计，创造空间丰富，亲近自然且具有人情味的居住环境。

Commercial Design 商业设计

The designers use the changes of the functional areas to transform the adverse aspects that the traffic main lines bring to the residences into the favorable aspects that enhance the commercial value of the community, and arrange supporting businesses at the northwest corner of the plot, which not only creates a relatively tranquil interior environment for the community, but also enriches the life experience of the neighborhoods.

利用功能分区的应变，把交通干线对住宅干扰的不利方面转化为增加社区商业价值的有利方面，在地块西北角布置配套商业，这样不但使小区获得相对宁静的内部环境，更丰富了社区的生活体验。

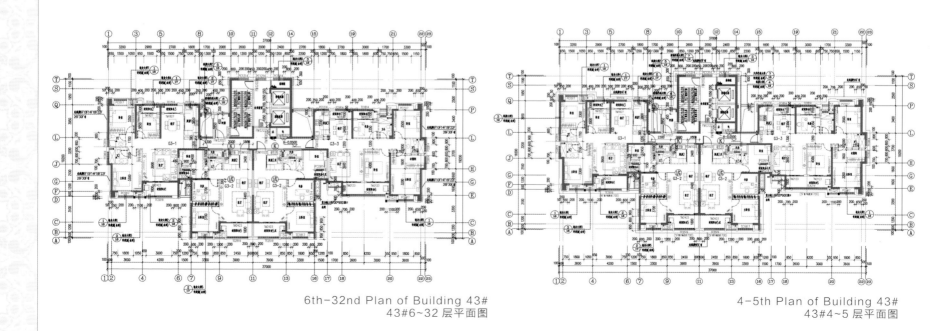

6th-32nd Plan of Building 43#
43#6~32 层平面图

4-5th Plan of Building 43#
43#4~5 层平面图

House Type Design 户型设计

The residences are designed in house types with different sizes, and the diversified house types are to meet different needs. Following the rule of each house facing south, the layout has fully considered the balance between the orientation and the landscape.

住宅产品设计了面积不等的户型，户型多元化，适应不同需求；遵循户户南向原则，布局充分考虑朝向与景观的均好性。

Landscape and Greening Design 景观与绿化设计

The buildings are built surrounding the community. From the community central garden to the group greening, overhead greening and then private home garden or private balcony, all these have formed the permeable landscape order of this project and achieved the effect of landscape in each house and each window. The approaches such as overhead greening, garden greening, roof greening and grass brick parking have enriched the greening levels of the community and made it full of green.

建筑围绕小区周边布置，"小区中心景观—组团绿化—架空绿化—入户私家花园/私家阳台"构成了本方案景观渗透序列，实现户户有景，窗窗有景。架空层绿化、花园绿化、屋顶绿化、植草砖绿化停车等手法丰富小区绿化层次，使小区绿意盎然。

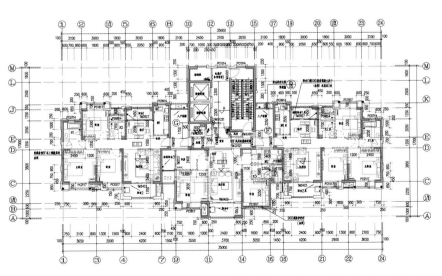

6th-32nd Plan of Building 22-23#
22-23#6~32层平面图

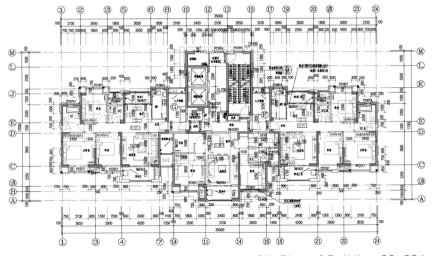

5th Plan of Building 22-23#
22-23#5层平面图

Brand Community with Comfortable, Modern & Noble Ambiance
居住舒适、集聚现代高尚气息的品牌社区

Urban Cradle Royal Mansion 万源御境

Location / 项目地点：Minhang District, Shanghai / 上海市闵行区
Developer / 开发单位：Shanghai Urban Development Group Co., Ltd. / 上海城开集团
Architectural Design / 建筑设计：Shanghai ZF Architectural Design Co., Ltd. / 上海中房建筑设计有限公司
Land Area / 占地面积：110 800 m²
Floor Area / 建筑面积：300 000 m²
Plot Ratio / 容积率：1.72

Keywords 关键词

Green Home; Ecological Building; One Family Shares One Lift on One Floor
绿岛家园 生态建筑 一梯一户

 "China's Most Beautiful Stylish Apartments" Finalist
入围"中国最美风格楼盘"奖

The facade design uses vertical lines to present the solemn, high quality and rich cultural Art Deco style. The wall of the residential building is mainly in stone, and the design uses the depiction of the material details to manifest the quality of this high-end residence.

立面造型设计以纵向线条表现出庄重、富有品质感和文化底蕴的ART DECO风格。住宅楼外墙材料以石材为主，通过材质的细部刻画，表现出高端住宅的品质。

Overview 项目概况

The site is located at the northwest corner of Urban Cradle residential community. The design uses central landscape water system and the sidewalks to divide the neighbor into four groups, and the central landscape water system runs through the underground driving way of the whole Urban Cradle. The whole community is planned to be high-end apartment, and the Wanyuan Road is equipped with commercial and related service supporting facilities such as kindergarten, bar and gym.

项目基地位于万源居住小区的西北角，通过中央景观水系及人行步道把街坊分为四个组团，中央景观水系贯穿整个万源城的地下行车廊。整个小区规划定位为高档公寓；沿万源路设置有幼稚园、酒吧、健身等商业设施及相关服务配套设施。

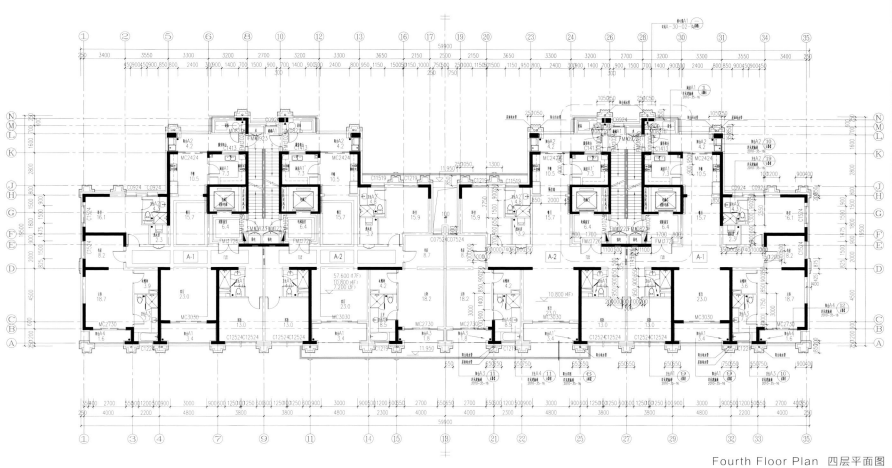

Fourth Floor Plan 四层平面图

Geometric Layout 几何形布局

The project design adopts the regular geometric layout that can manifest the quality of the luxury residential community to strive for the maximum distances between the residences. The others are four rows of townhouses arranged north-south at the site with a depth of 300m. By borrowing the landscape of split-level buildings, the open space between the floors broadens the sight and leads the advantaged landscape such as the southern open space into the community and becomes a part of the landscape itself.

项目设计采用能体现豪宅社区品质的规整几何形布局，并力求住宅间距的最大化。其余均为在基地南北向300多米的纵深内，安排四排建筑。通过建筑错列的借景手法，楼与楼间开阔的间距让视线开阔，将南侧的开放空间等得天独厚的景观引入社区环境，并成为自身景观的一部分。

Elevation 1 of 1# Building 1# 立面图 1

Elevation 2 of 1# Building 1# 立面图 2

Elevation 3 of 1# Building 1# 立面图 3

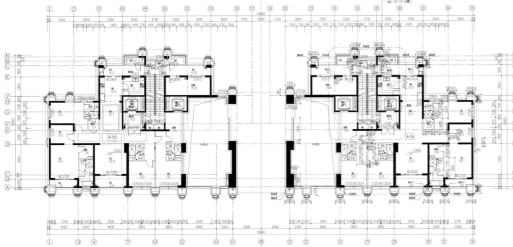

First Floor Plan 一层平面图

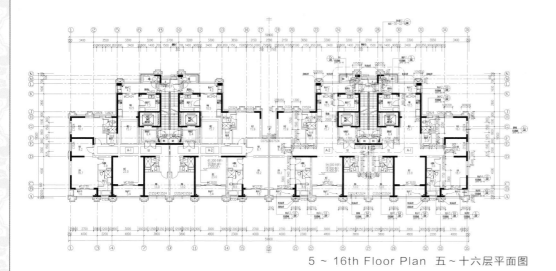

Second Floor Plan 二层平面图

5~16th Floor Plan 五~十六层平面图

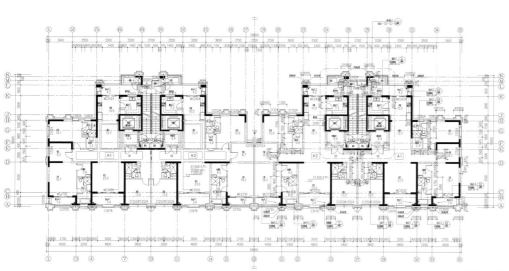

17th Floor Plan 十七层平面图

Design Concept 设计理念

Green Home

The site is arranged with water and planted with plenty of trees and grasses to make landscape and greenness run through this land and make the buildings to integrate into the green environment, making the whole site a green home.

Ecological Building

Firstly, it shows in its affinity with the natural environment. On the one hand, the design makes the building organically integrate into the natural environment and maximally reduces the destruction to the environment. On the other hand, it strives to lead the natural environment inside the buildings and arranges well-organized underground courtyard greening to endow the artificial environment inside the buildings with the quality of natural environment. Secondly, the ecology shows in energy conservation, which means it uses modern architectural technology to reduce the building energy consumption and adopts reasonable architectural layout to improve service efficiency.

High Quality Building

No matter the creation of the building form, organization of the traffic circulation or the design of the open space, all are studied with an insight of high quality design and view of grandness and magnificence. Based on the developing concept of "Noble Life, High Quality Living", the designers attempt to create this community into a brand community with comfortable, modern and noble ambiance.

绿岛家园

在项目基地中设置水体及大量种植林木、草地，以造景和绿化贯穿整个用地，让建筑融入绿化环境中，使整个基地成为绿岛家园。

生态建筑

这首先表现为与自然环境的亲和上，一方面通过设计使建筑能有机融入自然环境，最大程度上减少对环境的破坏；另一方面尽量将自然环境引入建筑内部，设置有序的地下庭院绿化，将自然环境的品质赋予建筑内部的人工环境。其次，生态表现为节能，即通过现代建筑技术来降低建筑能耗，依靠建筑合理布局来提高使用效率。

品位建筑

无论是建筑形态的塑造、交通流线的组织，还是开放空间的设计，都以品位设计的眼光，从宏观大气的角度进行研究。本着"尊贵人生，品质生活"的开发理念努力将本小区建设成为居住舒适、集聚现代高尚气息的品牌社区。

Axial Symmetrical Layout 中轴对称布局

The whole community adopts the architectural layout of axial symmetry, and emphasizes the symmetry between the courtyard groups on planning. The buildings are mainly plate-type high-rises with 16 to 22 floors. The central landscape water system surrounds the 22-floor point-mode residence, which locates at the vertices of the community and owns the solemn and gracious architectural form and good geographical location.

整个小区采用了中轴对称的建筑布局方式，在规划上强调组团院落间的对称，建筑以 16～22 层的板式高层为主，中央景观水系环绕的是小区的至高点 22 层的点式住宅，庄重大气的建筑造型与良好的地理位置。

Facade Design 立面设计

The project is a group of high-rises that is less than 100 m. The facade design uses vertical lines to present the solemn, high quality and rich cultural Art Deco style. The wall of the residential building is mainly in stone, and the design uses the depiction of the material details to manifest the quality of this high-end residence.

住宅为一组不超过 100 m 的高层住宅，立面造型设计以纵向线条表现出庄重、富有品质感和文化底蕴的 ART DECO 风格。住宅楼外墙材料以石材为主，通过材质的细部刻画，表现出高端住宅的品质。

Meiju Encyclopedia 美居小百科

"One family shares one lift on one floor" means that each house is equipped with a lift on the same floor. The owner takes the lift to his floor and then arrives to his own private space after getting out of the lift. The owner can open the door of his own floor by using the lift card while it's unavailable for the other owners, and to some extent, this has protected the privacy of the owner and is much safer. "One family shares one lift on one floor" can not only ensure the quality of the whole residence, but also improve the noble status of the owner.

  一梯一户，指在一层楼里，每家每户都配有一部电梯，业主乘坐电梯直达所在楼层后，出门便可以进入自己的私家空间，业主可以凭电梯卡刷开他所在的楼层这一户的单元门，而其他的业主是不可以进来的。这从某种程度上更保护了业主的私密性，而且安全性更高。一梯一户不但可以保障整个住宅的品质，而且可以提升主人尊贵身份的象征。

House Type Design 户型设计

One Family Shares One Lift on the Same Floor

The neighborhood has proposed the concept of "one family shares one lift on one floor" on the residential house type design. The sizes range from 200 to 500 m^2, and each house has the "five bright" layout, which means the bright living room, bright bed room, bright rest room and bright dining room. The living room and the dining room are north-south transparent design, and each house can enjoy the central landscape greenbelt and is ensured with good ventilation and lighting conditions.

High Standard Large Space

Each residential building is designed with entrance lobby, and the main functional rooms such as the living room and the bedroom are all with good orientations and landscape views to ensure the wellness of each house. Each suite is equipped with device platforms for placing terrestrial heat equipment, central vacuum cleaner and VRV outdoor air conditioning unit. Each house is designed with enough storage spaces, independent western style kitchen, living room and restroom, and ancillary spaces such as the restroom-equipped maid's room. Each bedroom is collocated with rest room, which means it's an independent suite design. The main bedroom is designed with independent cloakroom. The standard width of the living room ranges from 4.5 m to 8.55 m, and the standard width of the main bedroom ranges from 4.2m to 4.5 m. The height of each floor of this residential building is 3.6m.

一梯一户

街坊在住宅户型设计上提出一梯一户的理念，户型面积在 200～500 m^2，每户均有明厅、明卧、明厨、明卫、明餐厅的"五明"格局；客厅、餐厅为南北通厅设计，每户均能观赏到中心景观绿地且保证每户均有很好的通风采光条件。

高标准大空间

每幢住宅楼均设入口大堂；主要功能间如客厅、主卧等都有较好的朝向和景观视线，保证每户的均好性；每套户型均带有设备平台，考虑放置地热设备、中央吸尘器、空调 VRV 室外机组；每户均设置足够的储藏空间；每户均设有独立的西厨及客卫，并设有带卫生间的工人房等辅助空间；每间卧室均配置卫生间，为独立套房设计。主卧室均设置独立的衣帽间。客厅开间为 4.5～8.55 m，主卧室开间均为 4.2～4.5 m。住宅楼的每层层高为 3.6 m。

Apartment Design Approach Creates Urban Luxury Community
公寓设计手法塑造强烈都市感的豪华社区

Cixi Hengyuan Condo Residence　　慈溪恒元悦府公寓

Location / 项目地点：Cixi, Zhejiang / 浙江省慈溪市
Architectural Design / 建筑设计：CCDI Group / 悉地国际（CCDI）
Landscape Design / 景观设计：CCDI Group, Belt Collins / 悉地国际（CCDI）、贝尔高林
Interior Design / 室内设计：CCDI Group, PAL Design Consultants Ltd. / 悉地国际（CCDI）、PAL
Land Area / 占地面积：17 636 m²
Floor Area / 建筑面积：51 814 m²
Plot Ratio / 容积率：2.3

Keywords　关键词

Cordial & Pleasant; Elegant Life; Urban Luxury Residence
亲切宜人　优雅生活　都市豪宅

"China's Most Beautiful Stylish Apartments" Finalist
入围"中国最美风格楼盘"奖

The project breaks through the regular apartment design approach while designing this urban luxury community. It decorates both the exterior and interior with big opening and closing and gathers the high-end commerce and big residential units as a whole. The two blocks of hundred-meter luxury residences at the north and south stand face to face with unobstructed vision at the higher level, making the residents scan the prosperity of the city and enjoy the tranquil life as well.

项目将常规的公寓设计手法突破至具有强烈都市感的豪华社区，以大开大合之姿兼修内外，集高端商业和高档大户型住宅为一体，南北两栋百米华宅相峙而立，无遮挡高层视野，既可纵览城市繁华，亦能尽享静谧生活。

Overview　项目概况

Located in the east new district of Cixi city, the project is to the east of the 50 meter wide New Town road, to the south of the 34 meters' north second ring road, to the north of the Yingmochen new village at Hushan Street and to the west of the city water supply company, and it's only 200 m to the municipal administrative center. The development is commercial residences and is planned as 2 blocks of 30-floor high-rises and commercial podiums mainly in 2 floors and partially 3 floors.

　　基地位于慈溪市城东新区，西邻50 m宽的新城大道，北邻34 m宽的北二环路，南至浒山街道应莫陈新村，东至市自来水公司，距市行政中心200 m。用地性质为商业住宅用地，规划有两栋30层的高层住宅和2层为主、局部3层的商业裙房。

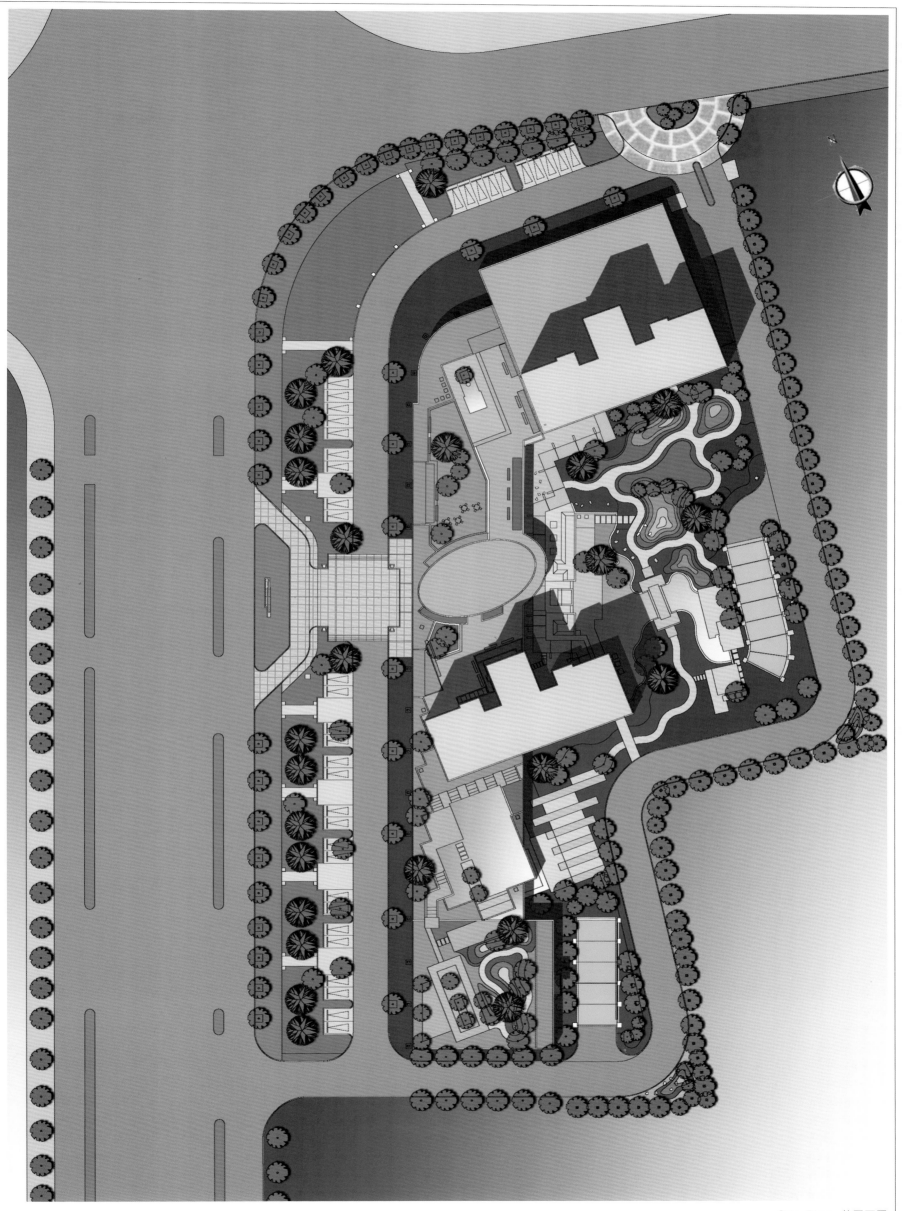

Site Plan 总平面图

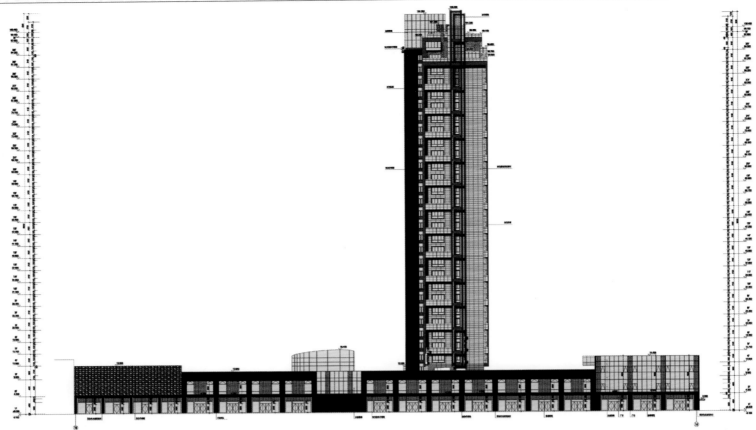

West Elevation of 2# Building 2号楼西立面图

Planning 规划布局

The designers make full use of the resources surrounding the base, and adopt the layout that adjusts to the local conditions and humanized structures to create an elegant living atmosphere and provide the residents with cordial and pleasant spatial sense. Studying the existing resources of the surrounding cities and grasping the entry point that is closely connected with the city is the basic principle to form the planning structure of this project.

充分利用基地周边资源，采用因地制宜的布局方式和人性化的结构特征，营造优雅的生活氛围，为居住者提供亲切宜人的空间感受，通过对周边现有城市资源的研究，抓住与城市紧密衔接的切入点，是本项目规划结构形成的基本原则。

Design Concept 设计理念

By overlaying and reorganizing the elements such as traffic, community supporting and landscape resources, the overall planning layout is formed to create pleasant activity scale and space and manifest the core idea of "people oriented". The designers have strengthened the transition sequence of "urban external – frontage transition – community internal space", defined some details to enhance the sense of belonging of the residents and paid attention to the continuation and transformation of space to promote the contact between the residents and create a friendly communicating interface.

通过对交通、社区配套、景观资源等元素的叠加、重组，构成总体规划格局，营造宜人活动的尺度和空间，体现"以人为本"的核心理念。强化"城市外部—沿街过渡—小区内部空间"的序列过渡，并进行局部限定以增强居者的归属感，注重空间的延续和转换，促进人与人的交往，塑造亲切的可交流界面。

North Elevation of 1# Building　1号楼北立面　　　　South Elevation of 1# Building　1号楼南立面图

Facade Design　立面设计

The front facade emphasizes the disposition of the wall and the balcony glass breast board, and the symmetrical aluminum frame design strengthens and enriches the linear sense. The back facade tends to be urbanized, and the entire equipments and tube wells are with integrated design and skillfully concealed inside the buildings. The facade is mainly in stone and combined with stone paint, glass wall, metal plate and blinds to make the building elegant and dignified, as well as light and graceful.

　　正立面强调幕墙与阳台玻璃栏板的处理，对称铝框设置让线条强烈而丰富；背立面趋于城市化，所有设备及管井进行整体设计，巧妙隐于建筑内。立面以石材为主结合仿石涂料、玻璃幕墙、金属板和百叶，使建筑典雅端庄而不失轻巧。

House Type Design 户型设计

Each unit is above 230 m², and with north oriented entrance garden as well as north-south ventilation. The double entrance lobby and 80% of astylar design have double enlarged the space. The science and technology such as drainage of the same floor, hollow floor slab, electric floor heating and natural balanced graft are implanted into this Hengyuan Condo Residence to manifest the orientation of urban luxury residence.

每户面积在230 m²以上，带北向入户花园，南北通透。双入口大堂、80%无柱设计将空间加倍放大。同层排水、空心楼板、电热地暖、自平衡通风，让科技智能"嵌入"恒元悦府，呈现都市豪宅的定位。

Even Level Plan of 1# Building 1# 楼偶数层平面图

The Flagship Residential Property of Xiaoshan District, Hangzhou
萧山区住宅地产的旗舰

Gemdale Tinyat Mansion　金地天逸

Location / 项目地点：Hangzhou, Zhejiang / 浙江省杭州市
Developer / 开发商：Gemdale Group Shanghai Real Estate / 金地集团上海地产
Architectural Design / 建筑设计：The Coast Palisade Consulting Group / 加拿大 C.P.C. 建筑设计顾问有限公司
Land Area / 占地面积：123 963 m²
Floor Area / 建筑面积：311 163 m²
Plot Ratio / 容积率：2.5
Greening Rate / 绿地率：30%

Keywords　关键词
Fashion Trend; Tranquil & Comfortable; Delicate & Elegant
时尚潮流　宁静舒适　细致典雅

"China's Most Beautiful Stylish Apartments" Finalist
入围"中国最美风格楼盘"奖

The delicate and elegant, charming and rich overall facade style fully shows the charm of the architectural style, tranquility of the residential community and the comfort atmosphere, and initiates the new fashion trends. The wall finishing materials is the combination of natural dry hanging granite stones and aluminum plates.

整体立面风格，细致中见典雅，丰富中见韵味，充分体现建筑的风格魅力与住宅小区的宁静、舒适气氛，创导时尚潮流的新气势。外墙装饰材料采用干挂天然花岗岩石材与铝板相结合。

Overview　项目概况

Located at the Duhu community of Chengxiang Road, Xiaoshan District, Hangzhou city, the project is a residential development. The base is to the east of Fengqing Road, to the north of Zhejiang-Jiangxi Railway, to the south of the Second Water Plant, and to the west of the urban planning road. According to the urban planning, the northwest corner of the base is arranged with subway station, which provides convenient and fast transportation resources.

项目位于杭州市萧山区城厢街道杜湖社区，属住宅开发项目。基地位于风情大道以东，浙赣铁路以北，二水厂以南，城市规划道路以西。根据城市规划，基地西北角设有地铁站，为本项目提供了便捷的交通资源。

Aerial View 鸟瞰图

Project Orientation　项目定位

The project is oriented to be a high-grade and high-quality luxury residence and a flagship residential property of Xiaoshan District. It will present a perfect lifestyle that combines leisure, culture, entertainment and living as a whole.

　　项目定位为杭州高档次、高品质的高档豪宅，萧山区住宅地产的旗舰，将体现休闲、文化、娱乐与居住相结合的完美生活方式。

Planning 规划布局

The development is planned to be a 30-floor high-rise residence which includes 17 blocks and 26 units and two main and subsidiary clubs. The plot is 50 meter away from the north and south boundaries and with an extreme manner of the king. As a cultivated land, the development shares the surrounding commercial and educational facilities such as the Bingjiang district living supporting, Xiaoshan downtown commercial clusters, Xiaoshan High School and Xiaoshan First Experiment Primary School.

规划开发为共 17 栋楼 26 个单元 30F 的高层住宅及两个主副会所。项目地块南北 50 m 退界，极具王者之气。且为"熟地"，周边共享滨江区城区生活配套、萧山中心城区商业集群及萧山中学、萧山市第一实验小学等商业、教育配套。

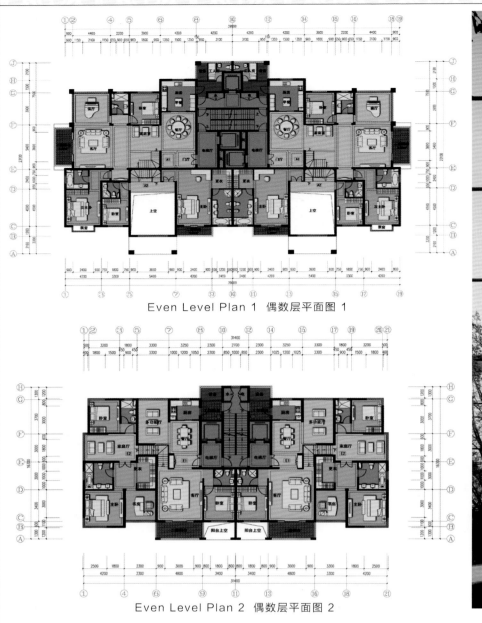

Even Level Plan 1 偶数层平面图 1

Even Level Plan 2 偶数层平面图 2

Facade Design 立面设计

The delicate and elegant, charming and rich overall facade style fully shows the charm of the architectural style, tranquility of the residential community and the comfortable atmosphere, and initiates the new fashion trends. The wall finishing materials is the combination of natural dry hanging granite stones and aluminum plates.

整体立面风格，细致中见典雅，丰富中见韵味，充分体现建筑的风格魅力与住宅小区的宁静、舒适气氛，创导时尚潮流的新气势。外墙装饰材料采用干挂天然花岗岩石材与铝板相结合。

Residential Design 住宅设计

The shape of the residence is unique, which creates the changeable outline and fully presents the charm of this modern building. The entire glass windows that reach half to the ground and the handrails are made as the facade shaping elements and the overall detailed patterns of the community residence. The double-layer hanging garden brings exquisite changes to the residential facade and extends the interior space as well. The broad and stretched windowsill for the interior makes the lightings and shadows diversified and changeful.

住宅造型独特，既创造出富有变化的轮廓，又充分强调了现代建筑的魅力。全玻璃的半落地窗户与栏杆作为立面的造型要素和小区住宅的整体细部格调。双层的空中花园，既创造了住宅立面的细腻变化，又使室内空间得到延展。提供给室内的宽大舒展窗台，使光影变化多端。

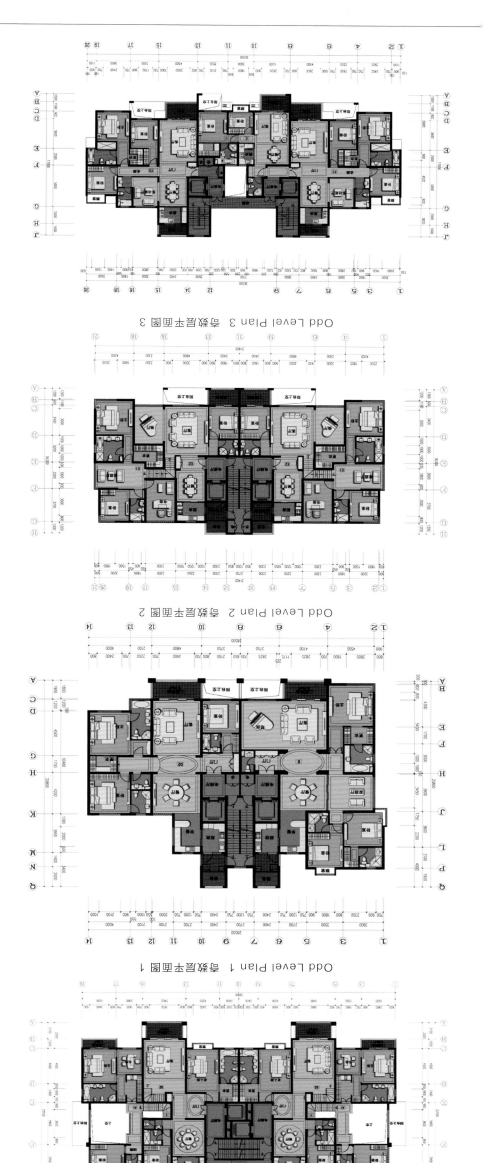

Odd Level Plan 1
Odd Level Plan 2
Odd Level Plan 3

Living Territory for Future Young Man of Chaoqing Plot
朝青板块的未来青年人居住领地

Zhonghong Pixel, Beijing
中弘·北京像素

Location / 项目地点：Chaoyang District, Beijing / 北京市朝阳区
Developer / 开发商：Zhonghong Real Estate / 中弘地产
Architectural Design / 建筑设计：SAKO Architects / SAKO 建筑设计工社
Land Area / 占地面积：209 000 m²
Floor Area / 建筑面积：698 000 m²
Plot Ratio / 容积率：2.38

Keywords 关键词
Duplex House Type; Rubik's Cube Facade; Windmill Shaped Planning
复式户型 魔方立面 风车式规划

"China's Most Beautiful Stylish Apartments" Finalist
入围"中国最美风格楼盘""家"

Inspired by the shape of honeycomb, the designers design the longitudinal facade with concave-convex walls and enrich the colors of the facade at the same time. The designers use the transition, volume increase and decrease of the building to form a split level of the building top so that to avoid the repression caused by the narrow space between the high-rise buildings.

以蜜蜂的家为灵感源，将建筑的纵向立面设计了几个凹凸墙面，并同时丰富立面色彩，通过建筑的转折、体块的加减变换，使建筑顶端错落，从而避免高层建筑由于间距过近造成的压抑。

Overview 项目概况

The project is divided into plot 1 and plot 2 with a gross land area of around 210,000 m² and approximately 10,000 residents. It's a complex that gathers commerce, residence and office as a whole. The development is quite close to the Caofang Station of metro line 6, 27 Diequan golf course and the Tonghui River waterfront ecological belt. The Chaoqing plot of this project enjoys the municipal planning benefits such as the CBD east outline and the media corridor.

项目划分为1号地、2号地，占地面积合计约21万 m²，约一万户入住的规模。住宅为以小为一体的商业建筑，项目紧邻地铁6号线草房站、27番翠园高尔夫、通惠河水岸生态带。项目所在朝青板块具有CBD东溢区作承载之属的规划利好等。

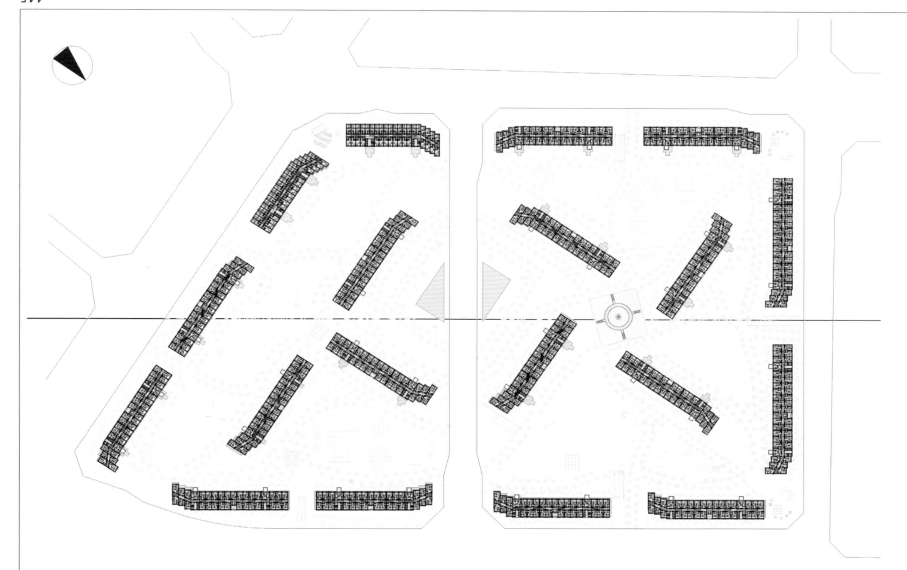

Planning 规划布局

In order to meet the essential function needs of the residence such as daylight and privacy, the arrangement of 19 residential blocks have been specially planned. 12 blocks of plate-type buildings are arranged along the red lines at the periphery of the plot, and the tilt axis along the Chaoyang North Road and its perpendicular are designed to arrange the residential buildings at the inner part of the plot.

The project planning breaks through the traditional row arrangement and uses windmill shaped planning instead. In this limited space, the designers fully considers the openness of the vision and breaks through the traditional row arrangement by using the naturally revolving windmill shape instead to ensure all the 31 blocks with daylight.

为了满足日照、居住私密性的要求，19栋住宅楼的布置进行了特殊设计。在地块外围沿红线排列着12栋板式楼栋，同时沿岔路朝阳北路的倾斜轴线以及与其相垂直的岔路轴线，共沿着该两条轴线排列地块内的住宅楼。

当项目建筑规划打破传统的排布方式，采用风车形状的规划。这样在有限的空间中，充分考虑视觉的开阔度，打破传统的建筑排布方式，采用以自然旋转的风车形式排列，使31栋建筑均可以采光。

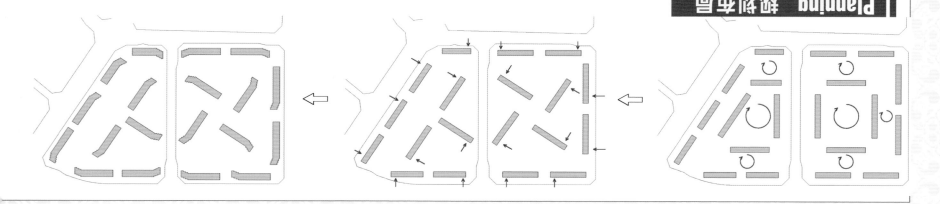

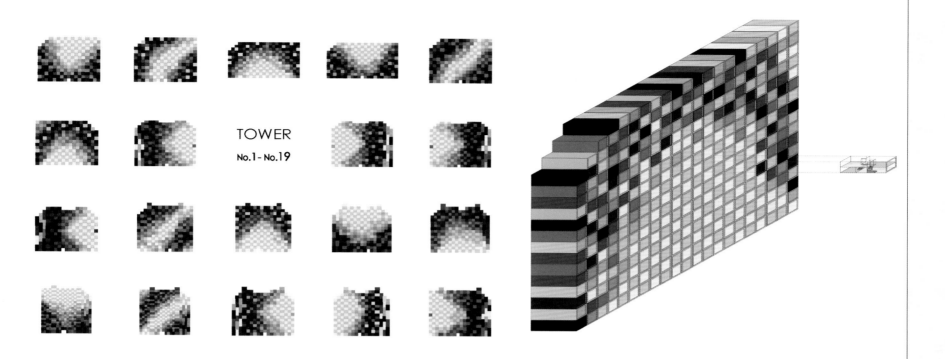

TOWER
No.1 - No.19

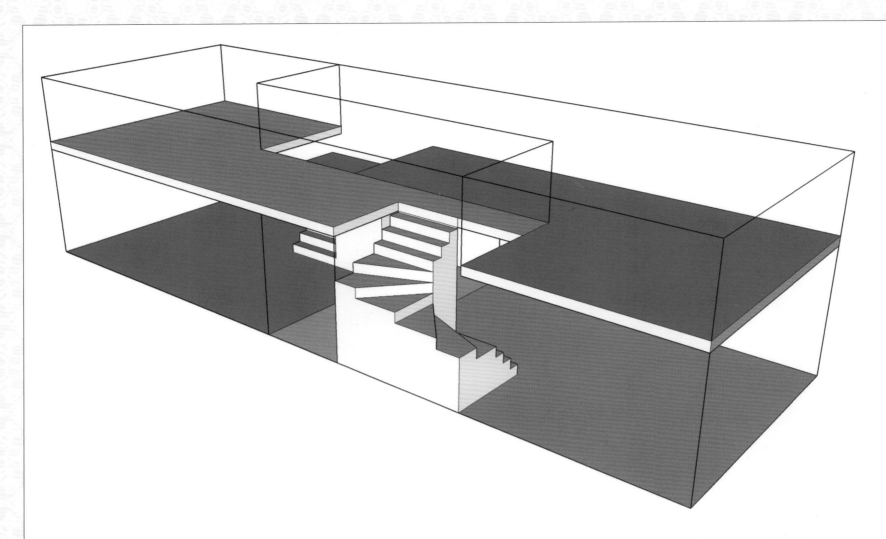

Rubik's Cube Facade 魔方立面

Inspired by the shape of honeycomb, the designers design the longitudinal facade with concave-convex walls and enrich the colors of the facade at the same time. The designers use the transition, volume increase and decrease of the building to form a split level of the building top so that to avoid the repression caused by the narrow space between the high-rise buildings.

以蜜蜂的家为创造灵感，将建筑的长向立面设计了众多凹凸墙面并同时丰富立面色彩，通过建筑的转折、体块的加减穿插，使建筑端部错落，从而避免高层建筑由于间距过近造成的压抑。

Facade Design 立面设计

Each residential unit with exterior wall coatings in four types of gray colors is formed by overlapping and accumulated volumes. By arranging the 12 plate shaped buildings at the periphery of the plot, the exterior wall facades jointly form a natural, smooth and integrated pattern. Otherwise, each block is designed with one theme color and the gradual change of the 10 colors goes through and creates the whole plot.

颜色深浅不一的四种灰色外墙涂料装饰的各住户的单元格，由重叠堆积的体块构成，通过位于地块外周的 12 栋板状建筑物的排列，12 栋的外墙立面共同形成了一个行云流水般整体的图案。另外，各栋分别设定一种主题彩色，用 10 种色彩的渐变贯穿并打造了整个地块。

House Type Design 户型设计

The main house type is divided into two kinds: the 4.75 m-high lofts in sizes of 40 m², 45 m², and 50 m², and the 5.48 m-high duplex in size of 56 m².

The 4.75 m-high and 5.5 m-wide space has reached a utilization of 90%. The designers have specially designed three types of empty space: empty space for double glazed windows, empty space for panoramic view and empty space for double hall.

The innovative 5.48 m-duplex loft design breaks through the traditional spatial layout, and the staggered design of the up and down layers which passes through the internal corridor has not only ensured the lighting passing through the two layers, but also created an extremely high utilization, as well as a width of 6 m.

主力户型分为两种：4.75 m 层高 40 m²、45 m²、50 m² 的 LOFT、5.48 m 高 56 m² 的对跃户型。

4.75m 层高的空间，5.5 m 的面宽，空间利用率高达 90%。设计师特别设计了三种不同形式的挑空：双层窗户挑空、选全景挑空、门庭双层挑空。

5.48 m 创新对跃式 LOFT 设计，突破传统 LOFT 空间布局，上下层之间跨过内走廊的交错设计，不仅充分保证了上下层通透的光线，且造就了超高利用率，更有阔达 6 m 的面宽。

Landscape Design 景观设计

The landscape planning of this project is base on the "integrated circuit plate". The designers use the core influence formed by the central building to sketch the road system and various thematic leisure space and build park landscape.

项目依托"集成电路板"构想景观规划，以中央建筑所形成的核心辐射，勾画出路线系统及各种主题休闲空间，打造园林小景。

Naturalistic Style Leads the Close-to-Nature Life
自然主义风格引领自然零距离生活

Gemdale River Town   **金地朗悦**

Location / 项目地点：Fangshan District, Beijing / 北京市房山区
Architectural Design / 建筑设计：Beijing Victory Star Architectural & CML Engineering Design Co., Ltd / 北京维拓时代建筑设计有限公司
Gross Land Area / 总占地面积：Approx 446 362 m² / 约 446 362 m²
Gross Floor Area / 总建筑面积：Approx 190 257 m² / 约 190 257 m²
Land Area (Residential) / 占地面积（住宅）：137 695 m²
Total Capacity of the Construction Area (Residential) / 总计容建筑面积（住宅）：227 817 m²

Keywords   **关键词**
Return to Nature; Humanized Living; Close-to-Nature
返璞归真　人居生活　自然零距离

 "China's Most Beautiful Stylish Apartments" Finalist
入围"中国最美风格楼盘"奖

The volume of the residence is diversified and without losing style and taste, making people touch the nature with their bodies for real and fully enjoy the sunshine, fresh air, rain and dew. The staggered volumes also form abundant facade texture.

住宅建筑体量丰富却不失格调品位，让人真真实实地用身体触摸着自然，充分享受着阳光、空气和雨露，体块的错落也形成了丰富的立面肌理。

Overview   **项目概况**

The gross residential land area of this project is 137,695 m², and the total capacity of the residential construction area is 227,817 m². The residential buildings are mainly 6 to 9-floor elevator garden houses and 11 to 15-floor high-rise apartments, meanwhile, a small number of commercial facilities are arranged at the frontage.

项目的住宅总用地面积为 137 695 m²，总计容建筑面积为 227 817 m²。住宅部分主要为 6～9 层电梯花园洋房和 11～15 层的高层公寓，同时在沿街布置少量的商业配套。

Aerial View 鸟瞰图

Planning 规划布局

The master plan of this project adopts the frontage image planning axis and the landscape planning axis inside the area to form the planning concept of the entire area, and it uses the artificial environment and natural environment to create a sense of order in the urban form and combines the natural landscape conditions to create a landscape axis that connects each group.

项目整体规划采用沿街形象规划轴和区内景观规划轴形成全区的规划理念，运用人造环境和自然景观两种处理手段在城市形态中产生了一种秩序感，并结合自然的景观条件来创造一个联络于各个组团之间的景观轴线。

Design Concept 设计理念

The architectural design returns to the ecologically rational space, re-pursues the unity of technological beauty and humanism and makes the residents' emotion return to tranquility and nature by using the pure naturalistic style in forms and details. The design always keeps the datum point of "humanized living", purses the comfort and taste of living and creates the uniqueness of the community, making the "people oriented, scientific living, healthy life" run through the whole process of the design.

建筑设计回归生态主义的理性空间，重新追寻技术美与人情味的统一，以造型、细节、纯粹的自然主义风格，使居住者情感回归于宁静与自然，设计中始终以"人居"为基准点，追求居住的舒适度与品位，同时建立社区的独特性，将"以人为本、科学居住、健康生活"贯穿于设计的全过程。

Facade Design 立面设计

The residential buildings are mainly 6 to 9-floor elevator garden houses and 11 to 15-floor high-rise apartments. The design uses neoclassical facade design which is vivacious, rich and full of style to meet the yearning and pursuit of romantic life of the clients in various ages. The design uses the top to provide rich details, manifest the close-to-nature and low-density life quality, and create warm, romantic, elegant, steady and return-to-nature life atmosphere.

项目住宅部分主要为6～9层电梯花园洋房和11～15层的高层公寓，采用新古典主义建筑立面风格，立面活泼丰富，富有风情感，以满足多种年龄群客户对浪漫生活品质的向往与追求。通过头部等提供丰富的细节，彰显与自然亲近的低密度生活品质，创造温馨、浪漫、典雅、稳重、回归自然的生活氛围。

Residential Design 住宅设计

The volume of the residence is diversified and without losing style and taste, making people touch the nature with their bodies for real and fully enjoy the sunshine, fresh air, rain and dew. The staggered volumes also form abundant facade texture. On the arrangement of the plan, the design uses transparent courtyard design while functionally combines the reasonable layout of different sizes, which not only meets the demands of private outdoor space but also naturally connects the public road greening and the centralized greenbelt, creating a harmonious relationship between the unit, neighborhood and community. It has ensured the privacy of the residents and made the residents finely enjoy the community life as well.

住宅建筑体量丰富却不失格调品位，让人真真实实地用身体触摸着自然，充分享受着阳光、空气和雨露，体块的错落也形成了丰富的立面肌理。在平面的处理上，功能结合不同面积合理布局的同时，利用通透型的庭院设计，在满足了私密的户外空间需要的同时，又自然地衔接公共的道路绿化、集中绿地，使得个体、邻里、社区之间层次关系十分融洽。在确保了居住者私密性的同时，又能很好的享受了社区生活。

Landscape Design 景观设计

The elegant landscape comes from the nature and this kind of attitude towards life runs across the entire community. The entrance of the community is the starting point of the landscape, and the rich plants scatter into the groups of the entire community along with the extension of the planning road and become the important components outside the groups.
Centralized greenbelts are arranged between each building, which provides the residents with a place to enjoy outdoor sunshine and realizes the closeness between living and nature.

优雅的风光源自自然，这种生活态度穿越了整个社区。景观中以小区的入口为起点，丰富的种植顺着规划路的延伸，分散到整个小区的组团中，形成组团外的重要构成元素。

每栋建筑间均布置了集中绿地，为住户提供了享受户外阳光的场所，也实现了居住与自然的零距离。

High-end Model of Industrialized Prefabricated Technology
工业化预制技术的高端典范

Tianjin Vanke Jin Lu Garden　天津万科锦庐园

Location / 项目地点：Binhai New District, Tianjin / 天津市滨海新区
Developer / 开发单位：Tianjin Vanke Real Estate Co., Ltd. / 天津市万科房地产有限公司
Architectural Design / 建筑设计：CAPOL / 华阳国际设计集团
Gross Floor Area / 总建筑面积：167 000 m²
Plot Ratio / 容积率：1.4
Project Type / 项目类型：Industrialized Residence / 工业化住宅

Keywords　关键词

Green & Ecological; Industrialized Prefabrication; Integrated Design
绿色生态 工业化预制 一体化设计

"China's Most Beautiful Stylish Apartments" Finalist
入围"中国最美风格楼盘"奖

As the high-end residence that first uses the industrialized prefabricated technology in North China, the project has reduced the on-site construction time of the alpine region, and this is the initial realization of the dream of prefabricated house.

作为华北地区首个应用工业化预制技术的高端住宅产品，项目缩短高寒地区现场施工时间，是预制装配房屋梦想的初步实现。

Overview　项目概况

Vanke Jin Lu Garden is located in the Sino-Singapore Tianjin Eco-City of Binhai New District, Tianjin, and the north of the plot is Central Road, the south is Heyun Road, the west is Hefeng Road, and the east is City Park and the commercial complex. High-rise residences are arranged along the east side of the project planning land, and the rest lands are all for multilayer houses. Among which, the external walls of the multilayer houses are built with industrialized prefabricated technology, and this is an attempt of industrialized design on the garden houses with complicated forms and diversified facade effects.

万科锦庐园位于天津滨海新区的中新生态城中，地块北侧为中心大道，南侧为和韵路，西侧为和风路，东侧为城市公园及商业综合体。项目规划用地东侧沿线设置高层住宅，其余用地全部为多层洋房。其中，多层洋房的外墙均采用预制工业化技术建造，是工业化设计在形体复杂、立面效果多样的花园洋房上的一次尝试。

Industrialized Design 工业化设计

Vanke Jin Lu Garden is an industrialized design based on the large-scale and complicated forms, and the facades of the prefabricated external walls are designed with various finish materials. The residence of the development is in the way of internal heat preservation, and the designers have systematically designed the utilization of semi-underground spaces, integrated design of the solar energy system and the architecture, rainwater collection and utilization, integration of civil and decoration engineering and interior natural ventilation. All the above design schemes have ensured the project with a three-star national green standard.

万科锦庐园是基于大规模的复杂形体进行的工业化设计,预制的外墙立面采用了多种饰面材料;项目中的住宅采用了内保温的方式,对半地下空间的利用、太阳能系统与建筑一体化设计、雨水收集及利用、土建与装修工程一体化、室内自然通风等进行了系统设计。上述诸多解决方案使本项目达到了国家绿色三星级别。

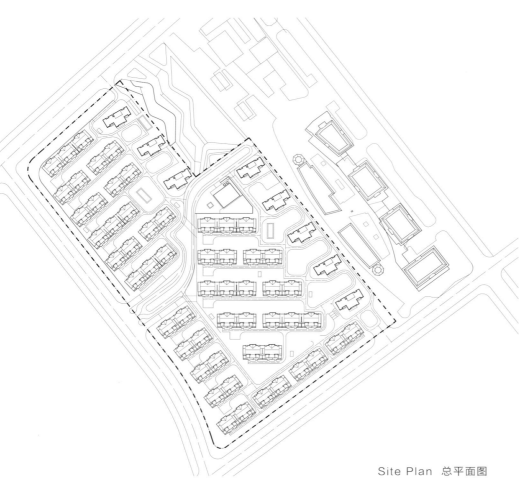

Site Plan 总平面图

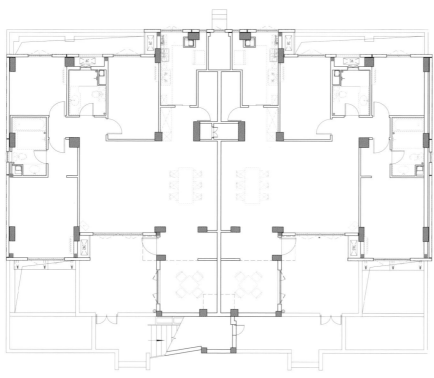

First Floor Plan 一层平面图

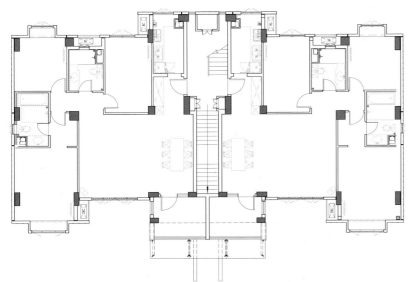

Second Floor Plan 二层平面图

Third Floor Plan 三层平面图

House Type Design 户型设计

The multilayer houses of this project use industrialized prefabricated external wall panels, and the designers have industrially divided the flat panels by combining the facade design at the initial stage of the scheme. On the premise of meeting the plan functional needs and facades' beauty, the designers have optimized the design of the location and size of the windows so that to reduce the template numbers and improve the economic benefit that the industrial production brings.

The standardized design of the house type shows in the several blocks of houses with the same storey height. There are three house types: the house type on the first floor with four rooms, two living rooms and two washrooms, the house type from the second floor to the fourth floor with three rooms, two living rooms and two washrooms, and the house type on the fifth floor with two rooms, two living rooms and one washroom.

项目中多层洋房采用的是工业化的预制外挂墙板，在方案初期就结合了立面设计对平面墙板进行了工业化拆分。在满足平面功能需求及立面美观的前提下，对开窗的位置及窗洞的大小进行优化设计，以便减少模板数量，提高工业化生产所带来的经济效益。

项目中户型的标准化设计体现在数栋同样层高的洋房中，共有三种户型，其中包括位于首层的四房两厅两卫户型、位于二层至四层的三房两厅两卫户型和位于五层的两房两厅一卫户型。

Eco-friendly and Environment-friendly　生态环保

Compared with the similar industrialized residential projects, the designers have entirely prefabricated the external walls and exterior finish materials to meet this high-end residence's precise demands of materials and construction details and to a large extent reduce the construction period of the alpine region to adapt to the local climate conditions. They have also added passive designs and the utilization of a series of green technology and equipments into the scheme, making energy-saving and environment-friendly a highlight of the project. And the 72% realization of the energy conservation target has also established the image of a garden house, a green eco-residential area with outstanding natural circulation characteristics in Sino-Singapore Tianjin Eco-City.

与同类型工业化住宅项目相比，设计师通过对外墙及外装材料的整体预制，实现了高端住宅产品对材质及施工细节的精准需求，同时较大程度缩短了高寒地区的施工周期以适应当地的气候条件；并在方案中增加了被动式设计及一系列绿色技术和设备的使用，使得节能环保成为项目的一大亮点。而72%的节能目标的实现，也树立了中新生态城区内颇具自然循环特性的花园洋房绿色生态住区形象。

New Urbanism City with Ecological Landscape
新都市主义的生态山水之城

East Shore International Zone B　东岸国际 B 区

Location / 项目地点：Hohhot, Inner Mongolia Autonomous Region / 内蒙古自治区呼和浩特市
Developer / 开发商：Inner Mongolia Ideal World Real Estate Development Co., Ltd / 内蒙古雅世春华房地产开发公司
Architectural Design / 建筑设计：Beijing Honest Architectural Design Co., Ltd / 北京奥思得建筑设计有限公司
Construction Drawing Co-design / 施工图合作设计单位：China Architecture Design & Research Group Residential Center / 中国建筑设计研究院住宅中心
Gross Land Area / 总用地面积：约 276 500 m²
Gross Floor Area / 总建筑面积：437 423.28 m²
Overground Floor Area / 地上建筑面积：304 166.52 m²
Underground Floor Area / 地下建筑面积：133 256.76 m²
Plot Ratio / 容积率：1.1
Green Coverage Ratio / 绿化率：41.3%

Keywords　关键词
Courtyard Space; Roof Garden; Flexible Connection
院落空间　屋顶花园　灵活串联

"China's Most Beautiful Stylish Apartments" Finalist
入围"中国最美风格楼盘"奖

The project is organized in the way of "center and functional groups", and it develops in the way of residential neighborhood enclosure. The natural, smooth and curve road and greenbelt flexibly connect the buildings, and multilayer residence and high-rise apartment have enclosed a clear group space, forming the courtyard space and the natural open and close of various spaces.

项目采用"中心—功能组团"的结构组织方式，以住宅邻里围合的形式层层展开。自然流畅的曲线道路和绿带将建筑灵活串联，多层住宅和高层公寓围合成明确的组团空间，形成院落空间，各种空间收放自然。

Overview　项目概况

The East International large-scale residential community is at the east bank of the East River of Ruyi District, Hohhot, and to the east of the city main road Xinhua Street. It's near the city's new and important building groups such as the Inner Mongolia Government Building, Inner Mongolia Museum and International Convention Center. The base is divided into three areas as A, B and C from north to south; and this development is the middle B area, which is a wasteland of the East River and it's with a square and smooth terrain.

　　东岸国际大型居住社区位于呼和浩特市如意区东河东岸，北接城市主要干道新华大街，临近内蒙古自治区党政大楼、内蒙古博物馆、国际会展中心等城市新建重要建筑群。地块自北向南分为 A、B、C 三区，本规划为中间的 B 区，原为东河河滩荒地，地形方正平坦。

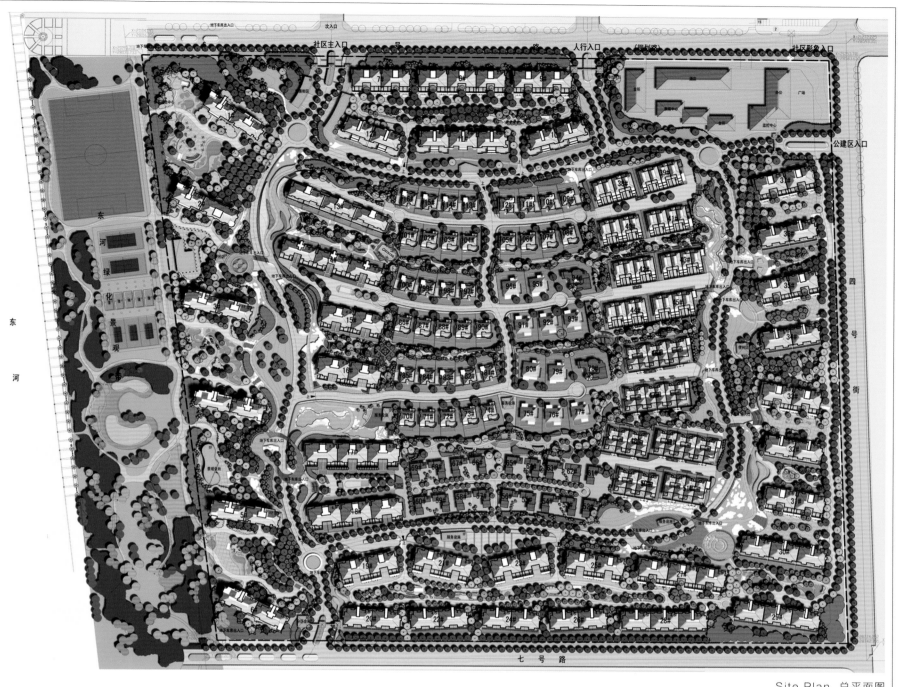

Site Plan 总平面图

Aerial View 鸟瞰图

Overall Layout 总体布局

The project is organized in the way of "center and functional groups", and it develops in the way of residential neighborhood enclosure. The natural, smooth and curve road and greenbelt flexibly connect the buildings, and multilayer residence and high-rise apartment have enclosed a clear group space, forming the courtyard space and the natural open and close of various spaces. The overall layout of the planning is in a freely grown leaf vein shape, and it fully uses the terrain and water flow to enhance the flexibility, random changes and interests of the space.

项目采用"中心—功能组团"的结构组织方式，以住宅邻里围合的形式层层展开。自然流畅的曲线道路和绿带将建筑灵活串联，多层住宅和高层公寓围合成明确的组团空间，形成院落空间，各种空间收放自然。规划总体格局采用的是一个自由生长形态的"叶脉状"布局形式，充分利用地形、水势，使空间的灵动性增加，随机性的变化多，趣味性比较强。

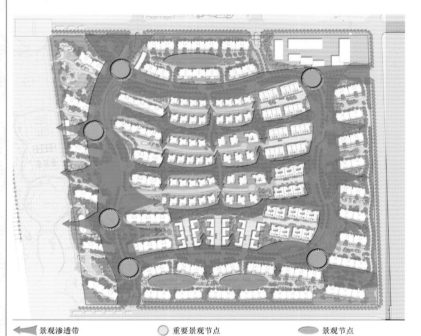

Landscape Analysis Drawing 景观分析图

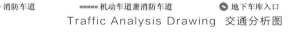

Traffic Analysis Drawing 交通分析图

Residential Design 住宅设计

Garden House

Garden house is the main product of this project, and in order to differentiate it from the traditional multilayer plate type residence, this product has tried innovation and adopted advanced methods regarding the plan distribution, functional organization, spring layers and equipment to truly meet its upgrading demands.

In order to highlight the two themes of green ecology and stereo landscape, the designers have specially designed a south oriented sunshine garden that is no less than 10 m² in each unit, lowered the floor to ensure the real soil planting, interlaced the green plant and the leisure deck chair and combined the living room and the main bedroom at the surrounding, providing the residents with interior environmental enjoyment that is totally different from the other residences. They combine the multilayer terrace whose facade retreats layer-by-layer to design the roof garden and bring the greenness into the air for real.

花园洋房

花园洋房是本项目的主力产品，为了区别于传统多层板式住宅，本方案在平面分区、功能组织、跃层空间及设备设施等多个方面尝试创新和采用先进的方式，真正满足项目升级的诉求。

为突出本项目绿色生态和立体景观的双重主题，设计中在每个户型均特别布置了不小于10 m²的南向阳光花园，其间通过降低楼板，确保实土种植，绿色植物和休闲躺椅交织，将起居厅和主要卧房结合在周边，提供给居住者完全不同于其它住宅的室内环境享受。结合立面造型层层进退的效果形成的多层露台设计成屋顶花园，把绿色真正带到空中。

Standard Floor Plan 1 标准平面图 1

Standard Floor Plan 2 标准平面图 2

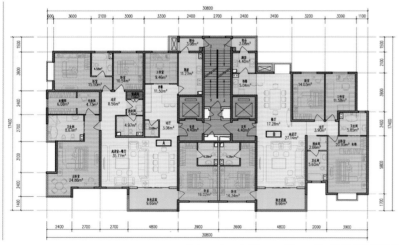

Standard Floor Plan 3 标准平面图 3

Standard Floor Plan 4 标准平面图 4

Single Family House

The single family house design of Zone B first adds the landscape space for public activities to meet the community activity demands of the residents. The interior space of the single family houses is organized in a clear, reasonable and interesting way. And the site elevation of this area has been raised to reduce the visual pressure of the surrounding high-rises and multilayer buildings and to further highlight the noble status of this core development in this core area.

独立别墅

B区的独立别墅设计首先增加了公共活动景观空间，满足了居住者社区活动的要求；同时在别墅内部空间组织上更简明合理，富有情趣；再者提高了别墅区域的场地标高，减少了周边高层和多层建筑的视觉压力，进一步突出了其核心区域核心产品的尊崇地位。

Townhouse

Townhouse is a brand new development of Zone B. Each townhouse is collocated with private sunken courtyard which well connects the public underground garage, enriches the architectural space and solves the lighting problem of the underground rooms. Two blocks of townhouses embrace the sunken courtyard to reduce the long feeling that's similar to the multilayer residence.

联排别墅

联排别墅是B区一种全新产品类型。每户联排别墅都配有各自独立的地下庭院，可以很好地与地下公共停车库串联，同时丰富了建筑空间，解决了地下房间的采光问题。两栋别墅建筑环抱地下庭院，弱化了联排别墅类似多层住宅的长度感觉。

Townhouse 1st Underground Floor Plan
联排地下一层平面图

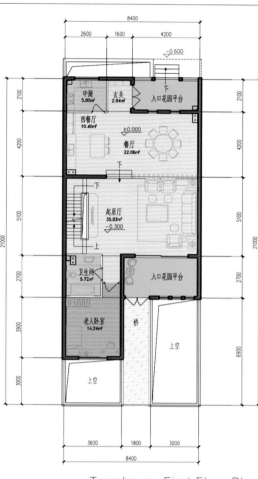
Townhouse First Floor Plan
联排一层平面图

Townhouse Second Floor Plan
联排二层平面图

High-rise Apartment

The plan design and house type arrangement develop on the basis of landscape theme. As it's the nearest to the East River and is near the most important internal landscape canyon corridor, the two house types at the east and west sides of two units are design into large-scale 5-room panoramic houses, and the two houses in the middle are with smaller 4 rooms. The 6 blocks are all designed in the type of "two family share one lift on one floor". According to different regional locations, the designers have adopted different approaches such as the split-level design, wide hall design and double main bedroom design, and each one has its own characteristics.

高层公寓

平面设计和户型安排围绕着景观主题展开，由于其位置最临近东河，同时又靠近最主要的内部景观峡谷走廊，因此两单元东西两端的两个户型设置为5居大面积的全景户型，中间两个户型为面积较小的4居户。6栋楼均为1梯2户，本方案根据其不同的区域位置，采用了不同的处理方式，有错层设计、宽厅设计、双主卧房设计等，每一种都各具特色。

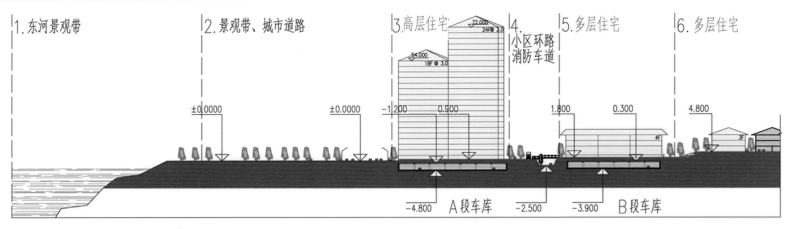

Vertical Analysis 1 竖向分析图 1

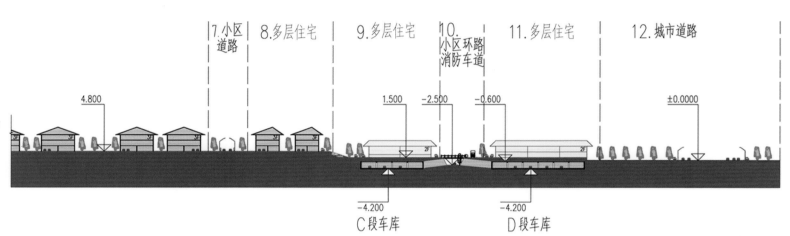

Vertical Analysis 2 竖向分析图 2

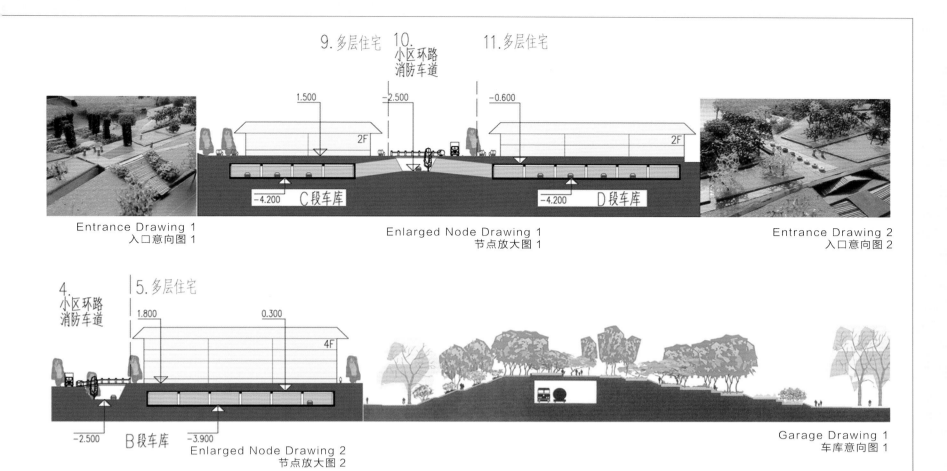

Entrance Drawing 1
入口意向图 1

Enlarged Node Drawing 1
节点放大图 1

Entrance Drawing 2
入口意向图 2

Enlarged Node Drawing 2
节点放大图 2

Garage Drawing 1
车库意向图 1

Landscape Design 景观设计

The project has built plenty groups of mutually interlaced linear greenbelt and combined with the winding water to closely run through each residential group, leading the landscape idea of the East River well into this community. It also highlights the waterfront residence concept, combines the city green landscape and culture of the community to improve the quality of green construction.

项目建立多组相互交织的流线型绿化带，配合蜿蜒环绕的水体，紧密地穿插在各组住宅群间，将东河的景观意念很好地引入到本社区内，突现本项目水岸住区地概念，把城市绿地景观与社区文化相结合，提升了绿化建设的品质。

The Elegant Ambiance and Aesthetic Culture of the Upper Class Create the Living Standard of Yinchuan

上流阶层典雅气息和美学文化缔造银川人居标准

Yinchuan Hailiang International Community | **银川海亮国际社区**

Location / 项目地点：Yinchuan, Ningxia Hui Autonomous Region, China / 中国宁夏回族自治区银川市
Developer / 开发商：Ningxia Hailiang Real Estate Co., Ltd. / 宁夏海亮房地产有限公司
Architectural Design / 建筑设计：United Architects & Engineers Co., Ltd. (UA Design) / 北京中联环建文建筑设计有限公司
Floor Area / 建筑面积：2 012 500 m²

Keywords 关键词

Hard and Soft; Elegant Ambiance; Extraordinary Manner
刚柔并济 典雅气息 非凡气度

"China's Most Beautiful Stylish Apartments" Finalist
入围"中国最美风格楼盘"奖

The project is composed of small high-rise, high-rise, semi-detached house and townhouse. The ultralow density and the extremely large landscape floor distance manifest the extraordinary life manner. Large-scale villa groups are quite rare in Yinchuan, and the high-rises are symmetric in forms with hard and soft lines.

项目由小高层、高层、双拼别墅、联排别墅组成，超低密度及超大宽景楼间距，丈量非凡人生气度，大规模的别墅群在银川也堪称稀缺，高层楼型对称，线条刚柔并济。

Overview 项目概况

Hailiang International Community, the first quality development that Hailiang Real Estate developed in Yinchuan, is located to the north of Helanshan Road, to the east of Tanglai Canal, and across the road to the North Tower Lake (Haibao Park). It's an important planning subordinated to the urban development strategy "North Development". The project shares various high-end living facilities such as the Municipal Government, City Cultural Center, gymnasium, district hospital and the new Secondary High School at the surrounding 6km inner race. Along with the improvement and perfection of the various facilities for the regional living, commerce and leisure, the livable value and appreciation value will be further proved by the market.

海亮地产在银川开发的第一个品质项目——海亮国际社区，位于贺兰山路以北，唐徕渠以东，与北塔湖（海宝公园）仅一路之隔，隶属城市发展战略"北拓"的重要规划。项目周边6 km内环伺市政府、市文化中心、体育馆、区医院及新二中等数个高端生活配套，随着区域内生活、商务、休闲等多项设施的相继完整和完善，此板块的宜居价值和升值潜力则得到进一步的市场证明。

Aerial View 鸟瞰图

Planning 规划布局

The project uses the different characteristics of the construction land to create different functional themes. Plot 1 is near the southern Aiyi River and will be created as high-end villa community; plot 2 will be created as high-end high-rise residential community with no residence at the northern part and the building height will not be restricted to sunshine; plot 3 faces the central landscape axis of this region and will be created as a luxury club and neighborhood center that gathers sales center and fitness as a whole.

项目利用建设用地的不同特性，打造不同的功能主题。1#地靠近南侧艾依河，将打造成高端的别墅社区；2#地北侧不再有住宅用地，建筑高度不受日照限制，将打造成高端的高层居住社区；3#地正对区域中心景观轴，将打造成兼顾售楼、健身功能于一体的豪华会所及社区邻里中心。

Design Concept 设计理念

Based on the geographical location and definition of the first party, the designers comprehensively create the atmosphere of high-end residential area, and bear this purpose on the design of function, plan layout, ancillary facilities and facade style to create high value-added product. The Hailiang International Community and the North Tower Lake high-end Eco-residential area are connected as a unity. The project is with water on three sides, enjoying the advantaged natural landscape and the strong and beyond compare cultural advantages. The British picturesque landscape, European-style townhouse, the Art Deco-styled high-rise and the magnificent luxury villa groups make the noble manner of this project reveal naturally. The multifunctional and integrative high-end club and the 360° international butler service concept guide the modernized living of Yinchuan.

根据项目所处地理位置及甲方的项目定位，设计上全方位营造高端居住区的氛围，无论在功能上、平面布置上、配套设施上，以及立面风格上均以此为目标，创造高附加值产品。海亮国际社区与北塔湖高档生态住宅区连为一体。项目三面环水，得天独厚的自然景观与厚重的人文优势无与争锋。英伦 CHAMBER 画意式园林景观，欧派建筑风格的联排别墅及 ART DECO 风格的高层，宏阔的奢贵别墅群，让项目的贵族气质不彰自显。多功能一体化的高档会所，360°全方位的国际管家服务理念，秉赋银川现代化人居版图。

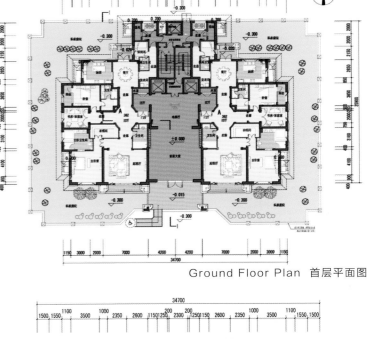

Ground Floor Plan 首层平面图

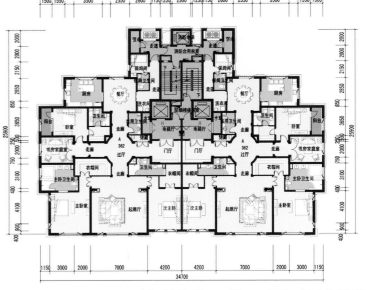

13~28th Floor Plan 十三~二十八平面图

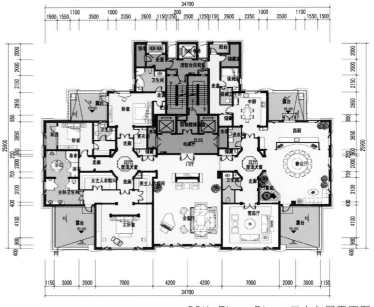

29th Floor Plan 二十九层平面图

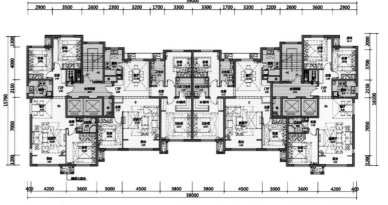

Standard Floor Plan of Compound Residence
组合住宅标准层平面图

Building Groups 建筑群落

The project is composed of small high-rise, high-rise, semi-detached house and townhouse. The ultralow density and the extremely large landscape floor distance show the extraordinary life manner. Large-scale villa groups are quite rare in Yinchuan, and the high-rises are symmetric in forms with hard and soft lines. The buildings are tall and straight, creating the elegant ambiance and aesthetic culture of the upper class and the living standard of Yinchuan.

项目由小高层、高层、双拼别墅、联排别墅组成，超低密度及超大宽景楼间距，丈量非凡人生气度，大规模的别墅群在银川也堪称稀缺，高层楼型对称，线条刚柔并济，建筑物高耸、挺拔，构筑出了上流阶层的典雅气息和美学文化，缔造银川人居标准。

Landscape Design 景观设计

With Aiyi River in the south and Tanglai Canal in the north, the project is with superior natural conditions. The designers make full use of the landscape at both sides on the planning of the high-rise area. The internal part of the project has reserved landscape spaces with different characteristics for different functional areas to remain some space for further in-depth landscape design.

用地南有艾依河，北有唐徕渠，自然条件优越，在规划上考虑高层区域对双侧景观的充分利用；同时，在项目内部为不同功能分区预留不同特色的景观空间，为进一步的景观深化设计留有余地。

The Youthful "Paradise Walk" Commercial Complex Group
洋溢青春气息的"天街"商业综合组团

Chongqing Longfor · U City　　**重庆龙湖 · U 城**

Location / 项目地点：Shapingba District, Chongqing / 重庆市沙坪坝区
Architectural Design / 建筑设计：Shenzhen Huahui Design Co.,Ltd. / 深圳市华汇设计有限公司
Land Area / 占地面积：198 700 m²
Floor Area / 建筑面积：658 600 m²
Building Density / 建筑密度：36.82%
Plot Ratio / 容积率：2.5
Green Coverage Ratio / 绿化率：30%

Keywords　关键词

Passive Energy Saving; Community Cohesion; Enclosed Courtyard
被动式节能　小区内聚性　院落围合感

"China's Most Beautiful Stylish Apartments" Finalist
入围"中国最美风格楼盘"奖

The "narrow in width, long in length" shape characteristics of the building can well accommodate itself to the terrain to form different spatial forms, bring people with changing landscapes, and ensure the comfortable vision and spatial feeling between the buildings.

"小面宽、大进深"的体形特点能很好地迁就地形组合不同的空间形态，为项目带来移步景迁的景观感受，同时又能保证楼栋之间有舒适的视野与空间感受。

Overview　项目概况

Located in the planning area of Chongqing University Town, Huxi Town, Shapingba District, the western Chongqing, the project is between the Zhongliang mountain chain and the Jinyun mountain chain; and near the Chongqing University, Sichuan Fine Arts Institute and Chongqing Normal University, the project enjoys a good culture and study atmosphere; the metro line 1 passes through the plot and the planned light rail station is near the development, this development is with quite convenient transportation and is the extended area of the majorly developed urban economic circle of Chongqing.

项目位于重庆市西部沙坪坝区虎溪镇重庆大学城规划区内，中梁山脉和缙云山脉之间，紧邻重庆大学、四川美术学院、重庆师范大学，拥有良好的人文学风氛围，地铁 1 号线穿过项目地块，规划的轻轨站点就在项目旁边，交通十分便捷，是重庆市重点发展的都市经济圈的扩展区域。

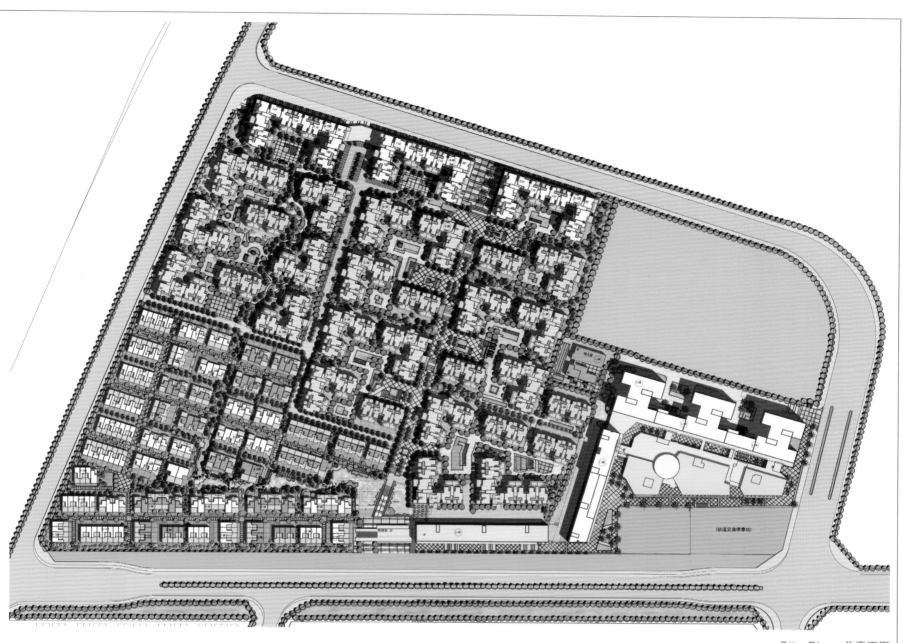

Site Plan 总平面图

Aerial View 鸟瞰图

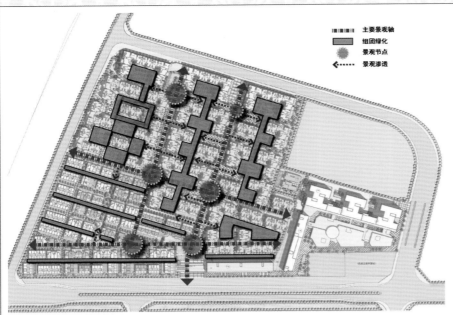
Landscape Analysis Drawing 景观分析图

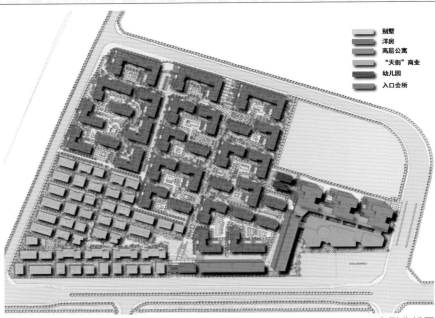
Housing Units Analysis Drawing 户型分析图

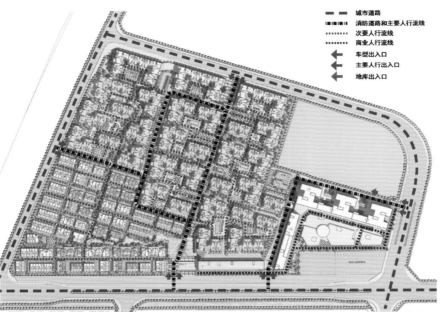
Traffic Analysis Drawing 交通分析图

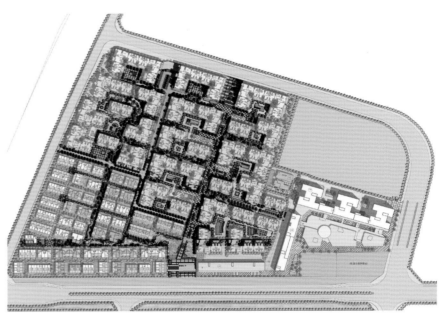
Landscape Planning Analysis Drawing 景观规划分析图

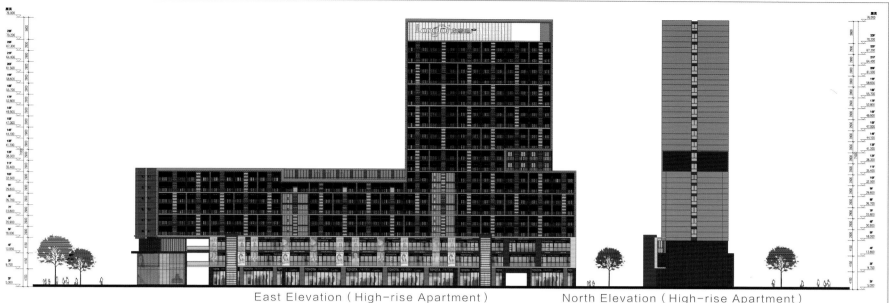

East Elevation (High-rise Apartment)
高层公寓东立面图

North Elevation (High-rise Apartment)
高层公寓北立面图

West Elevation (High-rise Apartment)
高层公寓西立面图

South Elevation (High-rise Apartment)
高层公寓南立面图

Planning 规划布局

Divided into three phases, the project is entirely developed by Chongqing Longfor Real Estate, and the overall layout is composed of three big groups, the large-scale commercial complex area, foreign-style house area and high-end villa area, which are with clear distribution and well arranged. The designers use the mutual penetration of the landscape axis and the inner courtyard landscape as well as the overall spatial form of the complex to organically organize the various groups, forming rich landscape sequences and lively community spatial experience. The community emphasizes the cohesion and enclosed courtyard of the groups to create a steady, safe and scenic community environment.

项目由重庆龙湖统一开发共分三期，整体布局由大型商业综合区、洋房居住区及高端别墅区三大组团组成，彼此分区清晰层次分明，通过景观轴线和内庭景观的相互渗透以复合型的整体空间形态，将多组团有机组织，形成了丰富的景观序列和生动的社区空间体验，小区强调内聚性和组团的院落围合感，营造稳定安全、景色优美的社区环境。

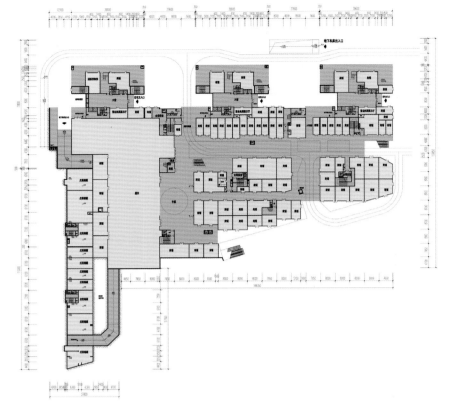

First Floor Plan of Commercial
商业一层平面图

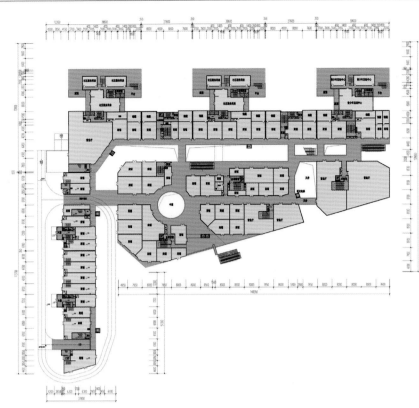

Second Floor Plan of Commercial
商业二层平面图

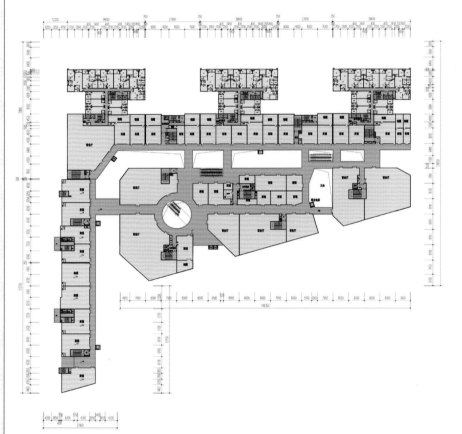

Third Floor Plan of Commercial
商业三层平面图

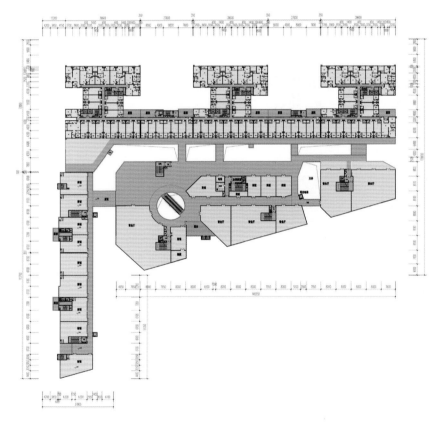

Fourth Floor Plan of Commercial
商业四层平面图

House Type Planning 户型规划

The featured development of the commercial complex group is SOHO small unit, and collocated with above 40,000 m² of centralized businesses of Longfor "Paradise Walk" series, the living supporting of this project has been improved to a higher level. The featured development of the foreign-style house group is the enclosed innovative unit that is specially designed for the planning form of this project. The "narrow in width, long in length" shape characteristics of the building can well accommodate itself to the terrain to form different spatial forms, bring people with changing landscapes, and ensure the comfortable vision and spatial feeling between the buildings.

商业综合组团特色产品为SOHO小户型，配上四万多平方米的龙湖"天街"系列集中商业使得项目的生活配套得到较高水平的提升。洋房组团特色产品侧是为了该项目的规划形态专门研发的围合式的创新户型。"小面宽、大进深"的体形特点能很好地迁就地形组合不同的空间形态，为项目带来移步景迁的景观感受，同时又能保证楼栋之间有舒适的视野与空间感受。

House Type Design 户型设计

The spatial manifestation of large space and small courtyards is expected to be created on planning, so the designers organically organize and combine 4 to 5 units to form semi-enclosed groups with different sizes and various forms; and they use the landscape approaches to organically connect the courtyard of each group and the landscape axis to form fishbone structured overall landscape system. The shape design of "narrow in width, long in depth" can enclose groups with different sizes according to the site situations and freely change the forms, and also can make full use of the width to ensure more residents with the best orientation and landscape angles to show the wellness of the unit. The group courtyards formed by combination create a "public living room" that belongs to the whole neighborhoods. And besides the public space of the community and the private space of the villas, a new spatial layer has been formed and a new living experience and state of life can be possible.

从规划上希望是营造一个大空间小院落的空间体现，将四到五个单元体有机的排布组合，从而形成大小不一，形态丰富的半围合组团；利用景观的手法把各个组团的庭院和景观轴线有机串联起来形成鱼骨架构的整体景观体系。"小面宽、大进深"的体形设计，一方面可以根据场地的情况围合出不同大小的组团并可较自由地变动形态，另一方面可充分地利用面宽使更多的住户有着最好的朝向和景观面，体现均好性。通过组合生成的组团庭院形成一个属于邻里间的"公共客厅"。从而在社区公共空间和别墅私家空间之外，一个新的空间层次由此生成，一种新的居住体验和生活状态成为可能。

Second Floor Plan of Garden Apartments
洋房二层平面图

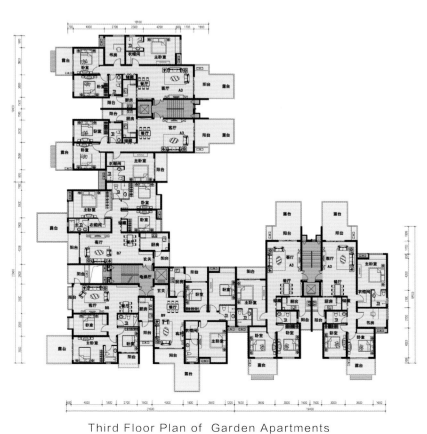

Third Floor Plan of Garden Apartments
商业三层平面图

East Elevation 东立面图

West Elevation 西立面图

South Elevation 南立面图

North Elevation 北立面图

Facade Design 立面设计

The facade design of the building begins with function, considers the basic requirement such as lighting, shelter and privacy, and combines the comprehensive control of volume, material, and color to realize the aesthetic integration on practicality and decorativeness.

建筑立面设计从功能本身出发，以采光、遮阳、私密性等方面的考虑为基本诉求，结合对体量、材质、色彩的综合把握，实现了实用性与装修性在审美上的统一。

Climate Adaptation 气候适应

The planning design fully considers the subtropical climate of the project's location, pays great attention to the natural ventilation, sunshade and summer heat insulation, makes the passive energy saving as starting point, and uses the architectural design itself instead of additional technical means to reach the ideal energy saving effect.

规划设计充分考虑了项目所处的亚热带气候的特征，充分强调自然通风、防晒遮阳及夏季散热保温等需求，以被动式节能为出发点，通过建筑设计本身，而不依赖附加技术手段达到了理想的节能效果。

Commercial Classic of Chinese Jade City
中国玉都的商业经典之作

Jieyang Wanyu Plaza, Phase I 揭阳万玉广场一期

Location / 项目地点：Rongcheng District, Jieyang, Guangdong / 广东省揭阳市榕城区
Developer / 开发商：Guangdong Jihai Investment Group / 广东吉海投资集团
Architectural Design / 建筑设计：Teamer Architectural Design and Consultants Co., Ltd. / 天萌国际设计集团
Landscape Design / 园林设计：Guangzhou Bonjing Landscape Design Co., Ltd. / 邦景（广州）园林绿化设计有限公司
Land Area / 用地面积：26 576.55 m²
Floor Area / 建筑面积：97 852.50 m²
Plot Ratio / 容积率：2.63

Keywords 关键词
Chaoshan Culture; Natural Interest; Neo-Chinese Simple Style
潮汕文化　自然情趣　新中式简约

"China's Most Beautiful Stylish Apartments" Finalist
入围"中国最美风格楼盘"奖

On the premise of remaining the pleasant street, and the commercial and residential complex of the traditional residential area, the project provides more residential spaces, larger central leisure spaces and richer business models to bring better core commercial area for the surroundings.

项目在延续传统住区的宜人街区、商住一体、宜商宜居的前提下，提供更多的居住空间，更大的中央休闲空间，更丰富的商业模式，为周边带来更好的核心商业区。

Overview 项目概况

Located in Yangmei, Rongcheng District, Jieyang, where is well known for jade, and the central core area where enjoys the reputation of "Chinese Jade City" and "Asian Jade City", Jieyang Wanyu Plaza phase I is to the north of Qiaonan jade center and to the west of Pandong Road. Wanyu Plaza phase I is undertaken in the west side and it's to the south of Qiaoxi white jade market, to the north of Qiaonan jade center and in the geometrical center of the three large jade markets. The project will be built as a diversified large-scale commercial center that gathers jade processing, business negotiation, transaction and appreciation as a whole.

揭阳万玉广场一期坐落于因玉而闻名遐迩的揭阳市榕城区阳美，并处在有着"亚洲玉都"与"中国玉都"美誉的中央核心地段上，位于乔南玉器中心北侧，磐东路西侧。西面承接万玉一期，北临乔西白玉市场，南接乔南玉器中心，处于三大玉器市场几何中心。项目将建设成为集玉器加工、商洽、交易、鉴赏于一体的多元化大型商贸中心。

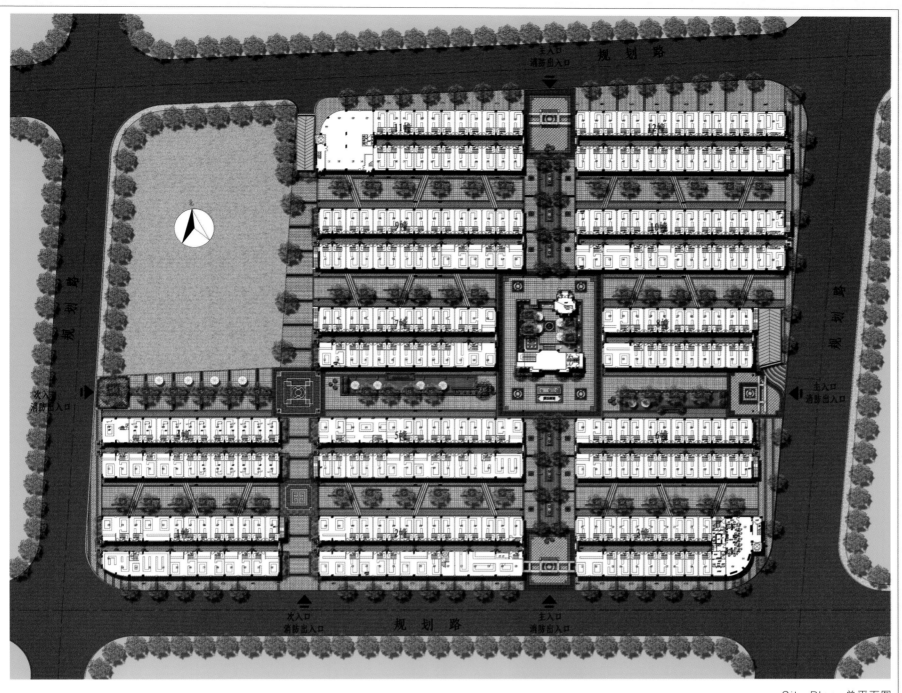

Site Plan 总平面图

Aerial View 鸟瞰图

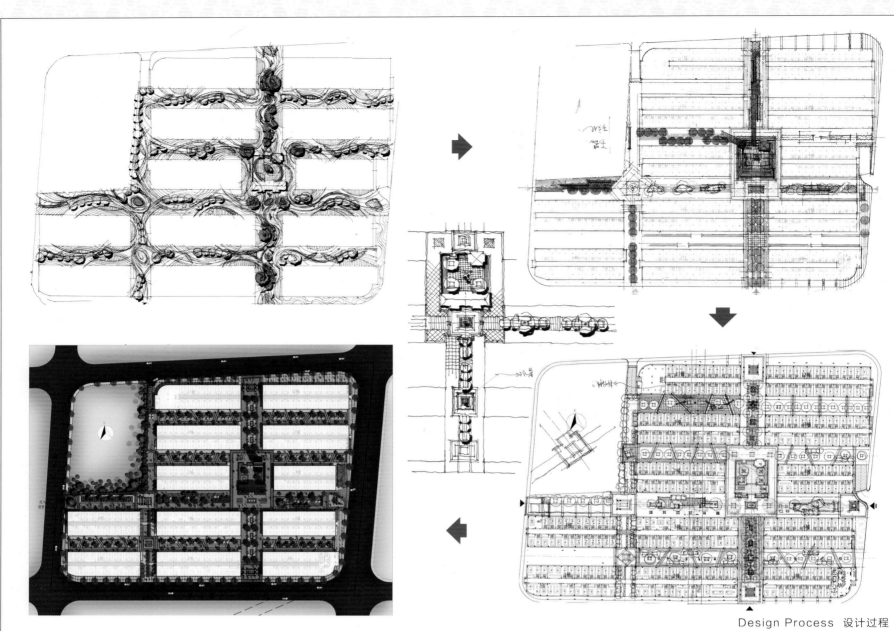

Design Process 设计过程

Planning 规划布局

The development is composed of two lands on the east and west with a gross land area of 72,936 m². phase Ⅰ is planned with a gross land area of 32,210.25 m² and a net land area of 26,576.55 m², and it's with a gross floor area of 97,852.50 m², among which the overground floor area is 68,200.65 m² and the underground floor area is 22,532.30 m². The project plans 12 blocks of single-family commercial villas with a totality of 322 units, and the first floor is designed as stores for jade ware and brand showcases. The second floor is designed as places for jade processing and private negotiations between the clients. The third to fifth floors are designed as functional areas for storages and residences.

项目用地由东西两块地组成，总用地为 72 936 m²。一期规划总用地面积 32 210.25 m²，规划净用地面积 26 576.55 m²，首期总建筑面积为 97 852.50 m²，其中地上建筑面积为 68 200.65 m²，地下建筑面积为 22 532.30 m²，项目规划 12 幢独栋式商业别墅，共 332 套，首层以商铺设计，为玉器展示、品牌展示区。二层可作为玉器加工场所，客户洽谈私密区。三至五层为仓储、居住功能区。

West Elevation Along Street 小区西向沿街立面图

East Elevation Along Street 小区东向沿街立面图

South Elevation Along Street 小区南向沿街立面图

North Elevation Along Street 小区北向沿街立面图

Section 1-1 小区 1-1 剖面图

Architectural Style 建筑风格

The architectural style continues the oriental feature and excellent quality of the modern Chinese-style which is simple and elegant, noble and steady, and pays great attention to the innovation of fashion. The facade design adopts the modern concise Hui-style form to creatively build a Chinese-style community which is pleasant in dimension, suitable for living and business, with Chinese-style appearance and white walls and black tiles, reflecting a modern concise, elegant and noble classic.

在建筑风格方面，继承了现代中式的东方特色及优秀品质，古朴高雅，贵气稳重，同时又非常注重时尚的创新，立面上采用现代简约徽派形式，创新地打造出一个具有宜人尺度，宜居宜商、中式外观、粉墙黛瓦的中式社区，反映出一副现代简约、雅致尊贵的上乘之作。

Architectural Analysis 建筑分析

The distinctive single-family commercial villa of this project can well coordinate the large commercial space with the small residential space, deal with the dimensions between the openness of the commercial space and the privacy of the residential space, separate the driveways and sidewalks of the commercial and residential areas and skillfully and organically combine the forms of the various spaces. This kind of property form has highly inspired the designers' passion.

项目特殊的独栋式商业别墅业态，很好地协调好商业空间的"大"和住宅空间的"小"，很好地处理了商业空间开放性和住宅空间私密性之间的尺度，并划分好商业及住宅的人车动线，巧妙地将各空间的形态有机地结合起来。这样的物业形态，极度激发了设计师的热情。

Architectural Layout 建筑布局

On the architectural layout, the designers have adopted the regular row to settle concise linear line for the landscape.

在建筑布局方面，采用了规整的行列式布局，为景观奠定了简约干练的线性空间。

Design Concept 设计构思

The design adopts the modern concise approach and symmetrical layout, which is in a fresh and elegant style. The color is mainly in cold tone, which is heavy and mature. The decorating detail advocates natural interest, and the texture of the finishing materials is integrated with modern elements and traditional styles to fully manifest the Chinese traditional landscaping charm. It organically combines the traditional Chinese garden and the modern arts to let people enjoy the concise and unique integration of the Neo-Chinese garden and modern arts and to manifest the Neo-Chinese simple style which possesses dual aesthetic interests.

设计采用现代简约手法，对称式布局方式，格调清新而高雅。色彩以冷色为基调，浓重而成熟。在装饰细节上崇尚自然情趣，饰面材料的质感肌理，融入时尚元素与传统风格，充分体现出中式传统造园神韵。同时将传统中式园林和现代艺术有机结合，品味新中式园林与现代艺术简约个性的融合，呈现出具备双重审美情趣的新中式简约风格。

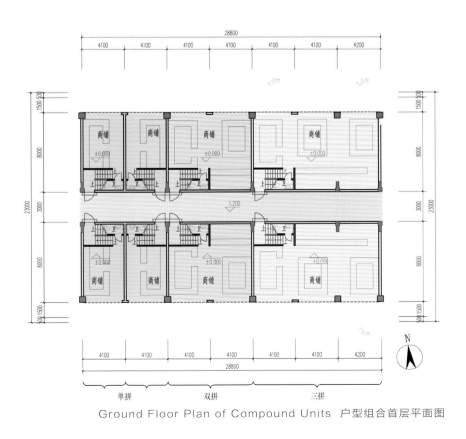

Ground Floor Plan of Compound Units 户型组合首层平面图

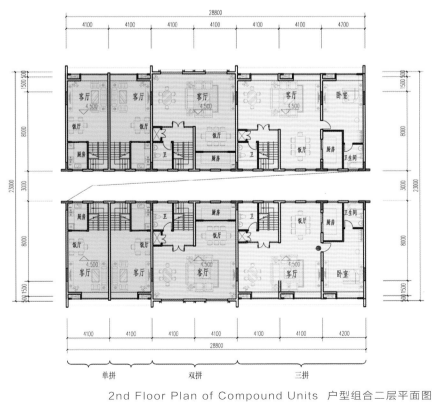

2nd Floor Plan of Compound Units 户型组合二层平面图

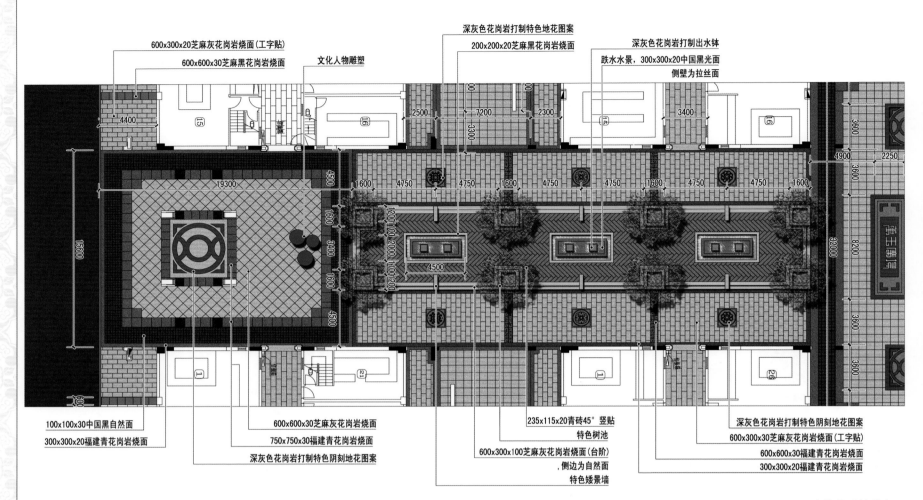

Jade Culture Landscape Node 1　玉石文化景观轴节点一

Landscape Design　景观设计

Based on the architectural planning layout, the landscape design has optimized the spatial structure of the place, completed the visual organization of the site, and divided the site into the landscape structure of "One Core, Two Axes and Multi-street".

One Core
Wanyu Jubao Core Plaza, the core landscape of the whole place, implies "gathering wealth from all directions".

Two Axes
As the main axis landscape of the site, the exhibition function and leisure function of the landscape is quite important, thus, the design mainly adopts the jade culture, together with the Chaoshan folk culture, and uses the trace of the landscaping elements to tell the story of the place, presenting people the quintessence of jade culture and the local conditions and customs of the Chaoshan region, and providing the one that saunters in it a leisure and comfortable public space.

1. Jade Culture Landscape Axis
It's a south-north oriented landscape axis which makes the jade culture as soul to present the thematic culture of Wanyu Plaza.

2. Folk Culture Landscape Axis (East-West Oriented)
It's an east-west oriented landscape axis which makes the Chaoshan folk culture as main part to promote the local conditions and customs of the Chaoshan region.

据规划建筑布局，景观设计优化了场地空间结构，完善场地的视线组织，将场地划分为"一核心、二轴线、多街道"的景观结构布局。

一核心
万玉聚宝核心广场，整个场地的核心景观，寓意为"纳四方财富"。

二轴线
作为场地的主要轴线景观，其景观的展示性功能与休闲性功能尤其重要，故此，设计以玉石文化为主，以潮汕民俗文化为辅，用造景元素本身的痕迹诉说着场地的故事，向人们展示玉石文化的精髓和潮汕民俗的风土人情，并给闲步其间的人们提供闲适的公共空间。

1、玉石文化景观轴
南北向景观轴线，以玉石文化为魂，体现万玉广场主题文化。

2、民俗文化景观轴（东西向）
东西向景观轴线，以潮汕民俗文化为主体，弘扬潮汕地区风土人情。

Multi-street

Each commercial pedestrian street of the place makes the ancient Chinese-style street as blueprint, presents the two different spatial dimensions of the "street" and "site", pays attentions to the detailed, humanized design concept and strives to provide people with a pleasant shopping environment.

多街道

场地各商业步行街，以中式韵味的百年老街为蓝本，体现"街"与"场"两种不同的空间尺度，注重细节化、人性化的设计理念，力求为人们提供一个愉悦的购物环境。

方案一 芝麻灰花岗岩打制石鼓透视图

方案二 芝麻灰花岗岩特色矮墙透视图

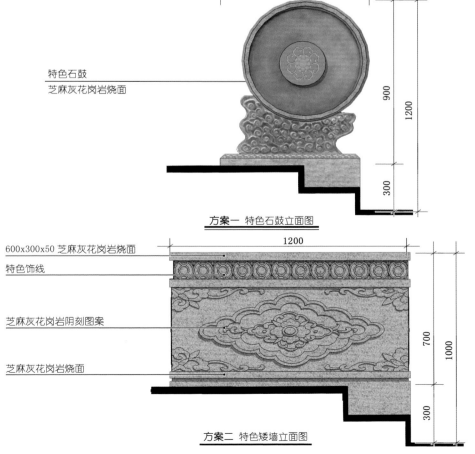

方案一 特色石鼓立面图

方案二 特色矮墙立面图

Parapet Style of Landscape Axis for Jade Culture and Folk Culture　玉石文化及民俗文化景观轴矮墙样式

Humanistic Residential Area that Situates in the Core Downtown Area of Bao'an
坐拥宝安核心旺地的人文住宅小区

Zhongzhou Central Park　中洲中央公园

Location / 项目地点：Shenzhen, Guangdong / 广东省深圳市
Developer / 开发商：Shenzhen Zhongzhou Baocheng Propertty Ltd. / 深圳市中洲宝城置业有限公司
Architectural Design / 建筑设计：Peddke Thorp Melbourne Pty Ltd. / 澳大利亚柏涛建筑设计有限公司
Land Area / 占地面积：90 836.24 m²
Floor Area / 建筑面积：406 000 m²
Green Coverage Rate / 绿化率：4.28

Keywords　关键词

Traditional Street; North-south Ventilation; Humanistic Ambiance
传统街区　南北对流　人文气息

"China's Most Beautiful Stylish Apartments" Finalist
入围"中国最美风格楼盘"奖

The residence emphasizes the north-south ventilation and regression to nature. The combination of static and dynamic interior spaces and the collocated Art Deco design fully manifest the elegant style. The traditional street with strong humanistic ambiance forms a delightful contrast with the residences.

项目住宅强调南北对流，寓于自然。室内动静空间相结合，配以装饰艺术派设计，尽显典雅风格。极具人文气息的传统街区也与住宅相映成趣。

Overview　项目概况

Located in Bao'an District of Shenzhen, the project is developed by Shenzhen Zhongzhou Baocheng Propertity Ltd and designed by Peddke Thorp Melbourne Pty Ltd. It boasts a land area of 90,836.24 m² and gross floor area of 406,000 m², including high-end residences, apartments and commercial buildings, and it's the rare high-end flagship urban complex project of Shenzhen.

项目位于深圳市宝安区，由深圳市中洲宝城置业有限公司开发，由柏涛设计完成。用地面积90 836.24 m²，总建筑面积406 000 m²，包括高尚住宅、公寓、商业等，是目前深圳罕有的高端旗舰城市综合体项目。

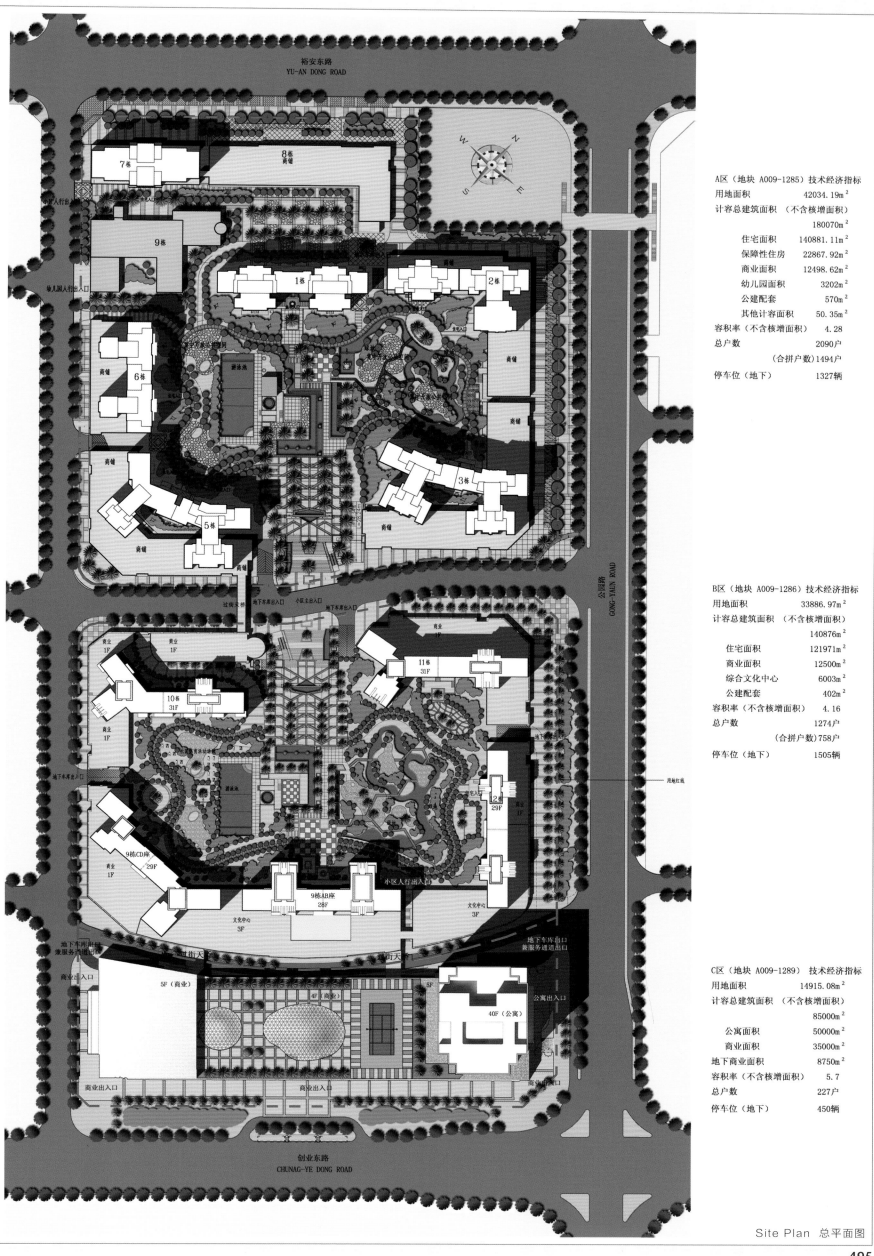

Site Plan 总平面图

Aerial View 鸟瞰图

Planning Design 规划设计

The project is defined as "urban complex" on planning. An interdependent and mutually beneficial relationship has been built between each part to form a multifunctional and high-efficient complex. During the planning design, the designers make the landscape as inner guidance of the site, use the approaches of landscape borrowing and landscaping, and make the greenness and waterscape as theme, forming a poetic imagery for the residential area and creating an elegant and tranquil living atmosphere for the successful people.

项目以"城市综合体"为规划定位。在各部分间建立一种相互依存，相互助益的能动关系，从而形成一个多功能、高效率的综合体。规划设计中在基地内部以景观为主导，采用借景与造景的手法，并以绿色与水景为主题，充分构筑了住宅小区内充满诗情画意的意境，为现代成功人士营造一个高雅、宁静的生活氛围。

Design Concept 设计理念

The project holds the intensive land-saving design concept. The buildings are arranged along the periphery of the plot to create comfortable and open courtyard greening space. The flexible layout provides the building itself with favorable conditions such as ventilation and lighting.

项目的设计理念为集约化节地型设计，建筑沿地块周边布置以创造舒适开敞的大庭院绿化空间。灵活的布局也为建筑自身的通风采光等提供有利的条件。

Facade Design 立面设计

Based on the residents' unique preferences of the traditional street with humanistic ambiance, the building is designed in Art Deco style. As the residential plot A and B have used the Art Deco style, this design of this plot considers from the perspective of urban landmark and facade of public buildings, adopts the modern elegant style which is more concise and with stronger commercial ambiance to highlight the simple, high-efficient, bright and fashionable theme.

The Art Deco-style facade design forms rich facade levels through the repeated scrutiny of the dimension of the volume and the mixed use of various materials such as aluminum plate, stone, face brick and coatings. The use of hanging stones adds a delicate luxury sense to the residence and manifests its quality.

基于此地区居民对于传统街区的人文气息有独特的偏好，本方案建筑的设计风格为 ART DECO，即装饰艺术派风格。在 A、B 块住宅采用 ART DECO 风格后，本地块设计从城市地标和公建化立面两个角度考虑，采用更简洁大方，商业气息更浓厚的现代典雅风格，突出简洁、高效、明快、时尚的主题。

ART DECO 风格的立面设计通过对形体尺度的反复推敲以及金属铝板、石材、面砖、涂料等多种材料的混搭运用，形成丰富的立面层次。干挂石材的运用更增添一份精致的奢华，彰显品质。

Residential House Type Design 住宅户型设计

The living rooms of all the house types face the main landscape, and each unit has south oriented bedroom which adopts large French door or bay-window to fully absorb the sunshine and beautiful landscape outside. Each unit has good ventilation for the interior spaces. It's with good north–south ventilation and comfortable interior environment and regresses to nature. The entrance garden and landscape balcony become the highlight of this house type design, which have organically led the outdoor greening into the residence and effectively improved the living quality of the residence. On the plan layout, the living room and the dining room are with dynamic spaces, and the bedroom is with static space, and the space of the kitchen is relatively independent, forming a linear, convenient and non-interfering interior space.

所有户型的厅均向主景观面，每户均有居室向南，采用大面积落地门窗或凸窗，充分吸纳外界阳光及美景。每户均有良好的室内空间对流条件。南北向通风良好，室内环境舒适，寓于自然。入户花园，观景阳台成为户型设计的亮点，将室外绿化有机地组织于住宅中，有效地改善了居室的生活环境质量。平面布局上，客饭厅的动空间，卧室的静空间，以及厨房后勤空间相对独立，交通流线便捷而互不干扰。

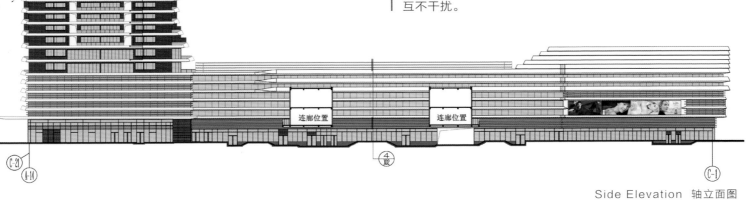

Side Elevation 轴立面图

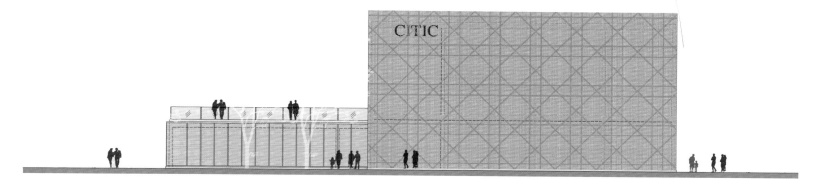

East Elevation 东立面图

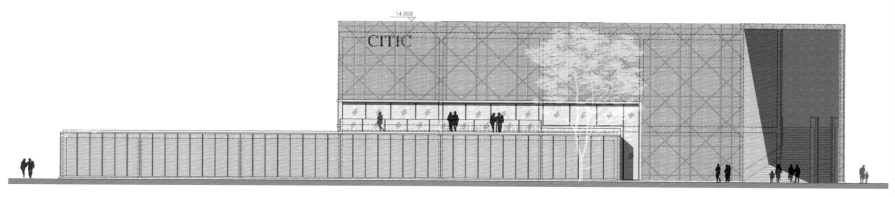

South Elevation 南立面图

Core Waterfront Mansion with Unique Garden of Qiandenghu Area
千灯湖区核心的别致花园水都

Comhope · Chief Lake Mansion　　创鸿 · 水韵尚都

Location / 项目地点：Foshan, Guangdong / 广东省佛山市
Architectural Design / 建筑设计：Guangzhou Hanhua Architects + Engineers Co., Ltd. / 广州瀚华建筑设计有限公司
Co-design / 合作单位：Shantou Building Design Institute / 汕头市建筑设计院
Land Area / 用地面积：64 400 m²
Floor Area / 建筑面积：239 365 m²

Keywords　关键词
Windward & Sunward; Circular Running Water; Open Space
清风迎阳　活水循环　开放式空间

 "China's Most Beautiful Stylish Apartments" Finalist
入围"中国最美风格楼盘"奖

The overall design concept uses "life" as original intention and aims to build a gardenesque life; and it integrates the Southeast Asian style, combines the master planning and architectural design and brings in the "circular running water" eco water system to create a "windward and sunward" comfortable living layout.

总体设计理念以"生活"为原点，以构造花园般的生活为设计宗旨，融入东南亚风情，结合项目总体规划和建筑设计，引入"活水循环式"生态水系，打造"清风迎阳"的舒适人居格局。

Overview　项目概况

Comhope · Chief Lake Mansion, with a land area of approx 65,000 m² and a gross floor area of approx 240,000 m², is composed of 10 blocks of north-south oriented super high-rise residences and 77 units of townhouses. The garden is with central waterscape and European hotel garden style, and the project would be a landmark of Foshan.

创鸿·水韵尚都占地接近6.5万m²，总建筑面积接近24万m²，由10栋南北向超高层住宅和77户联排别墅组成，园林风格为中央水景欧式酒店园林风情，项目建成之后将成为佛山的标志性建筑之一。

Location Advantage 区位优势

The "financial and business area" that the government strives to build is in the planning of Nanhai government, and the Qiandenghu area will be built as the backup financial industry cluster of Guangzhou-Foshan area and the future new engine of Nanhai's development, and it will focus on the development of the tertiary industry such as finance, commercial logistics, real estate and hotel. The development of financial business area is an important step for the north extension of Nanhai central urban area and also the "South Pole" of the future big urban area. Financial services as an advanced form of the urban economy will be the leading industry of Nanhai's tertiary industry.

政府着力打造的"金融商贸区"在南海区政府的规划中，千灯湖区域将打造成为广佛地区后援金融产业集群，未来南海城市发展的新引擎，重点发展金融、商贸物流、房地产、酒店等第三产业，开发金融商务区既是南海谋求中心城区北延的重要步骤，也是未来大中心城区的"南极"，金融服务业作为城市经济的一种高级形态，将成为南海区第三产业的龙头。

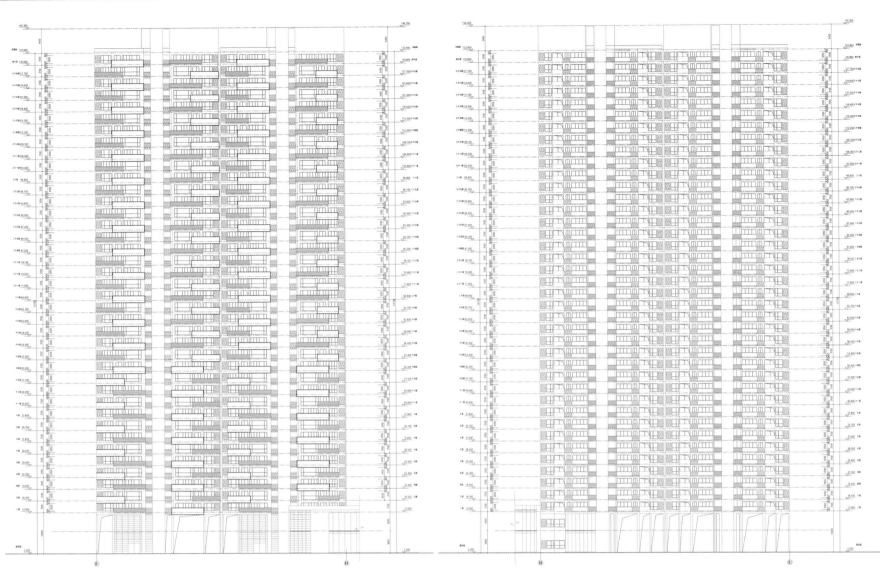

Side Elevation 1 轴立面图 1　　　　　　　　　　Side Elevation 2 轴立面图 2

Planning 规划布局

The designers absorb the oriental philosophy that man is an integral part of nature and adopt the design concept of "international texture, oriental connotation" to build the humanized high-end residential project. The entire project is composed of 4 blocks of 42-floor super high-rise residences and 77 units of waterfront townhouses, and it uses the windward and sunward geomancy planning which is lower in the south and higher in the south to creatively lead in the ecological and circular running water system, making the entire community surrounded by running water and ensuring each unit with beautiful scenery.

项目融入天人合一的东方哲学，采用"国际肌理、东方底蕴"的设计理念，打造出充满人性化的高端住宅项目，整个项目由 4 栋 42 层超高层华宅和 77 栋掬水联排别墅组成，采用南低北高的"清风迎阳"风水规划，创新引入生态活水循环系统，整个小区活水环绕，户户有景。

Design Concept 设计理念

The overall design concept uses "life" as original intention and aims to build a gardenesque life; and it integrates the Southeast Asian style, combines the master planning and architectural design and brings in the "circular running water" eco water system to create a "windward and sunward" comfortable living layout.

总体设计理念以"生活"为原点,以构造花园般的生活为设计宗旨,融入东南亚风情,结合项目总体规划和建筑设计,引入"活水循环式"生态水系,打造"清风迎阳"的舒适人居格局。

Residential Layout 住宅布局

The main landscape of this project is the Qiandenghu Park in the southeast, thus the designers arrange the high-rise residences centralized at the northwest of the plot and the low-rise residences at the southeast of the plot to form open living space and to make full use of the landscape resources of the park. The building floors above ground maximally counts to 42, and the shape tries to avoid the rigid barracks-style layout and uses different conformations to make the space rich and varied, natural and reasonable.

项目主要景观为东南面的千灯湖公园,故规划时将高层住宅集中布置于地块西北部,低层住宅布置于地块东南部,形成开放式的居住空间,并充分利用园景资源。建筑地面以上层数最高为42层,形态上尽量避免刻板的兵营式布局,通过不同的组织方式,令空间形体丰富多样、自然合理。

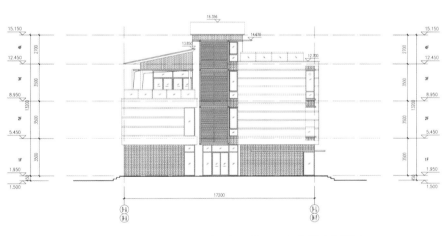

End Elevation 1 of 9# Building 9栋侧立面图1

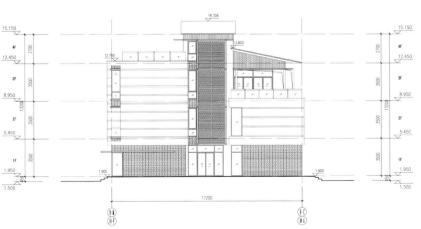

End Elevation 2 of 9# Building 9栋侧立面图2

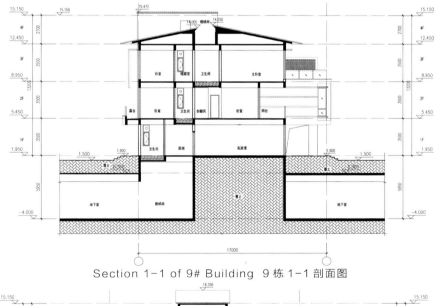

Section 1-1 of 9# Building 9栋1-1剖面图

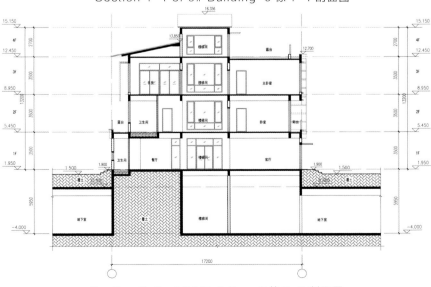

Section 2-2 of 9# Building 9栋2-2剖面图

Side Elevation 1 of 8# Building 8栋轴立面图1

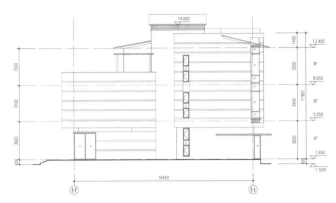

Side Elevation 2 of 8# Building 8栋轴立面图2

Side Elevation 3 of 8# Building 8栋轴立面图3

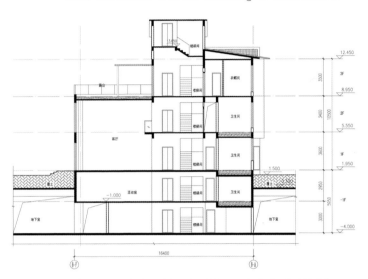

Section 3-3 of 8# Building 8栋3-3剖面图

Road Layout 道路布局

The main driveway and the buildings of the community are relatively independent, and the entrance paths to the residences are flexibly arranged and combined with the park. The basements are designed with different floor elevation to comply with the tableland, and some parts are higher than the roadbed to ensure partially natural ventilation, lighting and convenient transportation.

主要车行道与小区建筑相对独立，住宅入户的小路结合园林灵活布置。地下室顺应台地标高设计不同的底板标高，局部高出小区路面，可部分自然通风、采光并方便行车。

Fresh and Concise Seaside Style Manifests the Tropical Life Style of Sanya
清新简洁海派风格演绎三亚热带生活风情

Sanya Haitang Bay Linwang South Glamorous Small Town 三亚海棠湾林旺南风情小镇

Location / 项目地点：Sanya, Hainan / 海南省三亚市
Developer / 开发商：China Construction International Co., Ltd. / 中国对外建筑开发总公司
Architectural Design / 建筑设计：Beijing Honest Architectural Design Co., Ltd. / 北京奥思得建筑设计有限公司
Gross Land Area / 总用地面积：283 128.4 m²
Gross Floor Area / 总建筑面积：191 599.1 m²
Overground Floor Area / 地上建筑面积：184 099.1 m²
Underground Floor Area / 地下建筑面积：7 500 m²
Plot Ratio / 容积率：0.72
Green Coverage Ratio / 绿化率：38.1%

Keywords 关键词

Ocean Ambiance; Natural Harmony; Full of Glamour
海洋气息 自然和谐 风情洋溢

"China's Most Beautiful Stylish Apartments" Finalist
入围"中国最美风格楼盘"奖

The finishing materials of the building facade are mainly wood, paint and glass. The fresh and concise seaside style of the modern residence is manifested through the comparison of the facade, the rhythmic changes of the bay window, the interspersion of the structure and the emphasis of the building entrance.

建筑外立面装饰材料主要是木材、涂料和玻璃，通过立面构成对比、开窗的韵律变化、构架的穿插、建筑入口的强调，体现出现代住宅楼的清新、简洁的海派风格。

Overview 项目概况

Located in Haitang Bay, Sanya, Hainan, the project is to the north of Linnan No. 1 Road, to the south of Linnan No. 2 Road, to the west of Hebin Road and to the east of Linwang West Road. The base is divided into 6 square plots by the Linwang Central Road, Lingwang East Road and so on.

项目位于海南省三亚市海棠湾，南临林南一路，北临林南二路、东侧为河滨路、西侧为林旺西路，基地被林旺中路、林旺东路等分割成6个方形地块。

Base Status Analysis 基地现状分析

The base is in a rectangle shape with a north-south average length of approx 430 m, an east-west average width of approx 660m and a gross land area of 283,128.4 m² (approx 360 Mu). With a smooth terrain, the base is a resettlement housing plot of Linwang South village of Haitang Bay, Sanya city.

项目基地呈矩形状，南北平均长约430 m，东西平均宽约为660 m，总用地面积为283 128.4 m²（约360亩）。本项目基地地势基本平坦，基地为三亚市海棠湾林旺南村回迁安置房地块。

Site Plan 总平面图

Front Elevation Texture 正立面材质图

Planning Design 规划设计

Based on the concept of separating the driveway and the sidewalk and the people of different functions, the planning makes the residents flow and the commercial operation flow into independent and non-interfering areas; the parking and traffic of the commercial building form independent system which barely interlaces with the internal of the residential area. The residential buildings are arranged in an arc-shaped group which is rich in the moving feelings of the ocean ambiance; and the commercial pedestrian street and the landscape are integrated as a whole.

规划本着人车分流、不同功能的人群分流理念，将安置区居民流线和未来商业运营流线独立成区，互相不干扰；商业建筑的停车和交通独自成为系统，不与住宅小区内部交插。居住建筑成组团弧形排列，富有海洋气息的流动感；商业步行街与景观带融合为一体。

Back Elevation Texture 背立面材质图

Planning Principle 规划原则

1. Comprehensive Integration Principle: to consider the building, landscaping, natural resources and the ecological design that runs through the entire process as a whole.
2. Reasonable Function Principle: to fully consider the rationality of each functional area in the community, the convenience and smoothness of the traffic and their influence to the living environment; to fully improve the Eco-quality of the environment; and to fully consider the rationality and interest between the various spaces.
3. Ecological Priority Principle: to first develop the ecological construction, to collocate the land and environment resource for the most proper use, to promote the effective protection and use of the Eco-environment, making the resources high-efficiently used and the human and nature highly harmonious.
4. Market Economy Principle: to create the ordinary resettlement area into a glamorous tourist town, making the residents live and work in peace and contentment in this community.
5. Humanistic Culture Principle: to discover the Hainan local cultural connotations, enhancing the sense of belongs while living in this community and improving the living quality.
6. Residential Culture Principle: to try the best to create a high-quality living environment, leading the residential area into a model of higher level.

1. 全面整合原则：把建筑、景观营造、自然资源、以及贯穿始终的生态设计等内容作为一个整体考虑。
2. 功能合理原则：充分考虑区内各功能分区的合理，交通的便捷通畅，和对居住环境的影响；充分提高环境的生态质量；充分考虑各类空间的合理性和趣味性。
3. 生态优先原则：生态建设优先，将土地与环境资源配置以最适宜的用途，促进生态环境的有效保护利用，使资源得以高效利用，人与自然高度和谐。
4. 市场经济原则：把一个普通的安置小区打造成一处充满风情的旅游小镇，确保居民安居乐业于其中。
5. 人文文化原则：努力挖掘海南本地的人文内涵，增强社区居住归属感，提升品质。
6. 居住文化原则：竭力打造高质量的居住环境，引导住区向更高层次的模式演变。

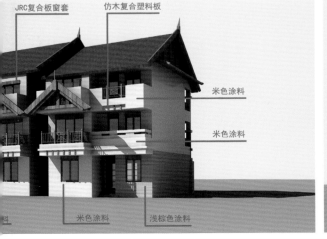

200 m² Ground Floor Plan
200 m² 首层平面图

200 m² Second Floor Plan
200 m² 二层平面图

150 m² Ground Floor Plan
150 m² 首层平面图

150 m² Ground Floor Plan
150 m² 首层平面图

Humanistic Theme 人文主题

"Flowers bloom in four seasons" is mainly manifested on the landscaping of the community. The designers use the local abundant and unique trees and flowers of Hainan and separately plant them in different plots, together with garden ornaments that are rich in local conditions and customs to create a unique community that are different in four seasons, color and humanity.

"Haitang Inn" is based on the houses of the villagers, and makes the villagers as management participants and the local cultural characteristics as supporting point to let the tourists participate and experience. As the affiliates of the Haitang Inn are quite familiar with their local culture and have their own unique understandings, so they can introduce the local culture to the tourists quite vividly and skillfully. Therefore, on the basis of maintaining the living and diet habits of each family of the villagers, Haitang Inn can form its own unique theme and connotation and build its own brand through innovating and promoting the essence of the local folk culture.

"花开四季"主要体现在社区景观的营建上，利用海南当地丰富的、特有的树木花卉分别规划在不用的区块内，搭配富有风土人情的园林小品，营造出四季不同、色彩不同、人文不同的独特社区。

"海棠客栈"是以村民家庭为依托、以村民为经营管理参与者、以当地文化特色为支撑点、以让游客参与体验为主要形式的，由于海棠客栈的加盟成员对自己的本土文化都非常了解，都有自己独特的见解，可以把本土文化很生动、熟练地介绍给游客，所以，海棠客栈能够在保持村民每户独特的住宿及饮食习性基础上，通过创新并发扬本土民俗文化精髓，从而形成自己独有的主题和内涵并树立品牌。

Architectural Design 建筑设计

Residential Building

The finishing materials of the building facade are mainly wood, paint and glass. The fresh and concise seaside style of the modern residence is manifested through the comparison of the facade, the rhythmic changes of the bay window, the interspersion of the structure and the emphasis of the building entrance.

住宅楼

建筑外立面装饰材料主要是木材、涂料和玻璃，通过立面构成对比、开窗的韵律变化、构架的穿插、建筑入口的强调、体现出现代住宅楼的清新、简洁的海派风格。

Commercial Building and Ancillary Building

Parts of the finishing materials of the commercial building facade are cultural stone, paint and glass and large French window, which is conducive to the exhibition of the products. The uniformly planned billboards are arranged above the entrance of each store, having achieved the effect of clear marking. The ancillary buildings emphasize the local features and that is the tropical building characteristics of the Southeast Asia. The arcade, open corridor and the sloping roof fully manifest the humanistic ambiance of the Haitang Bay of Sanya.

商业楼及配套建筑

商业楼部分立面装饰材料主要是文化石、涂料及玻璃。大面积落地窗，有利于商品的展示；统一规划的广告牌设于每户商铺入口上方，起到明确的标示作用；配套建筑注重地方特色，即东南亚热带建筑特征；骑楼、开敞的通廊、坡屋面，充分体现出三亚海棠湾的人文气息。

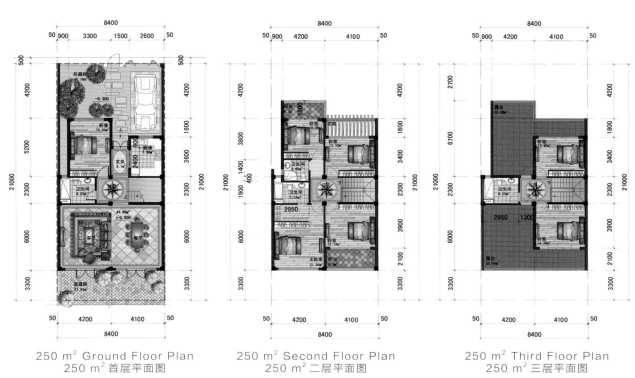

250 m² Ground Floor Plan
250 m² 首层平面图

250 m² Second Floor Plan
250 m² 二层平面图

250 m² Third Floor Plan
250 m² 三层平面图